P9-BIV-462

MICROFOSSILS

MICROFOSSILS

M. D. Brasier
University of Hull

London
GEORGE ALLEN & UNWIN
Boston Sydney

First published in 1980

GEORGE ALLEN & UNWIN LTD
40 Museum Street, London WC1A 1LU

British Library Cataloguing in Publication Data

Brasier, M D
Microfossils.
1. Micropaleontology
I. Title
560 QE719 79-40772

ISBN 0-04-562001-6
ISBN 0-04-562002-4 Pbk

Typeset in 10 on 12 point Times by Servis Filmsetting Ltd, Manchester
and printed in Great Britain
by Billing & Sons Ltd, Guildford, London and Worcester

To Cecilia

Preface

The past two decades have witnessed a dramatic expansion in the scope of the Earth sciences: geologists have probed the alien worlds of the Moon and the planets, fathomed the secrets of the ocean depths, and pondered over the strange and remote Precambrian Earth. As microfossils have played their part in the development of these and other new fields, it is not surprising that the scope and emphasis of their study, micropalaeontology, has also changed over this period. This subject now occupies a firm place in the teaching of most Earth science degree courses and is the basis of numerous postgraduate diplomas and degrees. The need for a more up to date but concise and inexpensive textbook has therefore become a pressing matter for both student and teacher.

This is a book about the nature of microfossils; hence its main concern is the specimen as seen down the microscope, considered as a once-living organism. The naming of parts, which is the first essential step, precedes the naming and classification of the specimen, in which a number of genera (chosen to give an outline view of morphological range) are described briefly and accompanied by labelled line drawings such as the student may prepare himself. More derivative data, as for the general history of a group, are broadly sketched but the all-important matter of their geological applications is covered for each group in the form of an 'animated' reference list, turning students towards books or articles of special interest.

Mindful of the value of involving the investigator in the collection and preparation of material, but also of the problems of overcrowded laboratories, of equipment expense and of stringent safety regulations, I have confined the preparatory methods to those that are safer, simpler and cheaper.

The matter in this book derives largely from the endeavours of other micropalaeontologists, both past and present and from many parts of the world. It has been a pleasurable education to draw this work together into a small volume which I hope will be helpful to other students of microfossils. I am particularly indebted to the following people for their valuable comments on various parts of the manuscript: Dr R. L. Austin, Prof. F. T. Banner, Dr R. H. Bate, Dr M. A. Butterworth, Dr E. N. K. Clarkson, Prof. C. Downie, Dr G. H. Evans, Dr R. Goldring, Dr R. Harland, Dr B. K. Holdsworth, Dr A. W. Medd, Prof. J. W. Murray, Dr R. Riding, Dr J. E. Robinson and Prof. F. M. Swain. Miss Linda Joseph helped to prepare the many versions of the typescript, Mrs Sharon Chambers prepared the line drawings and Mr Paul McSherry the diagrams, to whom I am most grateful.

Martin Brasier
Kingston-upon-Hull
1979

Acknowledgements

In addition to thanking all those authors and publishers who have kindly allowed the use of their illustrations, acknowledged throughout the text, the following permissions were also granted: *Treatise on invertebrate paleontology*, courtesy of the Geological Society of America and University of Kansas Press, for figures adapted from Campbell 1954, Campbell 1954a, Benson *et al.* 1961, Hass *et al.* 1962, Müller 1962 and Loeblich & Tappan 1964; Academic Press for figures adapted from Fogg *et al.* 1973 and Sarjeant 1974; Trustees of the British Museum (Natural History) for figures adapted from Marshall 1934, Harris 1938 and 1964, and Croft & George 1959; Society of Economic Paleontologists and Mineralogists for figures adapted from Colom 1948.

Contents

1 Introduction

Microfossils – what are they?

Perhaps as much as one sixth of the Earth's surface is covered by a thin blanket of soft white to buff-coloured ooze. Seen under the microscope this sediment can be a truly impressive sight. It contains countless numbers of tiny shells variously resembling miniature flügelhorns, shuttlecocks, water wheels, hip flasks, footballs, garden sieves, space ships and chinese lanterns. Some of these gleam with a hard glassy lustre, others are sugary white or strawberry-coloured. This aesthetically pleasing world of microfossils is a very ancient one and, at the biological level, a very earnest one.

Any dead organism that is vulnerable to the natural processes of sedimentation and erosion may be called a **fossil**, irrespective of the way it is preserved or of how recently it died. It is common to divide this fossil world into larger **macrofossils** and smaller **microfossils**, each kind with its own methods of collection, preparation and study. This distinction is, in practice, rather arbitrary and we shall largely confine the term 'microfossil' to those discrete remains whose study requires the use of a microscope throughout. Hence bivalve shells or dinosaur bones seen down a microscope do not constitute microfossils.

The study of microfossils is properly called **micropalaeontology**. There has, however, been a tendency to restrict this term to studies of mineral-walled microfossils (such as foraminifera and ostracods), as distinct from **palynology** which is the study of organic-walled microfossils (such as pollen grains and acritarchs). This division, which arises largely from differences in preparation techniques, is again rather arbitrary. It must be emphasised that macropalaeontology, micropalaeontology and palynology share identical aims: to unravel the history of the external surface of the planet. These are achieved more speedily and with greater reward when they proceed together.

Why study microfossils?

There are few sediments from which some kind of microfossil cannot be retrieved, the kind depending largely on the original age and the depositional environment of the sediment. Hence when a geologist wishes to know the age of a rock or the salinity and depth of water under which it was laid down, it is to microfossils that he will turn for a quick and reliable answer. Many geological surveys, oil companies and coal mines therefore employ a team of micropalaeontologists to learn more about the rocks they are handling. This commercial side to micropalaeontology has undoubtedly been a major stimulus to its growth. There are some philosophical and sociological sides to the subject, however. Our understanding of the development and stability of the present global ecosystem has much to learn from the microfossil record, especially since many microfossil groups have occupied a place at or near to the base of the food web. Studies into the nature of evolution cannot afford to overlook the microfossil record either, for it contains a wealth of examples. The importance of understanding microfossils is further augmented by recent discoveries in Precambrian rocks; microfossils now provide the main evidence for organic evolution through more than three-quarters of the history of life on Earth.

The cell

A great many microfossils are the product of single-celled (**unicellular**) organisms. A little knowledge of these cells can therefore help us to understand their way of life and, from this, their potential value to Earth scientists. Unicells are usually provided with a relatively elastic outer **cell membrane** (Fig. 1.1)

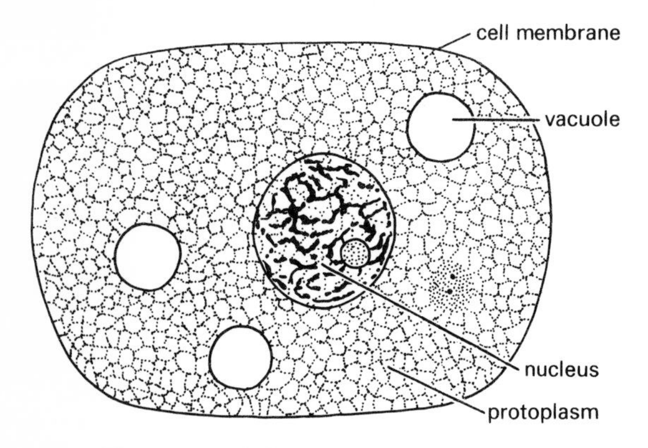

Figure 1.1 A living cell (diagrammatic).

that binds and protects the softer cell material within, called the **protoplasm** (or cytoplasm). Small 'bubbles' within the protoplasm, called **vacuoles**, are filled with food, excretory products or water and serve to nourish the cell or to regulate the salt and water balance. A darker, membrane-bound body, termed the **nucleus**, helps to control both vegetative and sexual division of the cell and the manufacture of proteins. Other small bodies concerned with vital functions within the cell are known as **organelles**. The whip-like thread that protrudes from some cells, called a **flagellum**, is a locomotory organelle. Some unicells bear many short flagella, collectively called **cilia**, whilst others get about by means of foot-like extensions of the cell wall and protoplasm, known as **pseudopodia**. Other organelles that can occur in abundance are the **chromoplasts** (or chloroplasts). These small structures contain chlorophyll or similar pigments for the process of photosynthesis.

Nutrition

There are two basic ways by which an organism can build up its body: by heterotrophy or by autotrophy. In **heterotrophy**, the creature captures and consumes living or dead organic matter, as we do ourselves. In **autotrophy**, the organism synthesises organic matter from inorganic matter by utilizing the effect of sunlight on CO_2 and H_2O in the presence of chlorophyll-like pigments, a process known as **photosynthesis**. Quite a number of microfossil groups employ these two strategies together and are therefore known as **mixotrophic**.

Reproduction

Asexual (or vegetative) and **sexual** reproduction are the two basic modes of cellular increase. The simple division of the cell found in asexual reproduction results in the production of two daughter cells with nuclear contents similar in proportion to those of the parent. In sexual reproduction, the aim is to halve these normal nuclear proportions so that sexual fusion with another 'halved' cell can eventually take place. Information contained in each cell can then be passed around to the advantage of the species. This halving process is achieved by a fourfold division of the cell, called **meiosis**, which results in four daughter cells rather than two.

The kingdoms of life

Living individuals all belong to naturally isolated units called **species**. Ideally, these species are freely interbreeding populations that share a common ecological niche. Even those lowly organisms that disdain sexual reproduction (such as the silicoflagellates) or do not have the organisation for it (such as the cyanophytes), occur in discrete morphological and ecological species. Obviously it is impossible to prove that a population of microfossils was freely interbreeding but, if specimens are sufficiently plentiful, it is possible to recognise both morphological and ecological discontinuities. These can serve as the basis for distinguishing fossil species from one another.

Whereas the species is a functioning unit, the higher taxonomic categories in the hierarchical system of classification are mere abstractions, implying varying degrees of shared ancestry. All species are placed within a **genus** that contains one or more closely related species. These will differ from other species in neighbouring genera by a distinct morphological, ecological or biochemical gap. Genera tend to be more widely distributed in time and space than do species, so they are not greatly valued for stratigraphic correlation. They are, however, of considerable value in palaeoecological and palaeogeographical studies. The successively higher categories of **family**, **order** and **class** (often with intervening sub- or super- categories) should each contain clusters of taxa with similar grades of body organisation and a common ancestor. They are of relatively little value in biostratigraphy and palaeoenvironmental studies. In 'animals' the **phylum** taxon is defined on the basis of major structural differences whereas in 'plants' the corresponding **division** is defined largely on structure, life history and photosynthetic pigments.

The highest category is the **kingdom**. In the nineteenth century it was usual to recognise only the two kingdoms: Plantae and Animalia. Plants were considered to be mainly non-motile, feeding by photosynthesis. Animals were considered to be motile, feeding by ingestion of pre-formed organic matter. Although these distinctions are evident amongst macroscopic organisms living on land, the largely aqueous world of microscopic life abounds with organisms that appear to straddle the plant/animal boundary. The classification of Whittaker (1969) overcomes these anomalies by recognising five kingdoms: the Monera, Protista, Plantae, Fungi and Animalia (Fig. 1.2).

The Monera are single celled but they lack a

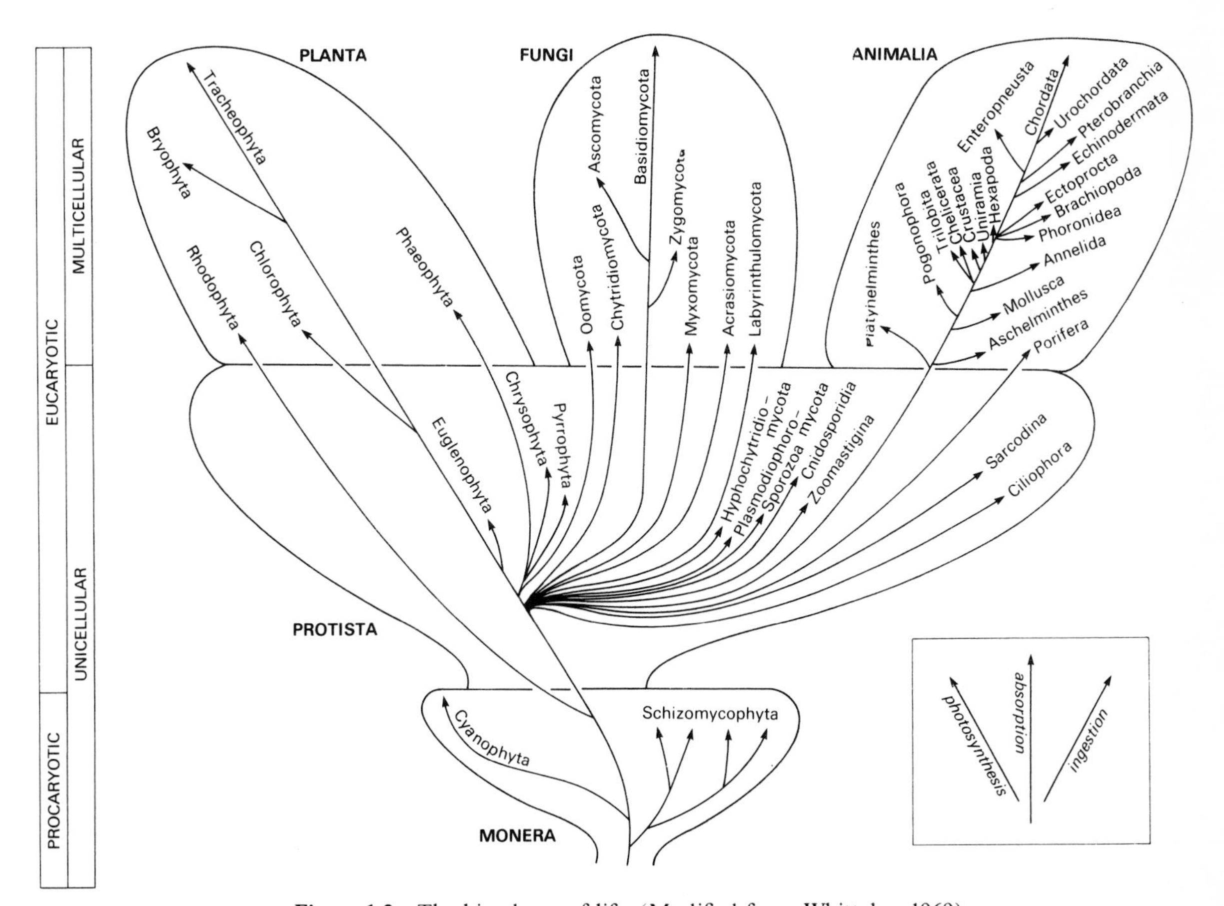

Figure 1.2 The kingdoms of life. (Modified from Whittaker 1969)

Figure 1.3 The stratigraphic column.

Millions of years before present	*Era/Erathem*	*System/Period*		*Series/Epoch*	
0	CAINOZOIC	NEOGENE		Pleistocene and Recent (Quaternary)	
1.8				Pliocene Miocene	
26		PALAEOGENE		Oligocene Eocene Palaeocene	
65	MESOZOIC	CRETACEOUS		Upper Cretaceous Lower Cretaceous	
135		JURASSIC		Upper Jurassic Middle Jurassic Lower Jurassic	
190–205		TRIASSIC		Upper Triassic Middle Triassic Lower Triassic	
225–45	UPPER PALAEOZOIC	PERMIAN		Upper Permian Lower Permian	
280–90		CARBONIFEROUS		Upper Carb. (Silesian)	Pennsylvanian
				Lower Carb. (Dinantian)	Mississippian
345–80		DEVONIAN		Upper Devonian Middle Devonian Lower Devonian	
395–430	LOWER PALAEOZOIC	SILURIAN		Downtonian Ludlow	Upper
				Wenlock	Middle
				Llandovery	Lower to Middle
430–60		ORDOVICIAN		Ashgill Caradoc	Upper
				Llandeilo Llanvirn	Middle
				Arenig Tremadoc	Lower
500–30		CAMBRIAN		Upper Cambrian (Merioneth) Middle Cambrian (St David's) Lower Cambrian (Comley)	
570–610	PROTEROZOIC	UPPER PROTEROZOIC (UPPER PRECAMBRIAN)	VENDIAN	Upper Vendian Lower Vendian	
c.700			RIPHEAN	Upper Riphean Middle Riphean Lower Riphean	
c.1700		LOWER PROTEROZOIC (MIDDLE PRECAMBRIAN)			
c.2500	ARCHAEAN (LOWER PRECAMBRIAN)				
c.4500					

nucleus, cell vacuoles and organelles. This primitive **procaryotic** condition, in which proper sexual reproduction is unknown, is characteristic of the Division Cyanophyta (blue-green algae) and the Division Schizomycophyta (bacteria). The other four kingdoms are **eucaryotic**. That is, their cells are nucleate with vacuoles and organelles, and are capable of properly coordinated cell division and sexual reproduction.

The Protista are motile unicellular organisms with rather varied body plans. Some, like the dinoflagellates (Division Pyrrhophyta), have whip-like flagella for locomotion, and photosynthetic pigments. Hence they resemble the true Plantae and are probably close to the ancestral line of that group. The foraminifera and the radiolarians (Phylum Sarcodina) engulf their food with the aid of mobile pseudopodia, whilst the tintinnids (Phylum Ciliophora) have a coat of bristle-like cilia and ingest their food through a mouth surrounded by 'tentacles'. These heterotrophic protists are therefore more akin to animals than to plants.

The multicellular and largely non-motile algae and the higher land plants both belong in the Kingdom Plantae. Their remains are, for the most part, the concern of **palaeobotany**, but the microscopic structures of green and red algae (especially the calcareous ones) and the spores and pollen of land plants are very common as microfossils.

The Fungi are heterotrophic, feeding by absorption of pre-formed organic matter. Unfortunately their reproductive spores and vegetative hyphae are rarely preserved and have received little study as fossils, so they will not be considered any further here.

The Kingdom Animalia comprises multicellular invertebrate animals and vertebrate animals that feed by ingestion of pre-formed organic matter, either alive or dead. Invertebrates which are microscopic even when fully grown, such as ostracods, may be considered as microfossils, but we shall be obliged to leave aside the microscopic elements of larger animals (such as sponge spicules, scolecodonts, echinoderm ossicles and juvenile individuals) with one exception: the conodonts. Although conodonts appear to have belonged to the apparatus of some extinct, macroscopic animal, the small size and stratigraphic value of conodont elements themselves gives them much in common with other microfossils. For more information on the macro-invertebrate fossil record the reader should consult Clarkson (1979).

Not all microfossils can be placed happily within the existing hierarchical classification. Those whose phylum or division remains uncertain (such as the acritarchs, chitinozoans and conodonts) are each accorded the informal and temporary status of a **group** in the following text.

The stratigraphic column

The succession of rocks exposed at the surface of the Earth can be arranged into a **stratigraphic column**, with the oldest rocks at the base and the youngest ones at the top (Fig. 1.3). Although the absolute ages have been determined from studies of radioactive isotopes, it is customary to use the names of stratigraphic units, mostly distinguished on the basis of differences in their included fossils. These units are arranged into a number of hierarchies relating to rock-based stratigraphy (**lithostratigraphy**), fossil-based stratigraphy (**biostratigraphy**) and time-based stratigraphy (**chronostratigraphy**).

Lithostratigraphic units, such as beds, members and formations, are widely used in geological mapping but will not concern us further here. The **biozone** is the fundamental biostratigraphic unit and comprises those rocks that are characterised by the occurrence of one or more specified kinds of fossil known as zone fossils.

Formal chronostratigraphic time units are also important and include, in ascending order of importance, the age, epoch, period and era. For example we may cite the Atdabanian Age, of the Comley Epoch, of the Cambrian Period, of the Palaeozoic Era. Rock units laid down during these times are properly referred to as stages, series, systems and erathems (i.e. the Atdabanian Stage, of the Comley Series, etc.). Less formal divisions are also widely used so that we may talk of the lower Cambrian rocks laid down during early Cambrian times. In the following text, these informal subdivisions are abbreviated as follows: lower (L.), middle (M.) and upper (U.).

The major subdivisions of the stratigraphic column are shown in Figure 1.3. Hedberg (1976) gives further information on stratigraphic terminology.

Part I
Procaryotes – the fundamental cells

2 Division Cyanophyta – Blue-green algae

The Cyanophyta are informally called blue-green algae on account of the colour imparted by the photosynthetic pigment phycocyanin. However, living cyanophytes may also be olive green or red in colour. These cyanophytes consist of small cells, mostly less than 25 μm in diameter, that have a unicellular or colonial organisation. Unlike other algae, higher plants or animals, they lack a number of cell features, especially a proper membrane-bound nucleus. They are therefore placed with the bacteria in the Kingdom Monera.

As might be expected these primitive cells have had an extensive geological history, ranging back to some of the oldest known Precambrian rocks (*c.* 3200 Ma before present (BP). The group has for much of this time been involved in the construction of organo-sedimentary structures (stromatolites and thrombolites), terrestrial calcareous crusts (travertine or tufa) and in the precipitation and degradation of carbonate sediments at sea and on land.

The cyanophytes deserve study because of their evolutionary significance and their potential as Precambrian and early Palaeozoic zone fossils. Stromatolites are also useful environmental markers.

The living cyanophyte

Cyanophytes have relatively small cells, between 1 and 25 μm in diameter. These may be spherical (coccoid), ovoid, discoidal, cylindrical or pear-shaped (pyriform) in outline. The cell has neither nuclear membrane, mitochondria nor contractile cell vacuoles. The phycocyanin and chlorophyll pigments are distributed in lamellae around the edges of the cell where they take part in photosynthesis.

Cells may be single (unicellular) or arranged in colonies bound by an outer mucilaginous **sheath** of cellulose fibrils. The arrangement of cells in a colony may be regular to irregular, e.g. flat, cuboid, spherical, uniseriate filamentous or branched filamentous (Figs 2.1, 2.2). The cells of a filamentous colony comprise the **trichome**.

Cyanophytes construct organic materials from inorganic ones by photosynthesis, evolving free oxygen in the process, as in higher plants. The need for light causes them to grow towards the sun. In filamentous forms this may be achieved by gliding upwards through the substrate, leaving behind the old sheath in the process. As these sheaths are of resistant cellulose, whilst the cell walls are mostly degradable amino acids and sugars, it is the sheath which has the better chance of preservation in the fossil record.

Cyanophyte life history

Cyanophytes are an extremely ancient and primitive group that have never developed the controlled division of cells by mitosis or meiosis. Sexual reproduction is therefore unknown, multiplication being entirely vegetative (asexual), usually effected by fragmentation, binary cell division, or the formation of endospores, akinetes or hormogonia. Cell division involves the splitting up of a cell into two new cells by inward growth of the wall (i.e. **binary fission**, Fig. 2.1a). The cell contents are randomly distributed between the new cells, unlike the orderly mitotic divisions of eucaryotes. **Fragmentation** is simply a breaking up of colonies into smaller ones. **Endospores** form by the internal subdivision of cells into two or more spores which are subsequently released to grow into new colonies (Fig. 2.1i). **Akinetes** are also spore cells, but these develop singly from vegetative cells by enlargement and the formation of a thick, often sculptured wall (Fig. 2.2c). After conditions of desiccation or chilling,

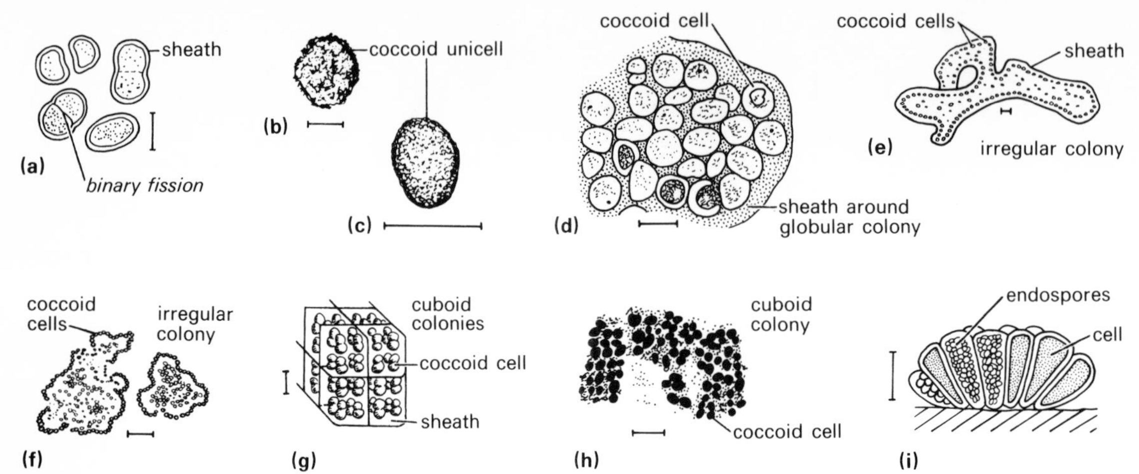

Figure 2.1 Order Chroococcales. (a) Recent *Synechocystis*; (b) Precambrian *Archaeosphaeroides*; (c) Precambrian *Huroniospora*; (d) Precambrian *Myxococcoides*; (e) Recent *Anacystis*; (f) fossil *Renalcis*; (g) Recent *Eucapsis* colony; (h) Precambrian *Eucapsis*-like colony; (i) Recent *Entophysalis*. Scale: bar = 10 μm. ((b) based on Schopf & Barghoorn (1967); (c) on Barghoorn & Tyler (1965); (d) & (h) on Cloud (1976); (g) on Fogg *et al.* (1973); (i) from Chapman & Chapman (1973))

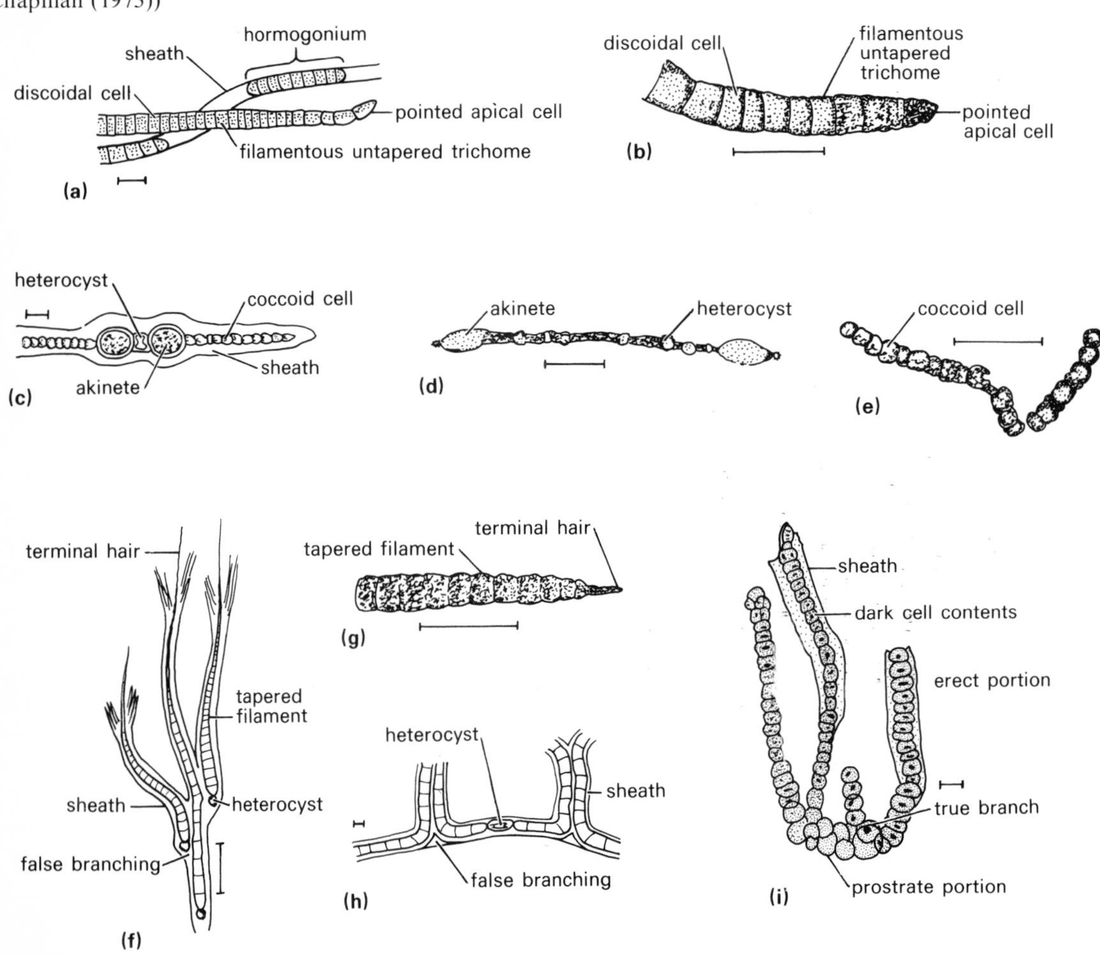

Figure 2.2 Order Nostocales. (a) Recent *Oscillatoria*; (b) Precambrian *Oscillatoria*-like filament; (c) Recent *Wollea*; (d) Precambrian *Gunflintia*; (e) Precambrian *Nostoc*-like filament; (f) Recent *Rivularia*; (g) Precambrian *Rivularia*-like filament; (h) Recent *Scytonema*. Order Stigonematales. (i) Devonian *Kidstonella*. Scale: bar = 10 μm. ((b), (e) & (g) based on Schopf (1972); (d) on Cloud (1976); (c), (f) & (h) redrawn from Fogg *et al.* (1973); (i) based on Croft & George (1959))

new filaments germinate from the akinete. **Hormogonia** are characteristic of filamentous forms. These are short detached pieces of the trichome which glide out of their sheath and develop separately (Fig. 2.2a).

Classification

Kingdom MONERA
Division CYANOPHYTA
Class CYANOPHYCEAE

Morphological classification of cyanophytes is based on the following characters in general order of importance: level of organisation, mode of reproduction, presence and type of branching, presence of heterocysts.

The most primitive organisation is found in the Order Chroococcales in which the cells are either solitary or loosely aggregated into colonies within a common sheath. In Recent *Synechocystis* (Fig. 2.1a) for example, the cells are solitary, as are the somewhat similar Precambrian unicells of *Archaeosphaeroides* (Fig. 2.1b) from the Fig Tree Chert and *Huroniospora* (Fig. 2.1c) from the Gunflint Chert. *Anacystis* is a living, irregular colony of unicells (Fig. 2.1e) with which fossil forms such as *Renalcis* (Fig. 2.1f, L.Camb.–Carb.) and *Myxococcoides* (Fig. 2.1d, Precamb.) may be compared. The cuboidal colonies of *Eucapsis* (Fig. 2.1g) also have their counterparts in the Precambrian (Fig. 2.1h) as do the clustered cells of *Entophysalis* (Fig. 2.1i).

The Order Nostocales comprises forms with a filamentous trichome. Where branching occurs it is of a false kind and does not give rise to distinct prostrate or upright filaments. Four families in this order are commonly met with: the Oscillatoriaceae, Nostocaceae, Scytonemataceae and Rivulariaceae. The Oscillatoriaceae are relatively primitive and amongst the commonest of cyanophytes. Their trichomes are uniseriate, untapered and sometimes arranged in bundles within a single sheath. The cells are generally cylindrical although the terminal ones are pointed. Fossil *Oscillatoria* from the Precambrian closely resemble Recent forms (Fig. 2.2a, b). The Nostocaceae also have uniserial (or rarely falsely branched), untapered trichomes in which the cells are commonly spherical or ovoid in shape. This and the following groups have enlarged cells with thick walls called **heterocysts**. These may occur along the length (intercalary, e.g. Recent *Wollea*, Fig. 2.2c) or be basal or terminal in position. Their function is debatable, but they are thought to be involved in nitrogen fixation. Heterocyst-like structures in the Precambrian *Gunflintia* (Fig. 2.2d) and the spherical cell-shape of the Precambrian filaments in Figure 2.2e both suggest nostocacean affinities. The Scytonemataceae have untapered filaments with false branching formed at points of breakage along the filament. Heterocysts are present and reproduction is by hormogonia (e.g. Recent *Scytonema*, Fig. 2.2h). The Rivulariaceae have filaments that are tapered into a fine terminal hair and sometimes have false branching (e.g. Recent *Rivularia*, Fig. 2.2f). Similar forms are also known from the upper Precambrian (Fig. 2.2g). Reproduction is by hormogonia and akinetes. Heterocysts are generally basal.

The Order Stigonematales is thought to be the most advanced cyanophyte group, bearing evidence of true branching with divisions into prostrate and upright filaments. Heterocysts are present (intercalary and terminal) and reproduction is largely by hormogonia or akinetes. *Kidstonella* (Fig. 2.2i) is a Devonian example.

Cyanophyte ecology

Cyanophytes are very self-sufficient organisms. They can tolerate extremely low oxygen concentrations and some can live anaerobically. They are the only organisms other than bacteria that can fix their own nitrogen, either with the aid of heterocysts in aerobic conditions, or without in anaerobic conditions. Cyanophytes as a group also have a wide resistance to high and low temperatures, ranging from polar climates to hot thermal springs. Their lack of cell vacuoles gives them great resistance to desiccation and plasmolysis, hence their presence in arid deserts, glacial regions, hypersaline seas and freshwater lakes.

Important limitations appear to be pH and light. They prefer neutral to alkaline environments and never more acid than pH 4·0. The photosynthetic pigment, phycocyanin, is sensitive to blue light and can work under very low light concentrations, cyanophytes having been found living some 300 mm below soil the surface on land and at depths of 1000 m or more in the oceans. They are also very resistant to ultraviolet light.

Of particular interest to geologists is the participation of cyanophytes in the accretion of organosedimentary structures called stromatolites, thrombolites and travertines (or tufa). Another important aspect is their role as rock borers and as generators of $CaCO_3$, both roles contributing to carbonate sedimentation in shallow marine areas.

Algal stromatolites. Stromatolite is the geological term for a fossil algal mat and dates from a time when their predominantly cyanophyte origin was uncertain. Algal mats form today in fresh, brackish and marine waters, generally under conditions unattractive to the grazing and burrowing activities of benthic invertebrates. A cohesive fabric of filaments, coccoid cells and sediment is produced by cyanophytes alone or with the assistance of eucaryotic algae. Commonly the organic and sedimentary constitutents are more or less disposed in thin **laminae**, each about 1 mm thick (Fig. 2.3a). Often the sediment is finer grained than in the surrounding habitats, being selectively trapped and bound by the mucilaginous sheaths. The laminations may represent a daily growth cycle (e.g. subtidal conditions) or tidal influxes of sediment (e.g. intertidal conditions). Littoral and supralittoral stromatolites commonly display desiccation features such as sun cracks, fenestral fabric and evaporite pseudomorphs, which are not seen in the sublittoral forms.

In fossil stromatolites the organic and trapped-sediment layers are usually seen as alternating or intergrading pale and dark **lamellae**. Preservation of filaments is rare but occurs in silicified stromatolites and in fine-grained carbonate ones. More usually, though, only the sheaths or the upward-gliding trails are preserved.

The gross form of a stromatolite is a combination of factors involving the environmental conditions and the species present. This form can be ascertained by serial horizontal and vertical sections. Many workers employ the binomial system of nomenclature in the description of stromatolite form. **Groups** (= genera) are based on general shape (planar, domed, columnar, oncolitic), mode of branching (if any), morphology of the 'wall' (marginal zone) and geometry of the laminae. **Forms**

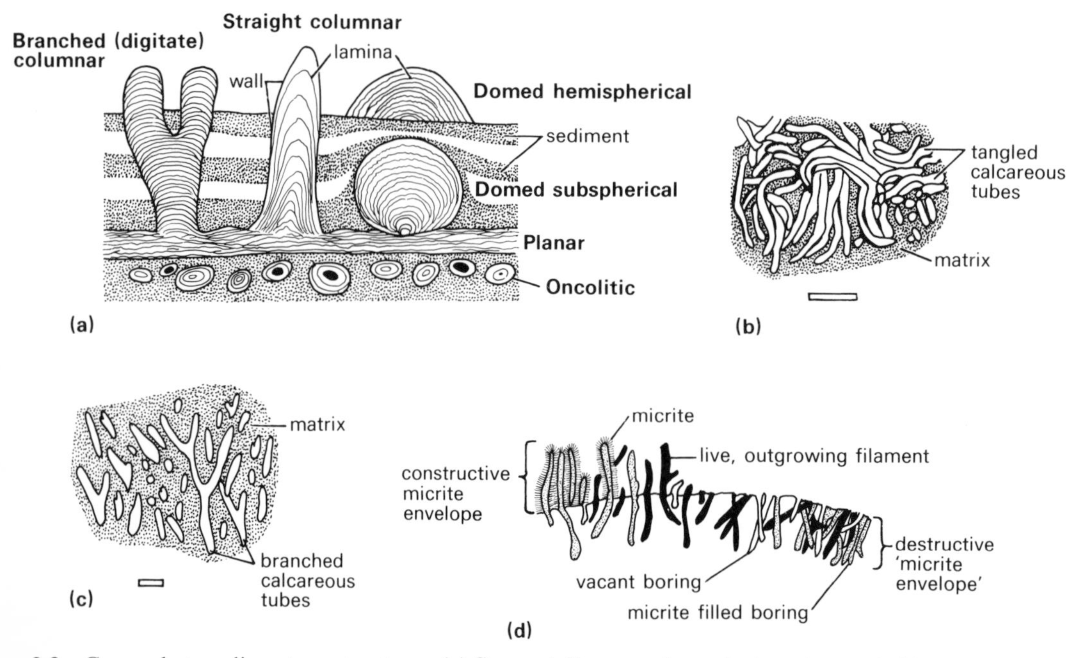

Figure 2.3 Cyanophyte sedimentary structures. (a) Stromatolite types in vertical section, × 1; (b) *Girvanella* tubes in skeletal oncolite; (c) *Ortonella* tubes in skeletal oncolite; (d) section through endolithic cyanophyte borings and skeletal envelopes (diagrammatic). Scale: double bar = 100 μm. ((d) based on Kobluk & Risk 1977)

(=species) are distinguished by microscopic textures and lamina geometry.

Skeletal stromatolites differ from the above **non-skeletal** stromatolites in that the form of the cells or sheaths has been moulded in $CaCO_3$, giving rise to micritic tubes with a micrite or sparry calcite internal filling. These tubes appear to represent calcification of the sheath, such as can occur during life in certain living algae because of CO_2 uptake during photosynthesis. Post mortem calcification is also possible. Such skeletal stromatolites and oncolites are known from both freshwater and marine waters. In *Girvanella* (Fig. 2.3b, L.Camb.–Rec.) the tubes are tangled and unbranched and occur in both oncolites and thrombolites. *Ortonella* (Fig. 2.3c, L.Carb.–Perm.) is an oncolitic form with branched tubes.

Thrombolites differ from stromatolites in lacking the internal laminations, having instead a mottled or clotted microtexture. They are built by coccoid cyanophytes (e.g. *Renalcis*) or by filamentous forms with wispy, tufted, branched or tangled growth rather than vertical filamentous growth. Thrombolites are found in sublittoral, often calcareous, facies in association with 'reef'-dwelling invertebrates.

Travertine develops in $CaCO_3$-supersaturated, usually fresh, waters in which coccoid and filamentous cyanophytes have become encrusted by physico-chemical precipitation of $CaCO_3$, again forming hollow tubes. In this case, however, the crystals greatly exceed the diameter of the original organic sheath. The moulds left by such fossilisation are of little taxonomic value.

Endolithic cyanophytes. A variety of marine cyanophytes bore into the surface of hard calcareous substrates such as shells and limestone by chemical dissolution (Fig. 2.3d). This **endolithic** boring is for protection rather than for food. Under conditions of $CaCO_3$ supersaturation the vacated borings are filled with micritic carbonate and the substrate thereby acquires an outer **micrite envelope**. Eventually, this boring may lead to destruction of the substrate and the formation of lime mud. If the filaments extend outwards from their borings and become calcified after death, however, a **skeletal envelope** may form (Fig. 2.3d). This constructive process requires relatively quiet conditions (see Kobluk & Risk 1977).

Cyanophyte borings are not easily distinguished from those of other algae or of fungi, but they are generally narrower than the former and broader than the latter (i.e. about 4–25 μm wide). The depth of water in which such borings may be found varies with water clarity and latitude, but is mostly shallower than 75 m.

Planktonic cyanophytes. Although usually benthic, certain coccoid and filamentous forms are known to lead a planktonic existence in tropical waters. Buoyancy is achieved either by the development of pseudovacuoles or by adherence to gas bubbles. Some filamentous forms float in bundles of up to 25 trichomes, forming mats at the surface of the ocean which can extend for many kilometres.

General history of cyanophytes and stromatolites

The history of cyanophytes is virtually the history of life for its first 3000 Ma. Knowledge of fossil forms is based on studies of preserved cells or sheaths in cherts and limestones and of stromatolites (Fig. 2.4). Nonetheless, the affinities of many Precambrian specimens still remain in doubt and some are even of dubious organic origin (see Golubić & Barghoorn, pp. 1–14 *in* Walter 1976, and Cloud 1976).

Coccoid cyanophytes are considered to be the most primitive type and examples of this group (e.g. *Archaeosphaeroides*, Fig. 2.1b) are amongst the oldest fossils known, being found in the carbonaceous Onverwacht and Fig Tree Cherts of South Africa (*c.* 3200–3100 Ma BP). Chroococcales are also known from the Transvaal stromatolites (*c.* 2200 Ma BP) and many younger horizons.

Although the first stromatolites are known from the Pongola System of South Africa (*c.* 3100 Ma BP), definite evidence for the participation of cyanophytes is not known for certain in rocks older than the Transvaal examples cited above, which contain nostocacean filaments. Nostocaceae and Oscillatoriaceae are also present in the Gunflint Cherts of the Canadian Shield (*c.* 2000–1600 Ma BP, e.g. Fig. 2.2d) and the Bitter Springs Cherts of Australia (*c.* 800 Ma BP, e.g. Fig. 2.2b, e). Branched filamentous forms appeared in the late Precambrian and flourished in the Phanerozoic (e.g. Figs 2.2g, 2.2i).

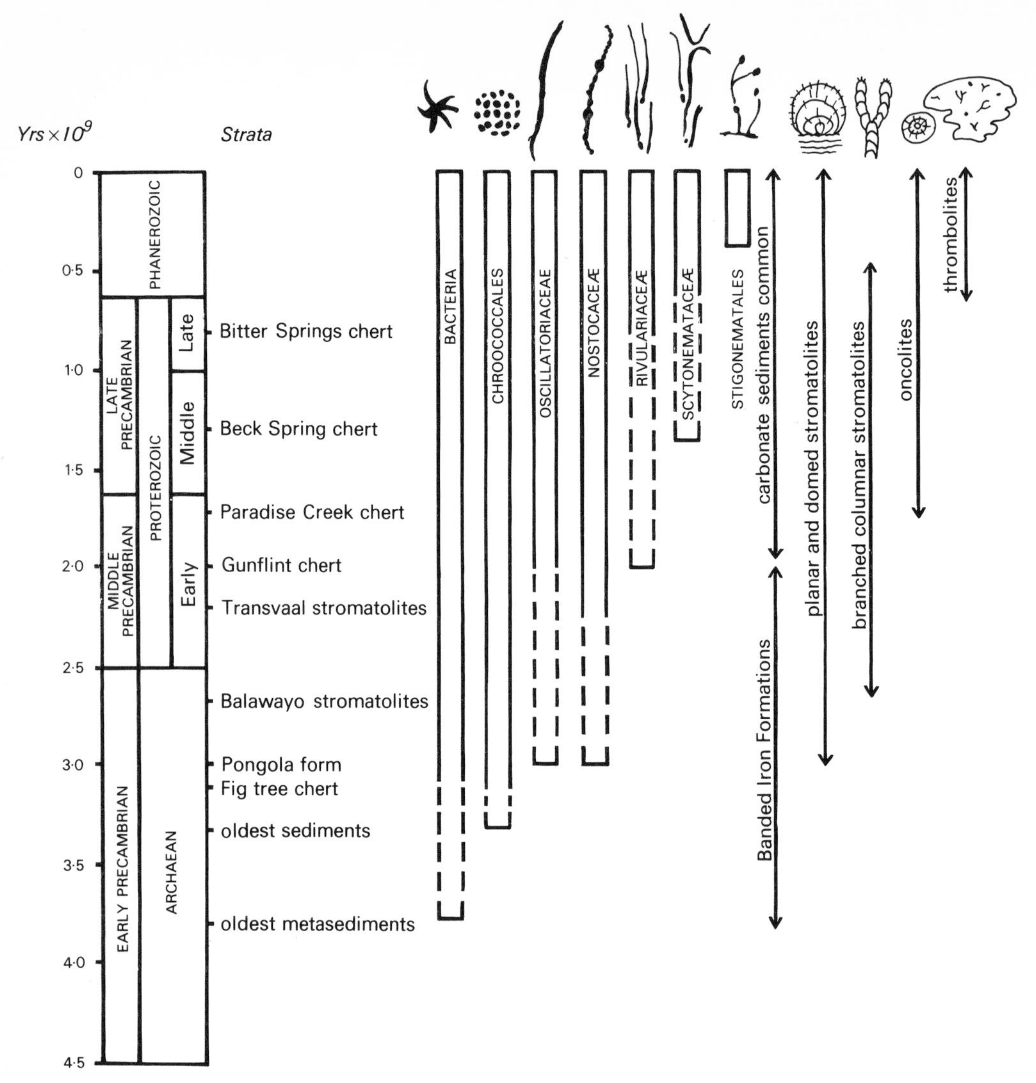

Figure 2.4 Distribution of the main procaryotic groups and stromatolites through geological time.

Stromatolites first became widespread about 2200 Ma ago when extensive carbonate platforms began to develop. Studies of Precambrian stromatolites suggest that changes of gross morphology through time may reflect the organic evolution of algal mat communities (Fig. 2.4). The acme of stromatolite diversity ranges from the upper Precambrian to the middle Ordovician (*c.* 1650–500 Ma BP), coinciding with a major period of shallow-water carbonate deposition and ending with exclusion from many habitats by the burrowing and grazing activities of benthic invertebrates. During this period, forms with erect, tall (> 150 mm) and branched columns were common and variable enough to allow their use for biostratigraphic subdivision of upper Precambrian (Riphean) rocks. The Vendian Stage (*c.* 700–600 Ma BP) marked the appearance of stromatolites with coarse vermiform microstructures, built perhaps by larger, eucaryotic filaments. The succeeding early Cambrian (*c.* 570 Ma BP) saw not only the radiation of shelled invertebrates but also of skeletal oncolites (e.g. *Girvanella*, Fig. 2.3b), thrombolites with *Girvanella*, *Renalcis* (Fig. 2.1f), *Epiphyton* (Fig. 10.4d) and endolithic algae.

The success of Precambrian cyanophytes may be attributed to their tolerance of low oxygen levels (which rose gradually due to continued release of photosynthetic oxygen), to their nitrogen-fixing abilities and to the lack of vigorous competition from plant and animal eucaryotes. Reviews of the fossil record of these microorganisms have been brought together by Schopf (1970, 1971) and Cloud (1976).

Studies of cyanophytes and stromatolites

Fossil cyanophytes are usually studied in one of three ways: as microfossils in stromatolitic cherts studied from thin sections; as microfossils in shales studied in isolated preparations after digestion of the rock, and sometimes referred to as 'acritarchs' (*q.v.*); as organo-sedimentary structures (oncolites, stromatolites, travertines) perhaps with sheaths, tubes or other traces of organic origin. Each method has its own advantages and disadvantages (see Schopf 1977).

Microfossils in stromatolitic cherts have been studied mainly from Precambrian strata and provide the bulk of our knowledge about life forms during that time (i.e. about 80% of the history of life on Earth). As more information becomes available it appears that the observed increase in the maximum diameter of coccoid and filamentous cells through time may provide a basis for Precambrian biostratigraphy (Schopf 1977).

Some assemblages appear to be time restricted and allow correlation between continents (e.g. Walter, Goode & Hall 1976). Where such microfossils closely resemble extant forms they may be used to reconstruct the conditions of deposition, such as the well illuminated freshwater facies inferred by Fairchild *et al.* (1973).

The gross form of stromatolites is now known to change through late Precambrian times, presumably because of biological evolution. This feature has been used for intra- and intercontinental correlation in the USSR and Australia (see Raaben 1969, Walter 1972, Preiss 1977 and various authors *in* Walter 1976, pp. 337–80). Pannella (1972) has studied the diurnal, tidal and seasonal laminae of such Precambrian stromatolites and inferred from them a gradual reduction in the number of days per Earth year with time. The ecological value of stromatolites as indices for supralittoral and littoral environments is generally overstressed, however. In each case the assumption needs to be tested against additional palaeontological and sedimentary criteria (see Ginsburg 1975, Monty, pp. 15–36 *in* Flügel 1977). The role of cyanophytes in carbonate genesis has been outlined by Golubić (pp. 434–72 *in* Carr & Whitton 1973).

Further reading

Aspects of the biology, ecology and classification of Recent cyanophytes are dealt with by Carr & Whitton (1973) and Fogg *et al.* (1973) whilst the fossil record and geological applications of stromatolites and cyanophytes are covered in Walter (1976), Flügel (1977) and Wray (1977).

Hints for collection and study

Recent cyanophytes can be collected from algal mats in almost any pond, lake or marsh. To prepare a temporary mount place a small amount of algal rich material on a glass slide with a pipette, cover with a drop of distilled water and view at over 400 × magnification with well condensed transmitted light.

Fossil cyanophytes can be found in petrological thin sections of carbonaceous cherts associated with oolites, oncolites and stromatolites. Their sheaths can also be found in thin sections or peels of freshwater and marine carbonates especially stromatolitic and oncolitic ones (see method N, in Appendix).

3 Division Schizomycophyta – Bacteria

The bacteria (Division Schizomycophyta, Class Schizomycetes) have extremely small cells, generally less than 1 μm in diameter. Bacterial cells lack a nucleus and are therefore placed with the cyanophytes in the Kingdom Monera. They may be single or colonial, the latter enclosed within a mucilaginous sheath called a **capsule**. Many bacterial cells bear a whip-like thread (flagellum) and some contain chlorophyll pigments for photosynthesis.

Although bacteria are rarely preserved in the fossil record as cells, the available evidence indicates that they were amongst the Earth's oldest inhabitants, occurring in Precambrian rocks about 3100 Ma old. Bacteria deserve further study as microfossils because of their involvement in the formation of stromatolites, iron and manganese ores, carbonate sediments and concretions, phosphorites, sulphide and sulphate minerals and native sulphur.

The living bacterium

The bacteria are an abundant and diverse group, closely resembling cyanophytes in most respects. However, their cells are usually smaller, (about 0·25–1·00 μm wide and 1–10 μm long), and of spherical, rod or corkscrew shape, collectively referred to as **cocci**, **bacilli** and **spirilla** respectively (Fig. 3.1a). These cells may be solitary or arranged in filamentous trichomes with or without branching of false or true type. Most of the bacilli and all the spirilla possess flagella (one or more per cell), but they are very thin and are rarely, if ever, preserved in the fossil state.

Most bacteria are aerobic, but many anaerobic forms are known. Bacteria may feed either on preformed organic matter (**heterotrophy**) or photosynthesise organic material from inorganic minerals (**autotrophy**). Unlike the cyanophyte case, however, no free oxygen is liberated by bacterial photosynthesis and many of the autotrophic forms are in fact anaerobic. As a group, the bacteria are relatively unaffected by salinity, and have a temperature tolerance of about 4–90°C. Nonetheless they dislike a pH outside of the range 6·0–9·0 and will die in bright sunlight. Their habitats range from the deep sea (planktonic and benthic) to terrestrial (including subterranean) and aerial.

Some geologically significant bacteria

The taxonomy of Recent bacteria is largely based on staining tests and aspects of biochemistry beyond the scope of palaeontology. A morphological classification would prove misleading because similar morphotypes occur in several different orders of bacteria. They have seldom been reported as microfossils, perhaps because of their small size and the difficulty of distinguishing them from fossil cyanophytes or fungi, or even from recent contamination, inorganic structures and artifacts formed during preparation of material. They have, however, been reported from a wide range of lithologies including limestones, cherts, phosphorites, iron and manganese ores (including deep-sea manganese nodules, pyrite nodules and banded iron ores), tonsteins, bauxites, oil shales, coal seams, plant tissues, coprolites and fossil animal remains. In cherts and phosphorites the cell wall may be preserved but more usually the wall, sheath or the whole structure has been replaced by minerals.

The Order Pseudomonadales contains most of the autotrophic bacteria, including the sulphur bacteria which liberate sulphur and sulphates from H_2S. Also included are the stalked bacteria (family Caulobacteraceae) whose fine stalks become encrusted with ferric hydroxide salts from oxidation

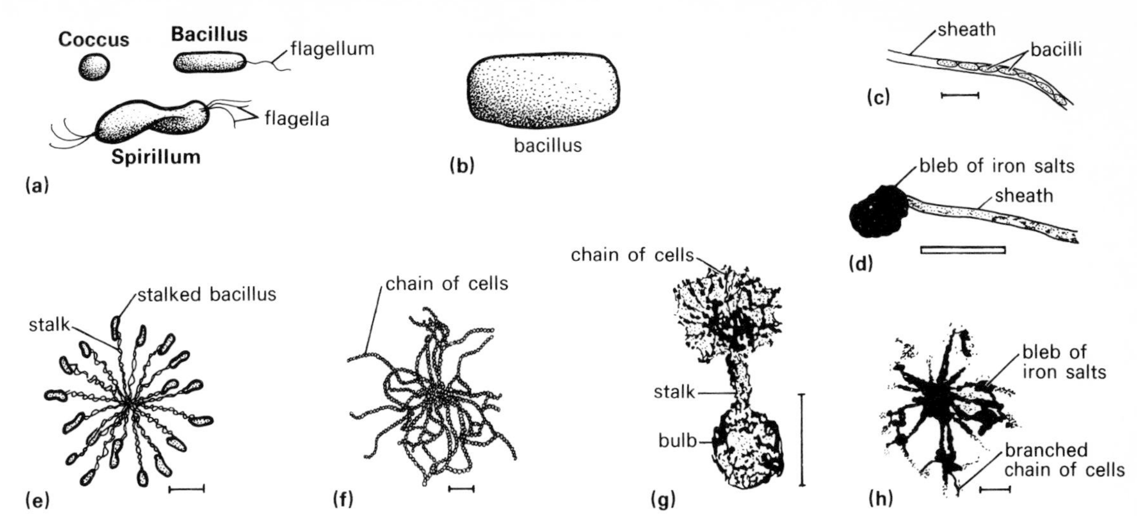

Figure 3.1 Bacteria. (a) Basic shapes of bacterial cells (schematic); (b) Precambrian *Eobacterium* (length 0·6 μm); (c) Recent 'sheathed' iron bacterium *Sphaerotilus*; (d) Precambrian *Sphaerotilus*-like form; (e) Recent 'stalked' iron bacterium *Caulobacter*; (f) Recent 'budding' bacterium *Metallogenium*; (g) Precambrian *Kakabekia*; (h) Precambrian *Eoastrion*. Scale: single bar = 10 μm; double bar = 100 μm. ((b) based on Bargoorn & Schopf 1966; (d) on Karkhanis 1976; (f) & (h) on Cloud 1976 and (g) on Barghoorn & Tyler 1965)

of dissolved ferrous iron (e.g. Recent *Caulobacter*, Fig. 3.1e). These organisms hence assist in the formation of bog iron ores. Carboniferous iron pyrites nodules have also yielded the somewhat similar genus *Gallionella* (Schopf *et al.* 1965).

The Order Chlamybacteriales, or sheathed bacteria, are also involved in iron ore formation. These have a trichome organisation with a sheath that can become encrusted with ferric or manganese oxides much as in the stalked bacteria (e.g. Recent *Sphaerotilus*, Fig. 3.1c). Similar bacteria may have participated in the formation of the world's most extensive iron ores, the early and middle Precambrian banded iron formations (Fig. 3.1d; see Karkhanis 1976) as well as in the formation of iron pyrite (Schopf *et al.* 1965).

The budding bacteria (Order Hyphomicrobiales) reproduce by budding; that is, threads grow out either from cells or other threads and themselves produce new cells. These may also be joined by threads, sometimes in aggregates connected to a common surface by stalks. One such Recent genus, *Metallogenium* (Fig. 3.1f), grows heterotrophically in low-oxygen environments, depositing crusts of manganese oxide around the filaments. Almost identical fossil bacteria, *Eoastrion* and *Kakabekia* occur in the Gunflint Chert flora, in association with banded iron formations (Fig. 3.1g, h; see Cloud 1976).

Examples of possible fossil 'true bacteria' (Order Eubacteriales) are reported from the 3100 Ma old Fig Tree Chert (*Eobacterium*, Fig. 3.1b). These are tiny bacillus-like structures discovered by electron microscopy of polished chert surfaces (Barghoorn & Schopf 1966). Bacilli may also be involved in the formation of lime mud (Maurin & Nöel, pp. 136–42 *in* Flügel 1977) and have been widely reported by Renault from various Phanerozoic rocks (see the review by Moore, pp. 170–3 *in* Harland *et al.* 1967).

The Beggiatoales are an order resembling unpigmented filamentous cyanophytes and thrive in H_2S-rich habitats. Hence, for example, the discovery of *Beggiatoa*-like remains in carboniferous iron pyrites (Schopf *et al.* 1965). Flexibacteria are an even more cyanophyte-like group with photosynthetic pigments and they dwell alongside cyanophytes in hot springs to be preserved, eventually, in sinters and stromatolites (Walter *et al.* 1972). Apart from the biological distinction of their not releasing free oxygen, it would be difficult to differentiate flexibacteria and cyanophytes except on the tenuous basis of cell diameter, the former rarely exceeding 2 μm in diameter and the latter usually exceeding this.

Unfortunately the fossil record is as yet too

incomplete to comment on the history or applications of the group. It seems probable that bacteria are older than cyanophytes, their ability to live in anaerobic conditions being a legacy from early and middle Precambrian times. The parasitic and saprozoic bacteria may, in part, be Phanerozoic developments, and their evolution could have had important consequences for ecosystem evolution in general.

The role of bacteria in the genesis of various kinds of ores has been reviewed in the book by Kuznetsov *et al.* (1963). Their ecological importance in the marine realm can be gathered from the beautiful portraits in Sieburth (1975).

Hints for collection and study

Living bacteria are easily cultured. Take a sample of water from decaying pond vegetation and place a drop on a glass slide with water and a cover slip. Allow the slide to dry out in a warm, dark place. When viewed with transmitted light at over 400 × magnification the slide will often be seen to contain clusters of minute bacilli. Fossil bacteria may well be encountered in thin sections of sedimentary ironstones, phosphatised faecal pellets, bauxites and evaporites, but the observer must be wary of the likelihood of more recent 'contamination'.

Part 2

Eucaryotes – success through variety

Origins of the eucaryotes

Eucaryotes are those organisms in which the cells have a proper membrane-bound nucleus, vacuoles and organelles. Reproduction may be either sexual or asexual, involving strictly controlled meiotic or mitotic cell divisions respectively. These features represent an evolutionary advance over the procaryote condition, but the nature

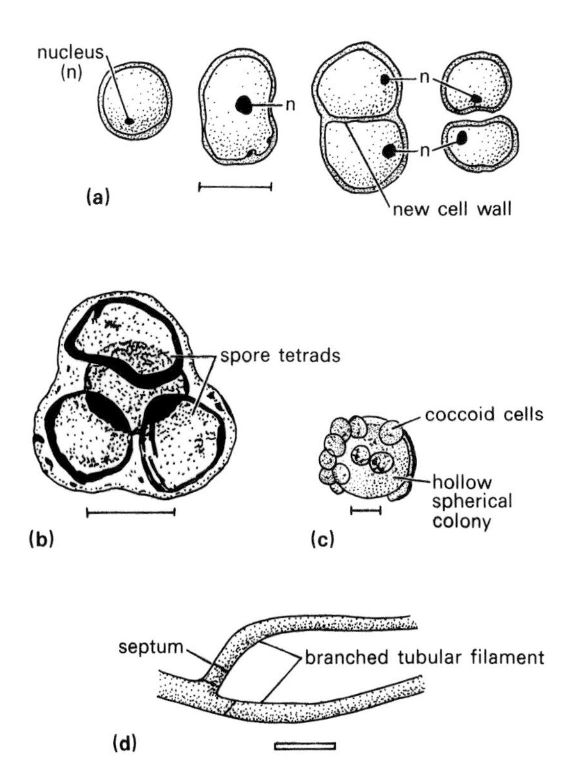

Early fossil 'eucaryotes'. (a) Sequence of Precambrian fossils indicating mitosis in *Glenobotrydion*; (b) meiotic spore tetrad of Precambrian *Eotetrahedrion*; (c) Precambrian *Eosphaera*; (d) Precambrian siphonalean-like filament. Scale: single bar = 10 μm; double bar = 100 μm. ((a), (b) & (d) based on Schopf 1972, and (c) on Cloud 1976)

of the changes that led to them is as yet uncertain. One theory suggests that eucaryotes arose from symbiotic association between formerly free-living procaryotes, each kind giving rise to mitochondria, to chromoplasts or to parts of the flagella (Margulis 1970).

The fossil record suggests that eucaryotes arose between 2000 and 1500 Ma ago, one ancient contender being *Eosphaera* from the Gunflint Chert (Fig. **c**). This possibly represents a *Volvox*-like, planktonic green algal colony (compare with Fig. 9.1c; see Kazmierczak 1976). Relatively large cells with diameters of 40 μm or more also become widespread in stromatolitic cherts and shales after about 1500 Ma BP (Schopf 1977). Cellular differentiation, root-like structures and the presence of nucleus-like spots are further indications of eucaryotic organisation in microfossils of late Precambrian age. There is even evidence for vegetative reproduction (i.e. mitosis, (Fig. **a**)) and sexual reproduction (meiotic spore tetrads, (Fig. **b**)) in the Bitter Springs Chert (about 800 Ma BP, Schopf 1971). The existence of branched cells like those of siphonalean green algae (Fig. **d**) and of macroscopic, ribbon-like algae about 1300 Ma ago (Cloud 1976, Walter, Oehler *et al.* 1976), suggests that sexual reproduction had evolved by this time. The microfossil evidence for the earliest eucaryotes is discussed more fully by Schopf (1971) and Cloud (1976).

The sexual eucaryote organisation has substantial selective advantages over the asexual procaryote way of life, particularly with regard to the exchange of genes and other forms of genetic recombination. For example, 10 genetic mutations in an asexual population can result in only 11 genotypes, that is the original type and those of the 10 mutants; the same number of mutations in a primitive diploid sexual population could be combined to produce up to 59 049 distinct genotypes (Schopf *et al.* 1973). Hence, in theory, the evolution of eucaryotic sexuality must have resulted in a prodigiously increased genetic variety of organisms, expressed by an increased rate of biological evolution in the fossil record.

4 Division Pyrrhophyta – Dinoflagellates and ebridians

The Pyrrhophyta, better known as dinoflagellates (meaning whirling whips), are single-celled organisms generally between 20 and 150 μm long, with both 'plant' and 'animal' characteristics. They are usually considered as 'plants', however, because of the presence of cellulose in the cell wall and chlorophyll pigments in the protoplasm. It is the carotenoid pigments dinoxanthin and peridinin, though, which give to these organisms the flame-like colours from which they derive their formal, botanical name of Pyrrhophyta (meaning fire plants). The majority of dinoflagellates are equipped with one whip-like and one ribbon-like flagellum for propulsion and have a prominent nucleus and a sculptured cell wall.

Both heterotrophic and autotrophic modes of nutrition occur, although the latter predominate and have formed an important contingent of oceanic phytoplankton since at least mid-Mesozoic times. Although these motile cells are abundant and wide ranging, it is the resistant resting cyst which leaves a fossil record. Dinoflagellate cysts (or **dinocysts**) have proved to be valuable tools for biostratigraphy and although they have contributed less to ecological and evolutionary palaeontology, these fields promise to become important.

The living dinoflagellate

Dinoflagellates of micropalaeontological interest exhibit two biological states with distinct morphology: a planktonic motile stage and a planktonic or benthic cyst stage. Only the cysts are preserved as fossils, but a knowledge of both stages is necessary for a proper understanding of the group.

The motile stage. Dinoflagellate cells range in size from 5–2000 μm. The cell wall may be either flexible and unarmoured or rigid and armoured. In the former case it comprises a proteinaceous envelope (**pellicle**) containing flattened cavities near the surface (Fig. 4.1a). In the armoured examples these cavities are occupied by plates of fibrous cellulose to form a closely fitting **theca** (Fig. 4.1b). The mode of arrangement of these plates, known as **tabulation**, is constant within each species.

Within the cell is a single, large nucleus and several fluid-filled vessels (**pulsules**) connected to the exterior via canals. Photosynthetic pigments, where present, are contained in round **chromoplasts** at the cell margins. Light sensory eye spots may also be present.

The two flagella arise either from pores at the anterior end or from the ventral surface (Figs 4.1c and 4.1d). The cell surface is generally traversed by two furrows, each of which bears a flagellum. One lies in a transverse furrow called the **cingulum**, which occupies a more or less equatorial position. The other lies in a longitudinal furrow called the **sulcus** and may trail behind like a tail. That half of the cell anterior to the cingulum is called the **epitheca** and that posterior to it is termed the **hypotheca** (Fig. 4.1g). The side bearing the sulcus is **ventral** (Fig. 4.1f), whilst the opposite side is **dorsal** (Fig. 4.1g). Many cells are dorso-ventrally compressed so that these two views are the ones usually illustrated.

The sulcus extends in a posterior direction and may terminate in a depression flanked by two **antapical horns**. The other, **apical**, end is often rounded, pointed or produced into an **apical horn**. Overall cell shape can be very varied even within one genus, but includes spherical, subspherical, ovoid, biconical, fusiform, rod-shaped, rectangular, polygonal, discoidal and 'peridinioid'.

Tabulation refers to the arrangement of plates in the armoured motile cells of the Order Peridiniales. In these, five plate series are found to encircle each

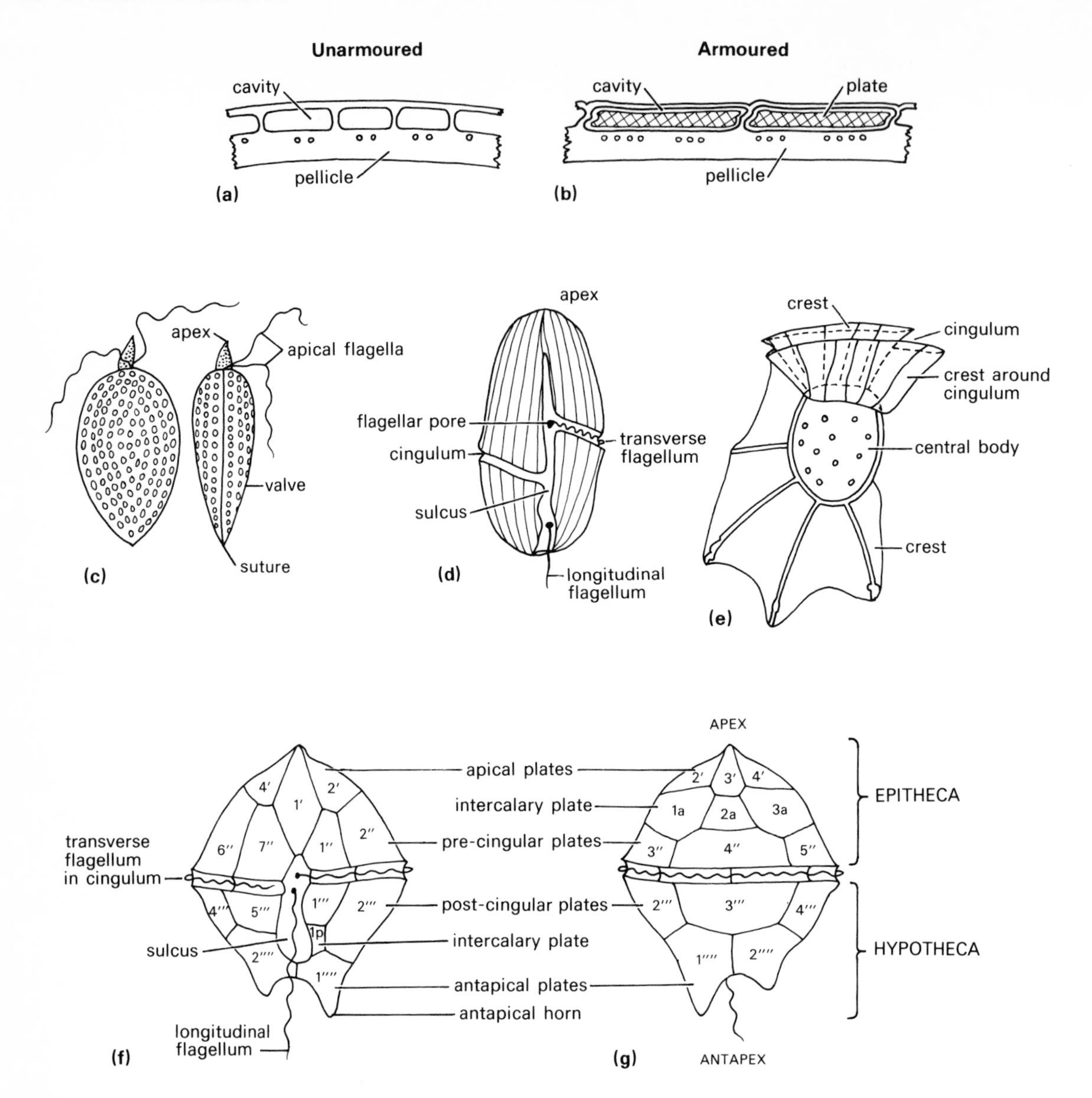

Figure 4.1 Dinoflagellate motile stages. (a) Schematic section through the wall of an unarmoured dinoflagellate; (b) schematic section through the wall of an armoured dinoflagellate; (c) Recent desmophycean *Prorocentrum*, about ×350; (d) Recent gymnodinialean *Gymnodinium*, about ×320; (e) Recent dinophysialean *Ornithocercus*, about ×275; (f) & (g) tabulation of a hypothetical peridinialean motile stage – ((f) ventral side; (g) dorsal side). ((a) & (b) modified from Dodge & Crawford 1970; (c) from Chapman & Chapman 1973; (d) from Kofoid & Swezy 1921; (e) after Barnes 1968 from Chatton; (f) & (g) from Evitt *in* Tschudy & Scott 1969)

cell, each plate being numbered for reference from the ventral area in a counterclockwise direction (Figs 4.1f and g). Around the epitheca occur the **apical** and **precingular** series. In the cingulum lie the **cingular** series whilst the **postcingular** and **antapical** series occur on the hypotheca. Additional **intercalary** plates may also develop at sites between the series, and the sulcus bears small **sulcal** plates that can be of taxonomic value.

The functional significance of cell shape and tabulation is little understood. As the planktonic forms maintain their position in the water by active

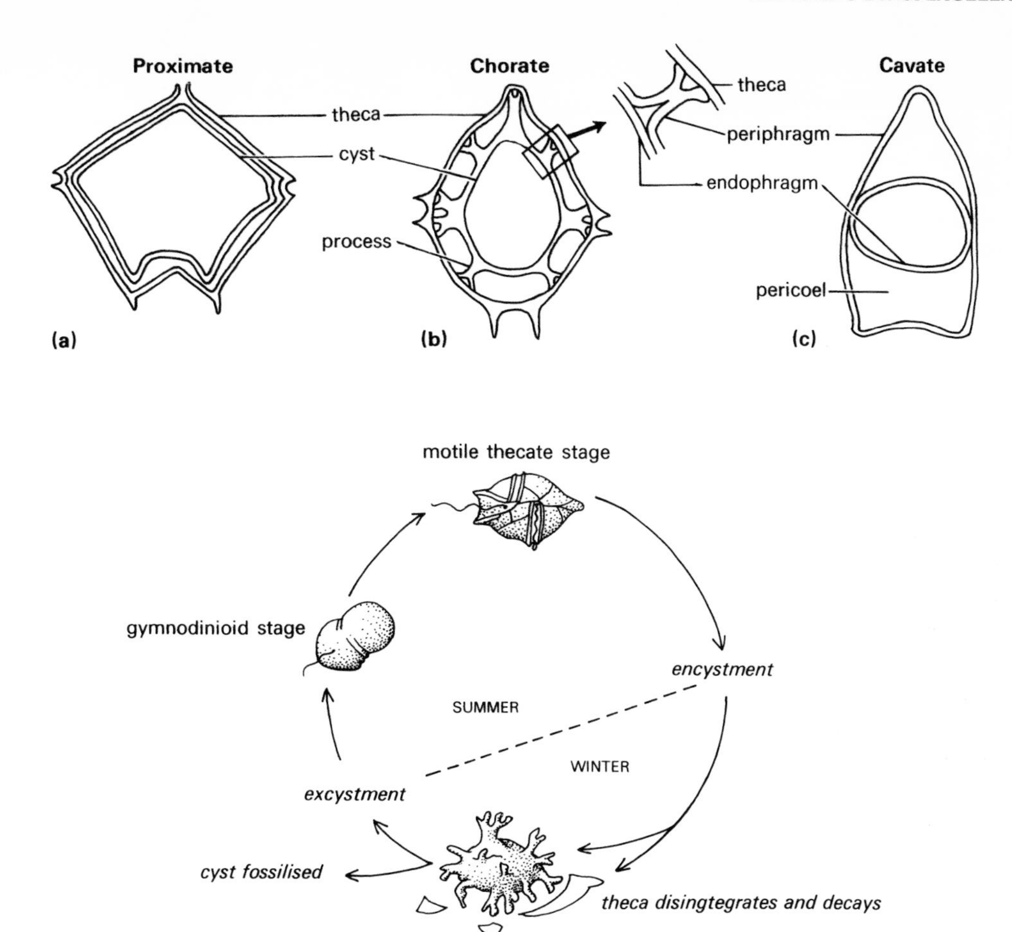

Figure 4.2 Dinoflagellate cysts. (a) Proximate cyst of *Peridinium* (axial section, about ×250); (b) chorate cyst of *Gonyaulax*, with detail of wall (axial section about ×250); (c) cavate cyst of *Deflandrea* (axial section, about ×220); (d) schematic life history of *Gonyaulax* sp., about ×600. ((a) & (b) based on Evitt *in* Tschudy & Scott 1969; (d) on Wall & Dale 1968a)

flagellar propulsion rather than by passive stabilisation, the cells tend to be streamlined. Nevertheless, the long horns of certain genera may serve to retard sinking.

The cyst stage. Although most, if not all, fossil dinoflagellates are cysts, only a few living genera are known to **encyst**, either in response to adverse environmental conditions or following sexual reproduction. The cyst is formed within the formerly motile cell and contains the same organelles. The cyst wall (**phragma**) is two-layered and consists of an outer **periphragm** and an inner **endophragm** (Fig. 4.2a–c). It is built of organic materials resistant to bacterial decay. Three basic kinds of cyst are recognised, termed proximate, chorate and cavate, although intergradations between these exist.

Proximate cysts develop with the phragma wall in contact with the wall of the enclosing motile cell (Fig. 4.2a). The tabulation, cingulum and sulcus are all reflected in the surface sculpture of proximate cysts.

Chorate cysts develop further within the original cell and are linked to it by spines (**processes**, Fig. 4.2b). These processes correspond in position with the plates or sutures of the thecal tabulation, but chorate cysts usually exhibit no traces of a reflected cingulum or sulcus.

Cavate cysts are a type in which the two layers of the cyst wall are partially separated, usually at the poles (Fig. 4.2c). The cavities thus formed (**pericoels**) may promote buoyancy in the cyst. Traces of the tabulation, the cingulum and the sulcus may also be seen, so it is doubtful whether such cysts formed far below the motile cell wall.

The surface of these cysts may be smooth or bear fine granules (granulate), irregular ridges (reticulate), short spines (spinose), indentations (punctate), raised crests (septate) or processes and horns. Where a reflected cingulum is present, that portion apical to it is called the **epitract** and the antapical portion is called the **hypotract** (Fig. 4.3f).

The cyst nature of such structures is demonstrated by the presence of an escape hole, called an **archaeopyle**. This is formed by the removal of one or several plates (thereby comprising an **operculum**) from either the apical series, the precingular series, an anterior intercalary plate or a combination of these. The form and position of the archaeopyle is constant within a species.

Further investigation is needed into the functional significance of cyst morphology. The processes of chorate cysts and the pericoels of cavate cysts may both be mechanisms to minimise the downward sinking of oceanic species. If cysts of such forms were to sink far below the photic zone before excystment, their chances of survival would appear to be slender.

Dinoflagellate life history

Sexual reproduction is known in very few Recent dinoflagellates. Asexual reproduction predominates and involves a division of the cell into two halves (by binary fission) without the aid of the guiding nuclear structures found in other eucaryote cells. Hence it may be that dinoflagellates represent an evolutionary condition intermediate between procaryotes and eucaryotes.

The cyst stage of the life cycle is known so far from only a few genera (e.g. *Gonyaulax*, Fig. 4.2d). Cysts form in the autumn with lowered temperatures, remaining dormant on the sea floor through the winter. During this period the surrounding thecal plates may drop away and begin to decay. With the amelioration of conditions in spring, the motile stage **excysts** through the archaeopyle to leave a resistant phragma for the fossil record. Before developing any armour, however, the new dinoflagellate must pass through a naked **gymnodinioid** stage (Fig. 4.2d).

Dinoflagellate ecology

Dinoflagellates currently form a major part of the ocean plankton, especially the armoured and autotrophic forms, and they play a prominent rôle in the food chains of the marine realm. The autotrophic forms thrive in areas of upwelling currents that are rich in nutrients such as nitrates and phosphates, whilst they are rarely found alive below 50 m depth because of their need for light. Flagellar locomotion is employed in bringing them to the surface at night and withdrawing them to greater depths in the day because they must avoid harmful ultra-violet light.

As a whole, the group has a wide temperature tolerance (1–35°C) with an optimum for most species of 18–25°C. Many dinoflagellates have geographic distributions which are broadly parallel to oceanic temperature zones (i.e. latitudinal) and hence may be used as indicators of climate oscillations through the Quaternary Era. Certain genera, such as *Gymnodinium* and *Peridinium*, are found in both fresh and salt water although the majority of species are marine and sensitive to changes in water mass, including salinity changes.

Sudden blooms of dinoflagellates, called 'red tides', may occur under optimal conditions but the build up of toxic excretions eventually kills off great numbers of fish and invertebrates. It is possible that some of the dinoflagellates avoid this fate by encystment.

Planktonic forms with a predatory or parasitic mode of life are usually unarmoured and belong mostly to the Order Gymnodiniales. Other orders of limited palaeontological interest contain immobile, benthic, colonial forms and the zooxanthellae which live symbiotically in the tissues of reef-building corals and larger foraminifera.

There are several inherent problems in interpreting the palaeoecology of fossil dinoflagellates. First, those of pre-Quaternary age are not easy to relate to taxa of known habit, although lineages can be traced back in a few cases. Secondly, some of the dinoflagellate flora may not encyst and therefore leave no fossil record. Thirdly, those cysts which are formed may sink and drift to be preserved at depths and conditions beyond the tolerance of the species. Recent studies, however, suggest that modern cyst assemblages from the sea floor in fact bear a strong resemblance to the main overlying water-mass distributions, so that the transport of cysts is probably not very great (see Wall *et al.* 1977). At present, dinoflagellate cysts are most abundant in sediments from the continental rise and slope, with from 1000 to 3000 cysts per gram. There is also a tendency for specific diversity to increase with distance from shore and to be greatest in low latitude waters, as with most other marine plankton. The distribution and ecology of Recent and Quaternary dinoflagellates are reviewed more fully by Williams (pp. 91–5, 231–43 *in* Funnell & Riedel 1971 and pp. 1288–92 *in* Ramsay 1977), by Wall (1971 and pp. 399–405 *in* Funnell & Riedel 1971) and in Wall *et al.* (1977). Palaeoecology is reviewed by Williams (pp. 1292–302 *in* Ramsay 1977). The factors affecting the preservation of dinoflagellate assemblages have been considered by Dale (1976).

Classification

Kingdom PROTISTA
Division PYRRHOPHYTA

Although on the boundary between 'plant' and 'animal' organisation, dinoflagellates are considered to be 'plants' by some because they contain cellulose and chlorophylls. Subdivision of Recent forms takes account of the following in order of importance: position of flagellar insertion, predominant habit (e.g. mobile and flagellate, mobile amoeboid, immobile solitary or immobile colonial), presence of armour, tabulation, shape and sculpture of the motile cell. Fossil dinoflagellate cysts are classified according to cyst type, reflected tabulation, archaeopyle position, general shape and sculpture (see Sarjeant 1974 and Williams pp. 1239–46 *in* Ramsay 1977).

At one time many dinoflagellate cysts were classed with the problematic 'hystrichospheres'. Evitt (1961, 1963) demonstrated that some of these were true dinoflagellate cysts, designating the remaining problematica to the group Acritarcha (*q.v.*).

Class Desmophyceae. These are thought to be the most primitive dinoflagellates. They have two flagella of equal length inserted at the anterior end of the motile cell, which is unarmoured. Cysts are unknown but may be included amongst the acritarchs. *Prorocentrum* is a living genus involved in red tide blooms. The theca is divided into two equal valves by a longitudinal suture (Fig. 4.1c).

Class Dinophyceae. Most dinoflagellates belong to this class. These have two flagella of unequal length arising from the ventral surface. There are three important orders: the Gymnodiniales, Peridiniales and Dinophysiales.

ORDER GYMNODINIALES. The Gymnodiniales are predatory and parasitic forms lacking armour but having a flexible pellicle. Their cells are commonly spherical, traversed by a deep equatorial cingulum and a shallow longitudinal sulcus. Although resting cysts are known, the lack of tabulation-related features makes it difficult to infer biological affinities. Such uncertain forms may therefore become classified as acritarchs.

The Recent genus *Gymnodinium* (Fig. 4.1d) has a motile cell with an equatorial cingulum. The common late Cretaceous genus *Dinogymnium* (Fig. 4.3a) is probably a proximate cyst and has longitudinal folds, a cingulum and an apical archaeopyle.

ORDER PERIDINIALES. The Order Peridiniales comprises forms with an armoured motile stage. In these the cingulum is equatorial with a slight spiral offset and there is a longitudinal sulcus. The plates are arranged into apical, precingular, cingular, postcingular and antapical series, with additional intercalary and sulcal plates.

Inevitably, classification of the Peridiniales has proceeded along two independent lines, one for the fossil cysts (the bulk of which belong here) and one for the living motile cells. In principle it would be best to combine this divergent information into a single, natural, classification. Unfortunately, however, cyst genera and motile genera do not always correspond, evolution having proceeded at different rates for these different stages of the life cycle.

The motile cell of Recent *Peridinium* (Fig. 4.3b) is laterally compressed and almost bilaterally symmetrical. The cyst stage is proximate with a peridinioid shape, clearly reflected tabulation and

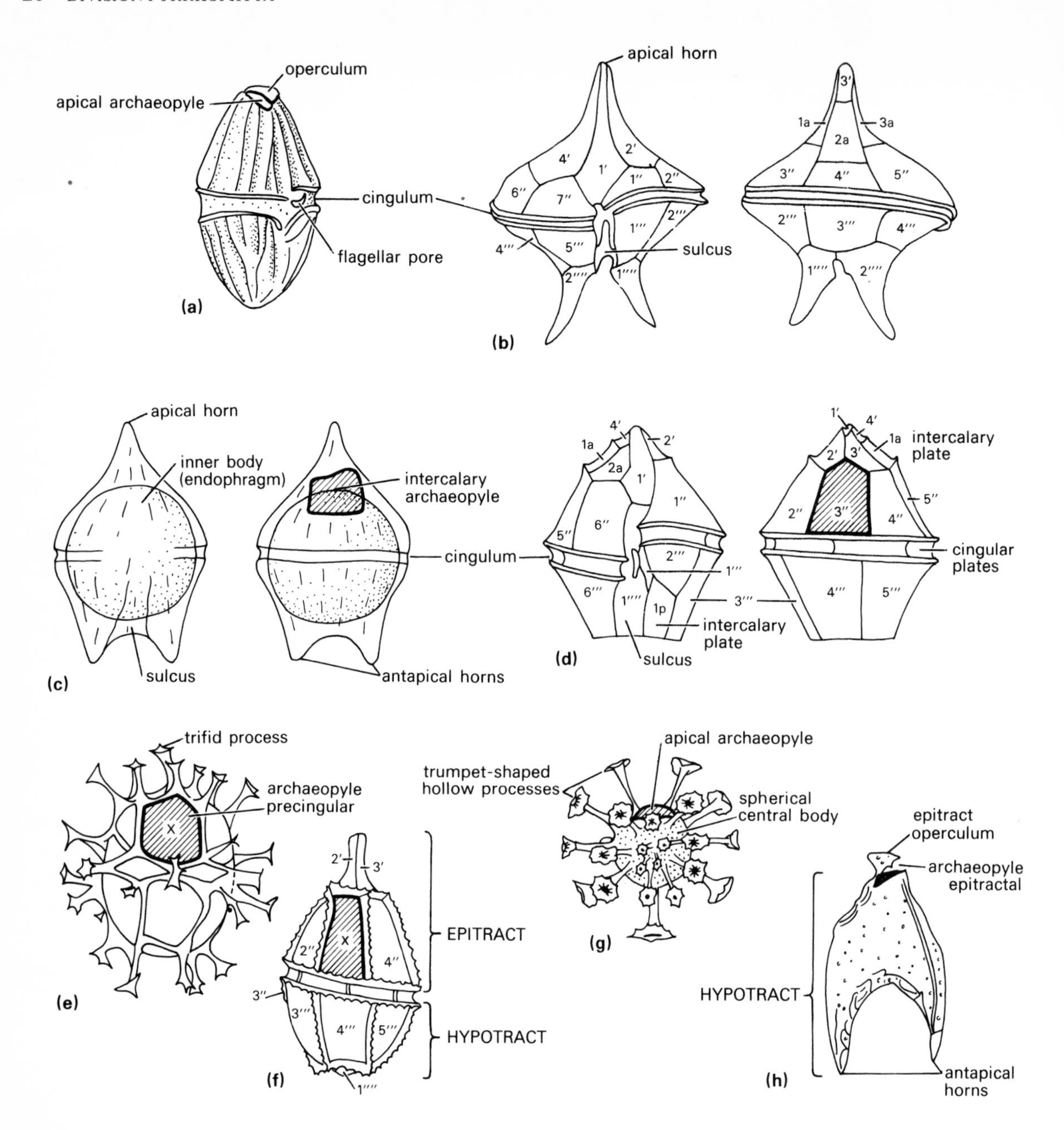

Figure 4.3 Dinoflagellates. (a) *Dinogymnium*, a fossil gymnodinialean cyst, about ×445; (b) motile cell of Recent *Peridinium*, about ×505; (c) cavate cyst of *Deflandrea*, left = ventral and right = dorsal views, about ×360; (d) motile cell of Recent *Gonyaulax*, about ×750; (e) proximochorate cyst of fossil *Spiniferites* ×465; (f) proximate cyst of fossil *Gonyaulacysta*, about ×405; (g) chorate cyst of fossil *Hystrichosphaeridium* ×400; (h) *Nannoceratopsis*, a fossil dinophysialean cyst, about ×680. ((a), (b), (c), (d), (f) & (h) based on Sarjeant 1974; (e) & (g) on Evitt *in* Tschudy & Scott 1969)

furrows. Both theca and cysts may bear two antapical horns (Fig. 4.2a). *Deflandrea* (L.Cret.–U.Olig., Fig. 4.3c) is a fossil cavate cyst of ellipsoidal shape, commonly with horns. The reflected tabulation is rarely visible but of *Peridinium* type with an intercalary archaeopyle.

In Recent *Gonyaulax* (Fig. 4.3d) the motile stage usually lacks horns and the tabulation is relatively asymmetrical. Its cyst is chorate or intermediate between proximate and chorate (**proximochorate**) with a precingular archaeopyle, being of the type once called *Hystrichosphaera* but now called *Spiniferites* (U. Jur.–Rec., Fig. 4.3e). The fossil proximate cyst *Gonyaulacysta* (M. Jur.–M.Mioc.) also has a reflected tabulation of *Gonyaulax* type with a precingular archaeopyle. The sutures are marked by crests and it bears an apical horn (Fig. 4.3f).

Hystrichosphaeridium (U. Jur.–M.Mioc.) is a fossil chorate cyst with a spherical body and radiating hollow processes, often with trumpet-like openings at the distal ends (Fig. 4.3g). Each process corresponds to the centre of a plate on the once enclosing theca. The archaeopyle is apical.

ORDER DINOPHYSIALES. Although the Dinophysiales are armoured, the plates lack a distinctive tabulation. The cingulum is very anterior in position and less spiralled than in the Peridiniales, uniting with the sulcus in a T- or Y-shaped junction. Both furrows are bordered by flange-like crests, as in Recent *Ornithocercus* (Fig. 4.1e). The cysts are proximate, the archaeopyle and operculum comprising the whole of the epitract (i.e. **epitractal**). Fossil examples are few but may include *Nannoceratopsis* from the Lower Jurassic (Fig. 4.3h). In this there are two prominent antapical horns and an epitractal archaeopyle.

General history of dinoflagellates

Although biologists hint at the primitive organisation and inferred great age of the group, the acme of peridinalean history appears to have been reached in the Mesozoic and Cainozoic Eras. The late Precambrian and Palaeozoic flowering of acritarchs may represent an earlier stage in dinoflagellate history, when non-tabulate forms thrived.

The earliest recorded peridinialean cyst is *Arpylorus*, a Silurian acritarch with tabulation, a cingulum and a precingular archaeopyle. The main radiation, however, began in the late Triassic Rhaetian Stage with proximate cysts. The latter type remained dominant throughout the Jurassic (e.g. *Gonyaulacysta jurassica*, Fig. 4.3f), although chorate, proximochorate and cavate types had all appeared by the middle Jurassic.

Dinoflagellate cysts of the Cretaceous were predominantly chorate (e.g. *Hystrichosphaeridium*, Fig. 4.3g) or proximochorate (e.g. *Spiniferites ramosus*, Fig. 4.3e) and it was at this time that the greatest diversity of cysts was reached (Fig. 4.4).

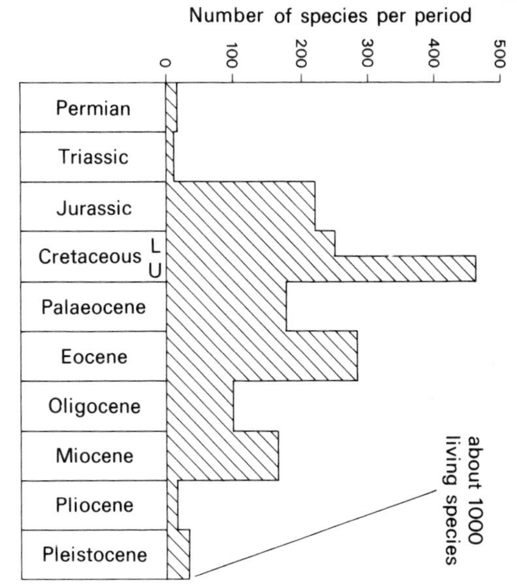

Figure 4.4 Specific diversity of described dinoflagellates through time, omitting the single Silurian record. (Based on Tappan & Loeblich 1973)

Cavate cysts began to flourish in latest Cretaceous times (e.g. *Deflandrea*, Fig. 4.3c) and dominated many Tertiary assemblages until the Oligocene, almost dying out in the Pliocene. Proximate and chorate cysts with complex processes occur in the Eocene and Oligocene, but simpler forms have prevailed since then. Dinoflagellate cysts first appeared in freshwater sediments during the Tertiary.

For fuller accounts of dinoflagellate history see Sarjeant (1967, 1974) and for a review of their evolution see Williams (pp. 1283–8 *in* Ramsay 1977).

Applications of dinoflagellate cysts

Dinoflagellate cysts have proved of considerable value in the biostratigraphy of marine Mesozoic and Cainozoic strata, where they may be used in conjunction with acritarchs, pollen grains and spores. Sarjeant (1967) and Harker & Sarjeant (1975) have outlined the stratigraphical distribution of dinoflagellate cysts. For examples of their biostratigraphic use we may take Bujak and Williams (pp. 321–9 *in* Swain 1977) for the Jurassic, Habib (1975 and pp. 341–67 *in* Swain 1977) and Aurisano and Habib (pp. 369–87 *in* Swain 1977) for the Cretaceous, and the papers by Costa & Downie (1976) and Manum (1976) for the Cainozoic. Williams (pp. 1246–83 *in* Ramsay 1977) reviews the biostratigraphic zones in use at present.

There is an increasing interest in the palaeoenvironmental value of dinoflagellate cysts. Manum (1976) for example, suggests that changing proportions of cysts to pollen, spores and indeterminate debris through the Cainozoic of the North Atlantic reflect widespread palaeogeographic changes associated with sea-floor spreading events. Dinoflagellate cysts and acritarchs reworked into younger sediments can be used to indicate the provenance of sediments and the directions of transport (see Stanley 1966). Different kinds of cyst may also help to distinguish nearshore from offshore sediments (e.g. Williams & Sarjeant 1967, Downie, Hussain *et al.* 1971, Wall *et al.* 1977). Wall and Dale (1968) found that Quaternary cyst assemblages change in response to probable climatic controls and are therefore of value in the correlation and interpretation of marine strata of this age.

Further reading

Useful introductions to the group may be found in the book by Sarjeant (1974), the chapter by Evitt *in* Tschudy and Scott (1969, pp. 439–68), by Williams *in* Ramsay (1977, pp. 1231–343) and the paper by Harland (1971). Many cysts can be identified with the assistance of the catalogue by Eisenack and Kjellström (1964-to-date) and the index by Norris and Sarjeant (1965).

Hints for collection and study

Dinoflagellate cysts are common in dark grey and black argillaceous rocks of post-Triassic age. They can be disaggregated by methods A to E (especially D) and sorted and concentrated by methods H or K (see Appendix). Temporary or permanent mounts on glass slides should be scanned with well condensed transmitted light at over 400 × magnification. More sophisticated methods of preparation and concentration are reviewed by Gray (pp. 530–86 *in* Kummel & Raup 1965) and by Sarjeant (1974).

EBRIDIANS

The ebridians are unicellular, marine and planktonic with an endoskeleton of silica, but unlike that of the similar silicoflagellates (*q.v.*), it is solid with a tetraxial or triaxial symmetry. Ebridians possess two flagella of unequal length and lack photosynthetic pigments, surviving instead by the ingestion of food (especially diatoms) with the aid of pseudopodia. Reproduction is mostly by asexual division.

Classification of ebridians is complicated by their uncertain biological status, resembling 'algal' groups such as the silicoflagellates and dinoflagellates as much as 'animal' groups like radiolarians. Generally regarded as 'algae' they are placed by some in the division Chrysophyta and by others in the Pyrrhophyta as a distinct class, the Ebriophyceae.

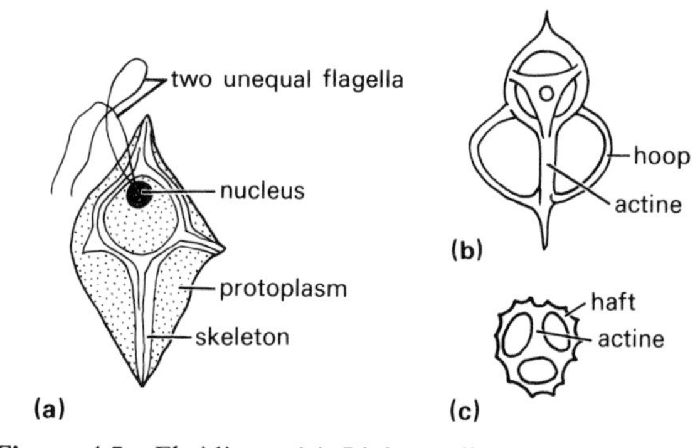

Figure 4.5 Ebridians. (a) Living cell and skeleton of *Hermesinum*; (b) *Hermesinum* skeleton × 500; (c) *Ebria* skeleton × 533. ((a) based on Hovasse 1934)

Genera and species are distinguished on the basis of endoskeleton morphology (see Loeblich *et al.* 1968). For example, *Ebria* (Mioc.–Rec., Fig. 4.5c) has three of four radiating bars (**actines**) with the ends joined by curved hoops called **hafts**. *Hermesinum* (Palaeoc.–Rec., Fig. 4.5a, b) consists essentially of four actines resembling a sponge spicule in their tetraxial arrangement, the ends of which are joined by a series of subcircular **hoops.**

Ebridians are known in rocks of Palaeocene age, the majority of genera thriving until the Pliocene when their diversity dropped sharply (Tappan & Loeblich 1972). The geological value of ebridians has been little exploited as yet, largely because they are neither abundant nor uniformly distributed and preserved. None the less, they have been used successfully with silicoflagellates in Cainozoic biozonal schemes, such as those in the north Pacific area (see Ling 1972, 1975).

5 Group Acritarcha

The Acritarcha are an informal rag-bag group to which any hollow, organic-walled unicellular vesicles may be assigned until their true affinities are discovered (Evitt 1963). Despite this, the majority of described acritarchs display a remarkable degree of similarity to one another. Most are 20–150 μm across and composed of a single-layered wall enclosing a central cavity. They may be entire or provided with an archaeopyle-like opening.

Ranging from late Precambrian to Recent times, acritarchs reached their acme in the early Palaeozoic and the Mesozoic. Like dinoflagellates, they are useful for inter-regional stratigraphic correlation but of more limited environmental value. Their great age, however, gives to them considerable evolutionary interest.

The vesicle

The acritarch vesicle consists of a **central body** enclosing a **central cavity** from which may arise spine-like **processes** and **crests**. These processes may be hollow and connected with the central cavity (open) or closed at the base, or solid. The tips of the processes can be simple, bifurcated or complexly branched.

Wall composition and structure. The acritarch wall consists of condensed fatty acids similar to those found in the spores of vascular plants (i.e. **sporopollenin**). Generally, this wall is single and homogeneous. Laminar walls with narrow radial pores occur in some types (see below). Double walls arise in *Visbysphaera* from the growth of an inner body of slightly smaller diameter inside the outer vesicle. The exterior surface of acritarch vesicles may be smooth, or finely ornamented with granules, short spines, indentations or pores.

The pylome. The escape hole in a vesicle is called a **pylome**, which, because it resembles a dinoflagellate archaeopyle, is likewise thought to have allowed release of a motile stage from the cyst (i.e. **excystment**). However, pylomes are present in only a minority of specimens and are not known from all genera, being especially scarce in those of Precambrian and Cambrian age.

The simplest form of opening is a **partial-split** of the vesicle (e.g. *Micrhystridium*, Fig. 5.1g). **Median-splitting** involves a complete division of the vesicle into two equal halves. If the splitting is less complete and proceeds along an arcuate fissure to leave a hinged flap, this is called an **epityche** opening (e.g. *Visbysphaera*, Fig. 5.1m). A circular or polygonal opening situated above the equator is called a **cyclopyle** (e.g. *Cymatiogalea*, Fig. 5.1d). Cyclopyles tend to form an **operculum** which may be hinged, released or fall inside the central cavity after excystment.

Classification

Group Acritarcha

Subdivision of the acritarchs has been hampered by a lack of biological information so that most classifications have been artificial. All described genera are form genera only, named and defined outside the jurisdiction of the Botanical and Zoological Codes. The following criteria may prove useful for a more natural classification, in order of importance: wall structure, pylome type, nature of processes and crests, and the form of the central body of the vesicle. Taking these criteria into account, about eight basic types of acritarch can be distinguished, named after well known genera (see Downie 1973).

Tasmanites group. These have a perforate wall with a cyclopyle or median-split opening prior to compaction. The vesicles were spherical and lacked spines or crests (i.e. **sphaeromorph**). Such forms (e.g.

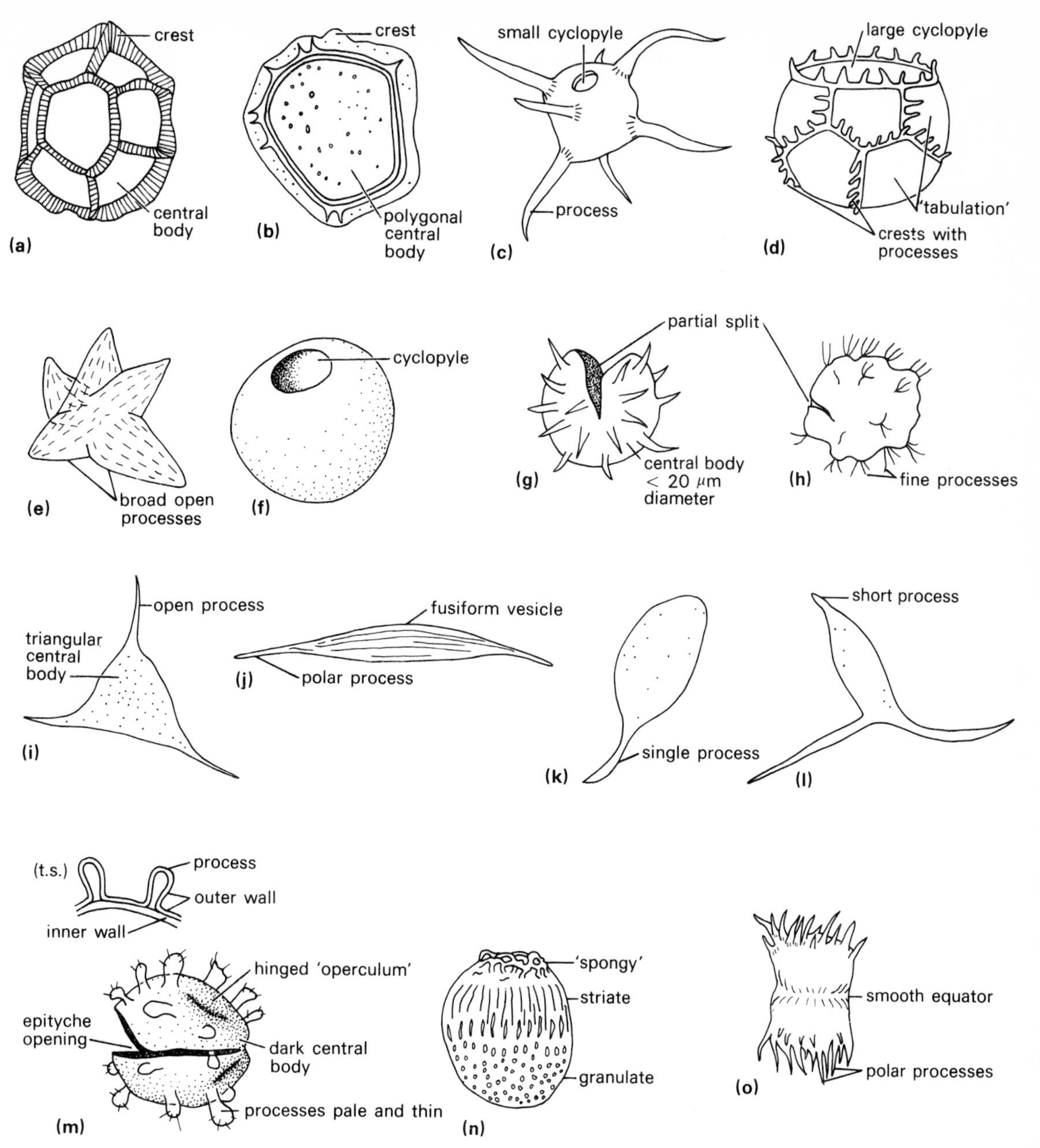

Figure 5.1 Acritarchs. (a) *Cymatiosphaera* ×900; (b) *Duvernaysphaera* ×800; (c) *Baltisphaeridium* ×250; (d) *Cymatiogalea* ×1100; (e) *Estiastra* ×400; (f) *Leiosphaeridia* ×400; (g) *Micrhystridium* ×1350; (h) *Vulcanisphaera* ×500; (i) *Veryhachium* ×400; (j) *Leiofusa* ×400; (k) *Deunffia* ×400; (l) *Domasia* ×400; (m) *Visbysphaera* ×670 with detail of transverse section; (n) *Ooidium* ×600; (o) *Acanthodiacrodium* ×400. ((a) & (o) based on Evitt *in* Tschudy & Scott 1969)

Tasmanites, Fig. 9.1a) range from Ordovician to Recent times. They may include *Pachysphaera*, a Recent unicellular green alga belonging to the Prasinophyceae (see Ch. 9).

Cymatiosphaera group. The wall is perforate and without known excystment openings. The vesicle was originally spherical or polygonal, divided into fields by crests. In *Cymatiosphaera* (L. Camb.–Rec.,

Fig. 5.1a) these fields are polygonal but in *Duvernaysphaera* (M. Sil.–U. Dev., Fig. 5.1b) the crests form an equatorial flange.

Baltisphaeridium group. In these the wall is perforate with a cyclopyle or median-split opening. The vesicle was spherical or polygonal but armoured with processes or crests. *Baltisphaeridium* (L. Camb.–L. Sil., Fig. 5.1c) has a spherical central body, over 20 μm in diameter, with simple hollow or solid processes with closed tips. In *Cymatiogalea* (M. Camb.–L. Sil., Fig. 5.1d) the vesicle is divided into polygonal fields with crests between, somewhat resembling a proximate dinoflagellate cyst, but it has a large cyclopyle opening. *Estiastra* (M. Ord.–U. Sil., Fig. 5.1e) is star-shaped with wide processes.

Leiosphaeridia group. These have a simple, imperforate wall with an irregular or cyclopyle opening. The vesicles were spherical and lacked processes and crests, corresponding with the group 'Sphaeromorphitae' of some classifications. The earliest acritarchs, such as *Chuaria* (U. Precamb.), may belong here, although this genus is exceptionally large (< 5 mm diameter). *Leiosphaeridia* (U. Precamb., Palaeozoic, Fig. 5.1f) may have had green algal affinities like *Tasmanites*.

Micrhystridium group. The *Micrhystridium* group has simple walls with partial-split or epityche openings. The vesicles are small and spherical, polygonal or elongate with hollow processes. For example, *Micrhystridium* (L. Camb.–Rec., Fig. 5.1g) has a spherical central body less than 20 μm in diameter with simple processes. *Veryhachium* (U. Camb.–Mioc., Fig. 5.1i) has a polygonal central body with from three to eight hollow pointed spines with closed tips. *Vulcanisphaera* (U. Camb.–U. Sil., Fig. 5.1h) has irregular protuberances which bear a bunch of from two to five spines of varying lengths.

Leiofusa group. This type has simple walls with either a median- or lateral-split or a C-shaped epityche opening. Typically, the vesicles are elongate with processes at the poles, as in fusiform *Leiofusa* (U. Camb.–U. Carb., Fig. 5.1j). *Deunffia* (Sil., Fig. 5.1k) bears a single process whilst *Domasia* has three processes (Sil., Fig. 5.1l).

Acanthodiacrodium group. The *Acanthodiacrodium* (or diacrodian) group has a simple wall which tends to split up into angular plates when damaged. The openings are of varying kinds but the vesicles are typically elongate with the sculpture concentrated at one or both poles. *Acanthodiacrodium* (M. Camb.–M. Ord., Fig. 5.1o) has small processes at both poles and a waist-like constriction. *Ooidium* (Camb., Fig. 5.1n) is more ovate with a granular sculpture at one pole and a spongy sculpture at the other, with fine striae between.

Visbysphaera group. These are characterised by a double-layered wall and an epityche opening. The vesicles may be elongate, triangular or spherical, as in *Visbysphaera* (L. Sil.–L. Dev., Fig. 5.1m) which bears processes that are produced from the outer wall.

Acritarch affinities and biology

Acritarchs resemble a wide range of organic structures, e.g. invertebrate egg cases, vascular plant spores, multicellular and unicellular algal zygospores, and prasinophycean and dinoflagellate cysts. Most of the similarities arise from convergent evolution, similar forms arising for similar functions in unrelated stocks.

Chemically the wall most resembles the sporopollenin wall of vascular plant spores, algal spores and cysts. In wall ultrastructure the *Tasmanites*, *Cymatiosphaera* and *Baltisphaeridium* groups compare closely with Recent prasinophycean cysts and cells. The *Leiosphaeridia* group compares with the spores of multicellular algae. The remainder may have affinities with the naked Dinoflagellates (Gymnodiniales) and the armoured Dinophysiales, which are known to develop non-tabulate cysts. Acritarchs differ from most peridinalean cysts in the absence of both reflected tabulation and of preformed excystment openings of definite form. At least one Recent peridinialean, however, is known to produce an acritarch-like cyst (Dale 1976).

Monospecific clusters of acritarchs have been found, especially in Precambrian and Cambrian rocks (see Combaz 1967). Although it has been suggested that these are the spores of multicellular algae this need not follow. Dinoflagellates are also known to aggregate in clusters.

Acritarch ecology

Acritarchs have mostly been found in marine strata, especially in shales and mudstones, but also occur in sandstones and limestones. Non-marine examples are first reported from Recent (Holocene) strata.

Acritarch abundance and diversity tends to increase away from the shoreline, those found in such facies being much abraded. Lagoonal facies are characterised by low diversity and monospecific assemblages of *Tasmanites, Leiosphaeridia* or *Leiofusa* type. Inshore facies may also contain abundant *Micrhystridium*, whereas quieter offshore facies are reflected in assemblages with longer, more delicate and elaborate processes and crests. Temperature range was probably a primary control of acritarch distribution for some assemblages have distributions which parallel the palaeolatitudes of the early Palaeozoic Earth. Even so, the group had a wide overall tolerance, being found from periglacial to tropical palaeoenvironments.

The geographical distribution of acritarchs suggests a partially or wholly planktonic mode of life. This is confirmed by their general morphological resemblance to dinoflagellate and prasinophycean cysts. Similar limitations may therefore exist for acritarchs and dinoflagellates in environmental studies.

General history of acritarchs

Acritarchs first became abundant in marine sediments about 1000 Ma BP, in the late Precambrian (or Riphean) Period. These ancient forms are largely spherical vesicles without obvious means of excystment and commonly occur in clusters.

In the early Cambrian appeared many spinose forms (e.g. *Micrhystridium, Baltisphaeridium*), crested forms (e.g. *Cymatiosphaera*), and the *Leiofusa* and *Acanthodiacrodium* groups. These last, plus *Cymatiogalea* and similar forms were at their acme in late Cambrian and early Ordovician times. *Baltisphaeridium* flourished throughout the Ordovician but died out in the early Silurian, a period dominated by *Micrhystridium*, *Veryhachium*, and similar genera.

Rich early Devonian assemblages of this kind were followed by a general decline in acritarch diversity and abundance (Fig. 5.2). The group then became scarce throughout the Carboniferous, Permian and Triassic. Certain genera made a limited come-back in the Jurassic, Cretaceous and Tertiary, e.g. *Tasmanites*, *Cymatiosphaera* and *Micrhystridium*.

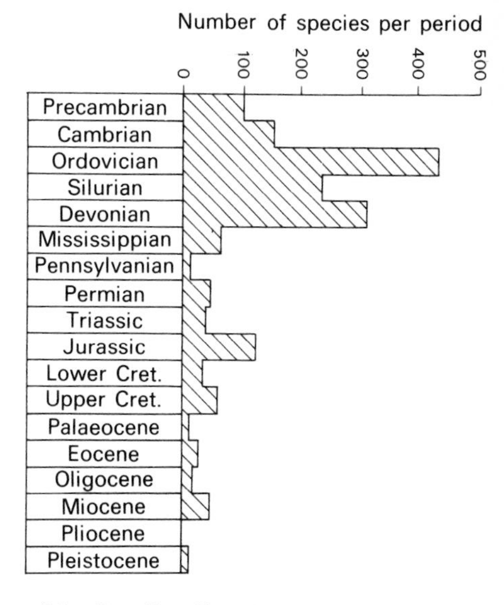

Figure 5.2 Specific diversity of described acritarchs through time. (Based on Tappan & Loeblich 1973)

We may note that the radiation of sculptured acritarchs in the early Cambrian coincides with the major radiation of invertebrate suspension feeders. It is therefore possible that acritarch evolution has played an important rôle in the development of life (see Brasier 1979).

The geological history of acritarchs is reviewed in more depth by Downie (1973).

Applications of acritarchs

Acritarchs have been used largely to correlate upper Precambrian and lower Palaeozoic rocks, especially those lacking other useful fossils. Papers by Downie, Lister *et al.* (1971), Downie (1974) and Vidal (1976) illustrate their potential in Precambrian and Cambrian rocks, whilst Wall (1965) examines their value in some Mesozoic strata. Swain (1977, pp. 137–49) provides a brief review of their stratigraphic use in Palaeozoic deposits from the Atlantic borderlands.

The abundance, diversity and degree of abrasion of acritarchs can be used to help distinguish near-shore from offshore sediments (see Smith & Saunders 1970, Riegel 1974), as may the sculpture of the vesicle itself (see Staplin 1961, Wall 1965). Their value as depth indicators in ancient sediments is reviewed by Williams and Sarjeant (1967). Geographically distinct acritarch provinces (see Vavrdová 1974, Cramer & Diez 1974) may assist the reconstruction of ancient ocean currents or climatic belts, especially if used in conjunction with other evidence.

Further reading

Helpful introductory reviews include those by *Schopf (*pp. *163–92) and Evitt (*pp. *439–68) in* Tschudy and Scott (1969) and that by Downie (1973). Their affinities and classification are discussed by Evitt (1963), Downie *et al.* (1963), Lister (1970) and Downie (1973). Acritarch morphology is clearly shown in the papers by Loeblich (1970) and Tappan & Loeblich (1971). The geological ranges of selected lower Palaeozoic taxa are portrayed by Diez and Cramer (1974) whilst identification should be assisted by reference to Eisenack and Kjellström (1964 to date).

Hints for collection and study

Acritarchs can often be obtained from dark carbonaceous shales, mudstones and clays disaggregated by methods A to E (see Appendix). Those occurring with dinoflagellate cysts in Mesozoic and Cainozoic rocks are usually more easy to extract. Acritarchs can be sorted and concentrated by methods H and K. Temporary and permanent mounts on glass slides should be scanned with well condensed transmitted light at 400 × magnification. For a fuller treatment of techniques see Gray (pp. 530–86 *in* Kummel & Raup 1965) and Sarjeant (1974).

6 Division Chrysophyta – Silicoflagellates and chrysomonads

The chrysophytes are protists with a distinctive golden colour imparted by their photosynthetic pigments (chlorophylls a and c, ß-carotene, fucoxanthin and carotenoids). Most are unicellular but there are some colonial forms. The motile cells may have either one or two flagella of unequal length inserted at the apex. Several groups with siliceous or calcareous skeletons in this division are of palaeontological interest: silicoflagellates, chrysomonads, diatoms and coccolithophores.

SILICOFLAGELLATES

Silicoflagellates have been minor components of marine phytoplankton since early Cretaceous times. They are only well preserved in siliceous rocks such as diatomites, though, and have been little used except in deep oceanic strata where they are now widely employed both for correlation and for estimation of palaeoclimatic conditions.

The living silicoflagellate

The unicellular organism is usually from 20–100 μm in diameter and contains golden-brown photosynthetic pigments, a nucleus, several pseudopodia and a single flagellum at the anterior end of the cell (Fig. 6.1a). The soft protoplasm is supported from within by a discoidal or hemispherical **skeleton** of hollow rods, built of opaline silica.

Reproduction appears to be predominantly asexual, beginning with the secretion of a daughter skeleton and followed by simple cell division. Silicoflagellates are mixotrophic; that is, they feed both by photosynthesis and by capture of prey with their pseudopodia. None the less, they are restricted to the shallow photic zone of the ocean (0–300 m), thriving in the silica-enriched waters associated with current upwelling in equatorial and high latitude waters along the western margins of continents. In these cooler waters they may also bloom seasonally. For these reasons the silicoflagellates are commonest as fossils in biogenic silica deposits formed during cool periods with marked seasonality or with strong upwelling. They are unknown in freshwater habitats.

The silicoflagellate skeleton

The basic skeleton is built upon a **basal ring** which may be elliptical, circular or pentagonal with from two to seven **spines** on the corners of the outer margins (Fig. 6.1b). This basal ring is usually traversed by one or more **apical bars** arched upwards in an apical direction and connected to the basal ring by shorter **lateral bars**. In some genera these features are elaborated into a complex hemispherical **lattice** (Fig. 6.1g).

The skeleton is thought to function as a mechanism for improvement of buoyancy, spreading out the protoplasm and increasing resistance to downward sinking. To reduce weight further, the skeleton is also hollow. In life, the domed apical portion is orientated upwards towards the light.

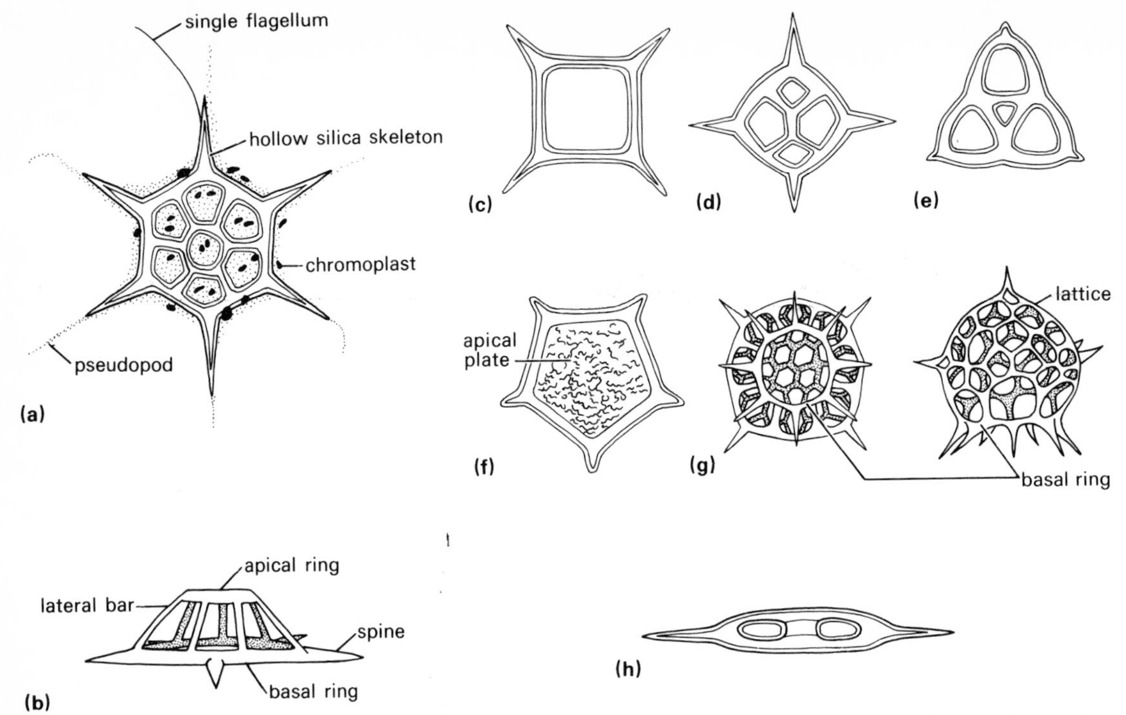

Figure 6.1 Silicoflagellates. (a) Living cell and skeleton of *Distephanus* ×267; (b) Side view of *Distephanus* skeleton ×267; (c) *Mesocena* ×533; (d) *Dictyocha* ×400; (e) *Corbisema* ×533; (f) *Vallacerta* ×446; (g) *Cannopilus* ×500; (h) *Naviculopsis* ×373. ((a) modified from Marshall 1934)

Classification

Kingdom PROTISTA
Class CHRYSOPHYCEAE
Order CHRYSOMONADALES

Silicoflagellates (Suborder Silicoflagellinae) belong to the above class and order on account of their unicellular organisation and their simple flagella. The siliceous endoskeleton distinguishes the Silicoflagellinae from other true chrysophyceans and only a few genera (< 20) are recognised, distinguished on the basis skeletal symmetry and architecture (see Loeblich *et al.* 1968).

Mesocena (Palaeoc.–Rec., Fig. 6.1c) comprises a simple polygonal basal ring, often with spines but without bars. In *Dictyocha* (Cret.–Rec., Fig. 6.1d) the quadrate basal ring has corner spines and a diagonal apical bar with bifid ends (lateral bars). In *Corbisema* (Cret.–Rec., Fig. 6.1e) the symmetry is trigonal whilst in *Distephanus* (Palaeoc.–Rec., Fig. 6.1a, b) it is hexagonal, both genera having an inner apical ring connected by lateral bars to an outer basal ring. *Vallacerta* (Cret., Fig. 6.1f) has a pentagonal basal ring with corner spines and a convex, sculptured disc (**apical plate**) of silica. *Naviculopsis* (Palaeoc.–Mioc., Fig. 6.1h) has a long and narrow ring with an arched cross bar and a spine at each end. More complex is *Cannopilus* (Olig.–Rec., Fig. 6.1g) which resembles a radiolarian but has a hemispherical lattice with spines both on the basal ring and on the lattice.

General history of silicoflagellates

Although silicoflagellates first appeared in early Cretaceous times, many of the fossil genera are still extant. *Dictyocha* is probably the stock from which the other forms arose by skeletal modification.

Silicoflagellates have been both numerous and diverse during periods of climatic cooling, i.e. the late Cretaceous, late Eocene, Miocene and Quaternary (Fig. 6.2). At these times oceanic current

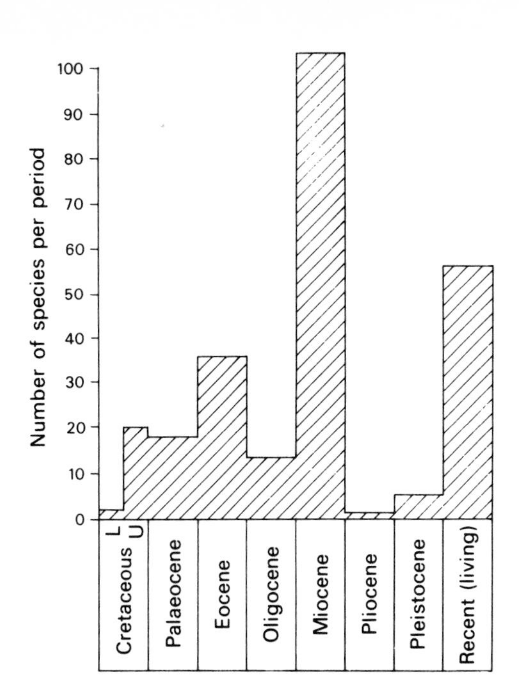

Figure 6.2 Species diversity of described silicoflagellates through time. (Based on Tappan & Loeblich 1973)

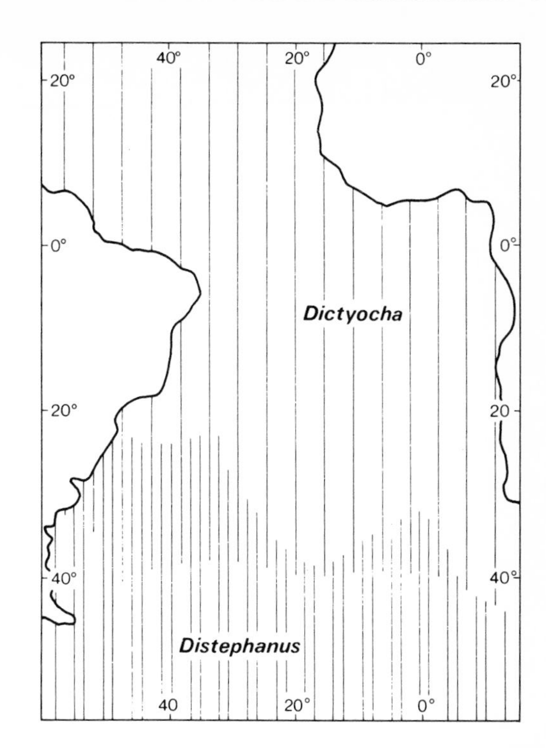

Figure 6.3 Distribution of Recent *Dictyocha* and *Distephanus* in the South Atlantic waters. (Based on Lipps 1970)

circulation is thought to have been more rapid leading to more vigorous upwelling of mineral-rich waters and blooms of siliceous phytoplankton (see Lipps 1970).

Applications of silicoflagellates

The value of silicoflagellates has been well tested by microfossil studies on cores of sediments collected during the JOIDES Deep Sea Drilling Project. Because they have evolved slowly, silicoflagellate biozones are comparatively few and long-ranging, with reliance on the relative abundance of species (or assemblages of species), but workable biostratigraphic schemes are being used (see Martini & Muller 1976). This work is briefly reviewed by Casey (pp. 545–52 *in* Swain 1977) and Martini (pp. 1335–7 *in* Ramsay 1977). Emphasis has also been placed on their value as palaeoclimatic indicators, especially from ratios of the warm water *Dictyocha* to the cool water *Distephanus* in sediments (Fig. 6.3; see Bukry & Foster 1973, Ciesielski 1975 and Bukry 1976). Warm and cool water species of *Dictyocha* have been used by Cornell (1974) to indicate fluctuations in the Miocene climate of California. Similar aspects of palaeoclimatology are covered *in* Funnell and Riedel (1971) by Jousé (pp. 407–21) and Muhina (pp. 423–31). Although only a minor fraction of biogenic silicates, their role in sedimentology has also been outlined in the above volume by Kozlova (pp. 271–5).

CHRYSOMONAD CYSTS

The remaining non-skeletalised taxa in the Chrysomonadales are for the most part unicellular, non-marine and phytoplanktonic, although marine forms do occur. Others can be colonial with a coccoid or filamentous habit but most of them may at some time form benthic resting cysts called **statospores**, especially after reproduction.

Chrysomonad cysts are from 3–25 μm across, each comprising a subspherical, silica-impregnated cellulose cell wall enclosing the dormant protoplasm, as in recent *Chrysocapsa* (Fig. 6.4a). An apical pore, often borne on a neck, is closed off by a silicified cellulose plug. The outer surface sculpture, form of neck and pore and the overall shape may be used to distinguish cyst genera and species (e.g. fossil *Archaeomonas*, Fig. 6.4b). These form-genera bear little relation to those usually recognised by biologists on the basis of cellular ultrastructure.

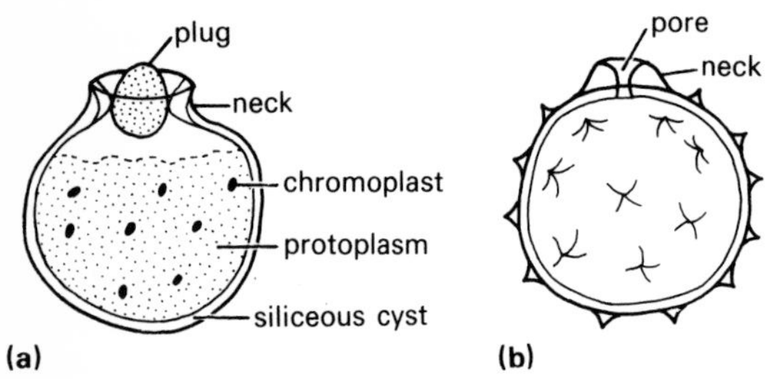

Figure 6.4 Chrysomonad cysts. (a) Recent statospore of *Chrysocapsa* ×2670; (b) fossil *Archaeomonas* ×3330.

Fossil chrysomonad cysts, usually called **archaeomonads**, are mainly known from late Cretaceous and younger diatomites, shales, and silts (Cornell 1970, Tynan 1971) but similar structures are reported from the late Precambrian Beck Spring Chert, about 1300 Ma old (Cloud 1976). It appears that some of the more distinctive archaeomonad species may have potential as guide fossils in deep-sea strata (Gombos 1977).

Hints for collection and study

Silicoflagellates and chrysomonads are most readily obtained from marine diatomites and prepared and studied in the same way as diatoms (*q.v.*). Disaggregated residues in water can be smeared on a glass slide and covered with a cover slip for viewing in transmitted light. For more permanent mounts allow the residue to dry on the slide, add a drop of Caedax or Canada Balsam to the cover slip and place over the residue. Allow to dry before examining with transmitted light.

7 Division Chrysophyta – Diatoms

The Class Bacillariophyceae, better known as diatoms, are unicellular algae with chrysophyte-like photosynthetic pigments but they differ from other chrysophytes in lacking flagella. The cell wall is silicified to form a **frustule**, comprising two **valves**, one overlapping the other like the lid of a pillbox or an agar dish. Diatoms occupy a very wide range of habitats and may occur in enormous numbers in **diatomites**. Fortunately, a century of careful botanical research into living forms has resulted in a relatively clear cut taxonomy and some knowledge of their biology and ecology. Because high proportions of fossil genera and species are still extant, diatoms can be valuable tools in palaeoenvironmental studies. Increasing research into deep-sea history has also encouraged their use as biostratigraphic zone fossils.

The living diatom

The diatom cell ranges in size from 5–2000 μm in length, although most species encountered are in the size range 20–200 μm. The cell may be single or colonial, the latter attached together by mucous filaments or by bands into long chains. Each cell possesses yellow, olive or golden-brown photosynthetic granules (chromoplasts), a central vacuole and a large central nucleus, although it lacks flagella and pseudopodia. Pennate diatoms (Fig. 7.3) can glide over the substrate by the production of a stream of mucus between the frustule and the sediment, but the planktonic centric diatoms (Fig. 7.4) are non-motile. To avoid sinking below the photic zone, the latter are therefore provided with low density fat droplets or occasionally with spines.

Reproduction is primarily asexual, by simple division of the parent cell into two. Because each daughter cell takes one of the parent valves for its own larger dorsal valve, or **epivalve**, and adds a new ventral **hypovalve**, there is a gradual diminution of size in the diatom stock with each generation. This trend is eventually reversed by sexual reproduction.

The frustule

About 95% of the cell wall in diatoms is impregnated with opaline silica. The region of overlap between the epivalve and hypovalve is called the **girdle**, and a study of the two valve views and the girdle view aids identification (Fig. 7.3a). Frustules are usually either circular (**centric**) or elliptical (**pennate**) in valve view, these kinds also comprising the two orders of diatoms (Centrales and Pennales). From 10–30% of the valve surface is perforated by tiny **punctae**, the arrangement of which is also significant for classification. These punctae, which allow connection between the protoplasm and the external environment, can either be simple holes or are occluded by thin transverse plates with minute pores, referred to as **sieve membranes** (Fig. 7.3d). Arrangement of the punctae in lines gives rise to **striae**, usually separated by imperforate ridges called **costae**.

Diatom distribution and ecology

Most diatoms are autotrophic and form the basis of food chains in many aqueous ecosystems. Different species occupy benthic and planktonic niches in ponds, lakes, rivers, salt marshes, lagoons, seas and oceans whilst some even live in the soil or attached to trees.

Pennate diatoms dominate the freshwater, soil and epiphytic niches although they also thrive in benthic marine habitats. Centric diatoms thrive as

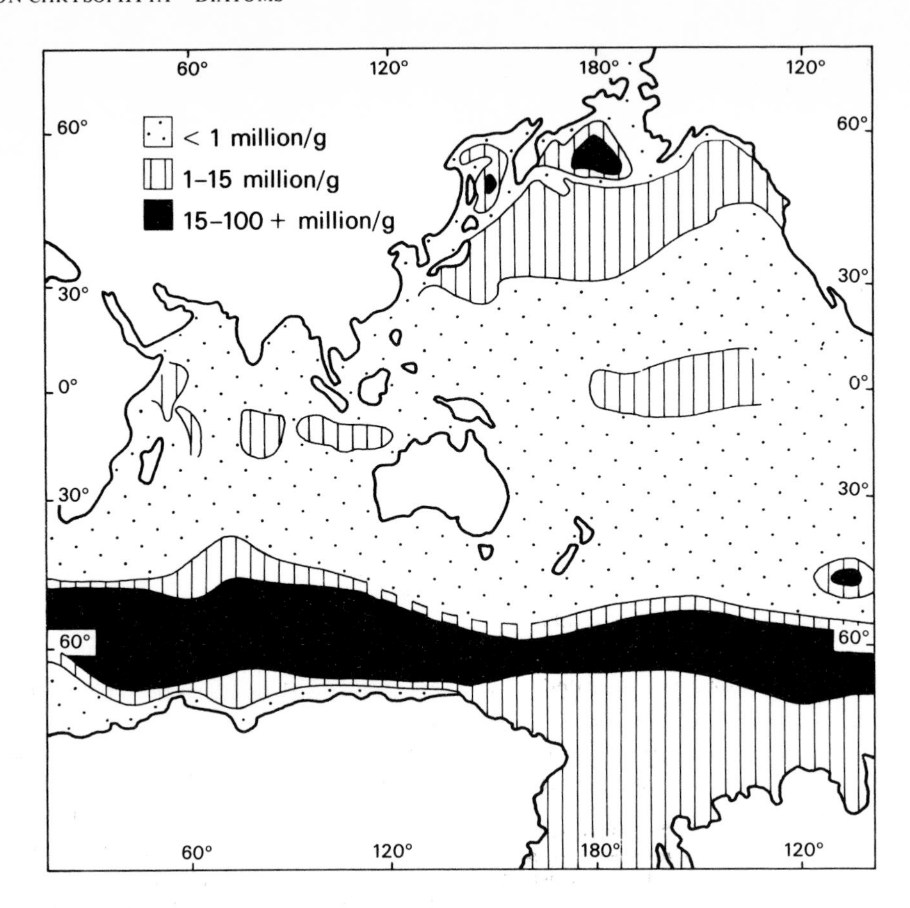

Figure 7.1 Distribution of diatom frustules in surface sediments of the Indian and Pacific Oceans, in millions per gram of sediment. (Based on Lisitzin, Fig. 10.11, *in* Funnell & Riedel 1971)

plankton in marine waters, especially at subpolar and temperate latitudes. Distinct planktonic assemblages are known to dwell in nearshore, neritic and oceanic environments. They can also occur as plankton in freshwater bodies.

Diatoms require light and are therefore limited to the photic zone (< 200 m) during life. Each species tends to have distinct requirements for temperature, salinity, acidity, oxygen and mineral concentrations. Seasonal fluxes in these factors at high latitudes may lead to spring and late summer blooms, especially amongst the plankton where diatoms may number as much as 1000 million cells per m^3 of water. Diatoms are also abundant in regions of oceanic upwelling caused by current divergences, as in those of the Antarctic divergence. These waters are favoured because of their high silica, phosphate, nitrate and iron content. Cool waters are also more dense and, together with the ascending currents, they pose the minimum of buoyancy problems for these non-motile organisms.

Diatoms and sedimentology

Under optimal conditions of productivity such as noted above, diatoms can accumulate in lakes and oceans to form silica deposits called diatomites. These are presently forming on the deep ocean floors beneath sub-arctic and sub-antarctic waters where diatom productivity is high, terrigenous influx is low and $CaCO_3$ solubility relatively great (Fig. 7.1). These sediments may contain over 400 million valves per gram. Diatomaceous oozes (largely of *Ethmodiscus* sp.) some 4–6 m thick are also forming at abyssal depths in equatorial regions in the Indian and Pacific Oceans because of solution of the other, more abundant, calcareous microfossils. Nevertheless, diatoms are themselves prone to dissolution by pressure at depth or under alkaline conditions, especially the less robust or more weakly silicified forms (see Fig. 7.2). This selective solution of the living flora in marine environments and the non-preservation of many freshwater forms leads to fossil assemblages which are rarely representative. Furthermore, planktonic diatoms may travel far before coming to rest on the sea bed; even freshwater diatoms are not uncommon in deep-sea sediments. The latter are apparently blown off the land by strong winds or brought in by turbidity currents.

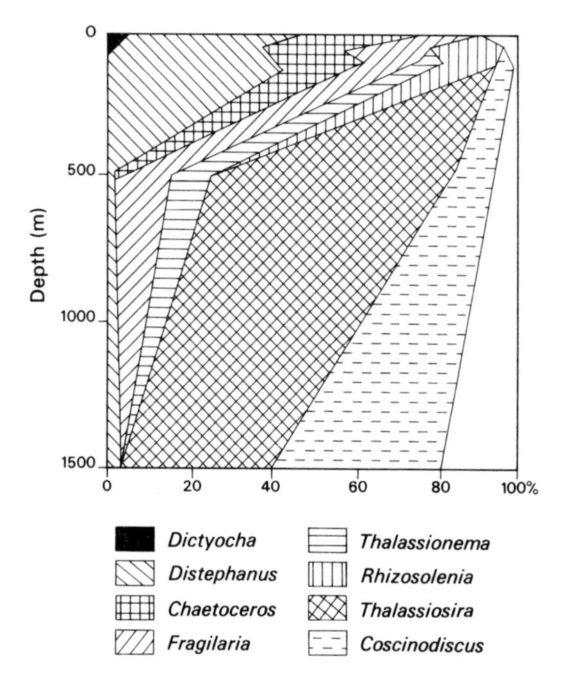

Figure 7.2 Changes in the silicoflagellate and diatom flora with depth, mainly through solution. *Dictyocha* and *Distephanus* are silicoflagellates; the rest are diatoms. (Based on Lisitzin, Fig. 10.8, *in* Funnell & Riedel 1971)

Classification

Kingdom PROTISTA
Division CHRYSOPHYTA
Class BACILLARIOPHYCEAE

The classification of diatoms is based largely on frustule form and sculpture. Hustedt (1930) gave the group the status of a division, Bacillariophyta, but Hendey (1964) and many others regard diatoms as a class within the Chrysophyta. Two orders are recognised by Hendey, namely the Pennales and the Centrales.

The Pennales, or pennate diatoms, have frustules which are elliptical or rectangular in valve view, with bilaterally symmetrical sculpture. The latter includes a longitudinal unsilicified groove called a **raphe** down the middle of each valve face, with rows of punctae arranged at right angles on either side (Fig. 7.3a). Some do not have a groove but merely a similar, silicified area clear of punctae, called a **pseudoraphe** (Figs 7.3b, c). A **central nodule** in the mid point of the valve face divides the raphe into two, and similar **polar nodules** may occur at the extremities (Fig. 7.3a). The raphe or pseudoraphe can occur on one or both valves. Such features are used for further taxonomic subdivision of pennate diatoms. The Suborder Araphidineae have only a pseudoraphe and generally occur attached by mucilage pads at the apex of the cell. For example, *Fragilaria* (Fig. 7.3b) is a benthic, freshwater genus with a very narrow frustule, rectangular in girdle view and commonly united on the valve faces into long chains. The punctae are arranged in striae without intervening costae. In the Suborder Monoraphidineae, a raphe is present on the hypovalve and a pseudoraphe on the epivalve. *Achnanthes* (Fig. 7.3c) for example, is solitary or united in chains and has boat-shaped (naviculoid) valves with punctae arranged in striae. The figured example is a brackish-water species but freshwater and marine species occur. The Biraphidineae have a true raphe on both valves, such as in the common freshwater genus *Pinnularia* (Fig. 7.3a).

The Centrales, or centric diatoms, have frustules which are circular, triangular or quadrate in valve view and rectangular or ovate in girdle view. Being mostly planktonic and non-motile, they lack the raphe and pseudoraphe. *Melosira* (Fig. 7.4a) thrives in freshwater and brackish-water habitats, its

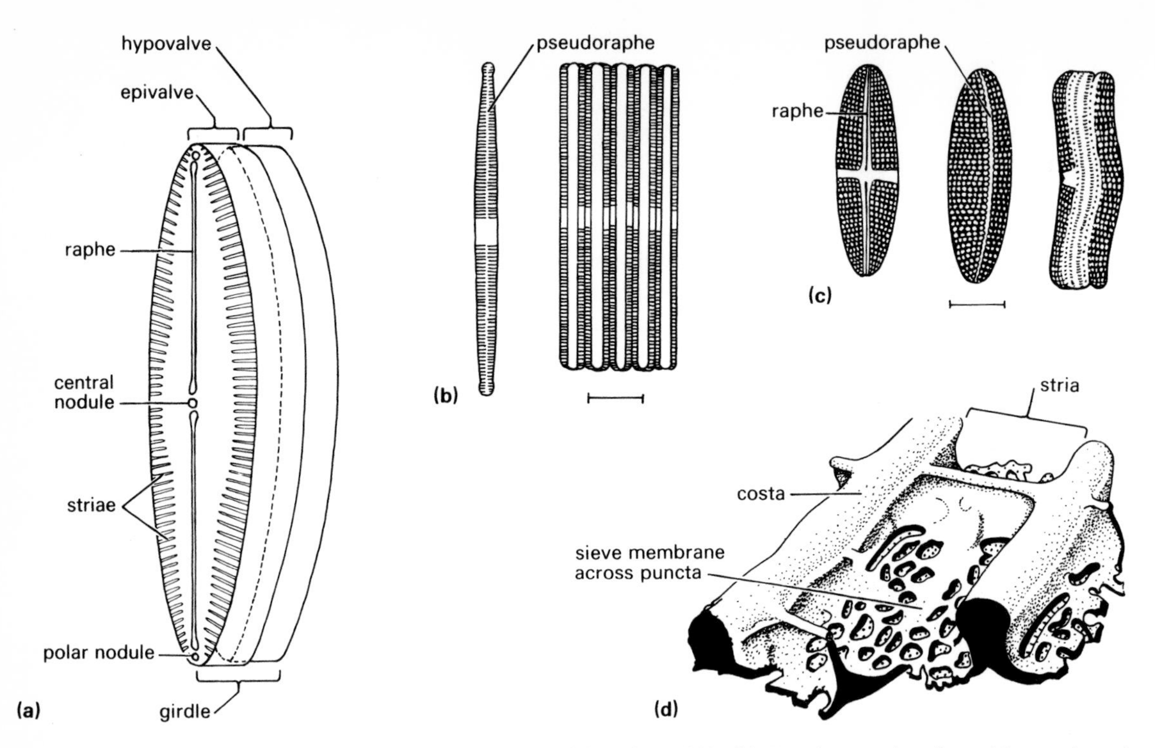

Figure 7.3 Pennate diatoms. (a) *Pinnularia*, oblique view with raphe ×320; (b) *Fragilaria*, valve view with pseudoraphe (left) and girdle view of colony, (right, about ×545); (c) *Achnanthes*, hypovalve view with raphe (left), epivalve view with pseudoraphe (centre) and girdle view (right, all about ×545); (d) detail of diatom punctae. Scale bar = 10 μm, ((a) after Scagel *et al.* 1965; (b) and (c) after van der Werff & Huls 1957–63; (d) after Chapman & Chapman 1973 from Fott.

pillbox-like frustules united into long filaments. The punctae are small and arranged in numerous fine striae radiating from a central region of fewer punctae. In *Coscinodiscus* (Fig. 7.4b) the frustule is also discoidal but with very large radiating punctae. This genus is typical of many inshore and outer shelf planktonic assemblages. *Actinoptychus* (Fig. 7.4c) has the valve face divided into compartments, alternatively elevated and depressed with punctae of different size and shape. It thrives in the near-shore plankton. *Thalassiosira* (Fig. 7.4d) is an open-ocean planktonic form with radial punctae and small submarginal spines, the frustules united in chains by a delicate mucus filament.

General history of diatoms

The earliest recorded diatoms are centric (*cf. Coscinodiscus*) from Cretaceous strata, although there are dubious reports of Jurassic specimens. A major radiation took place amongst centric diatoms in the Palaeocene when the first pennate types also appeared (Fig. 7.5), expanding their numbers gradually throughout the Tertiary. Although the Centrales thrived during the Miocene, their diversity has dwindled somewhat since then. The high number of living species in Figure 7.5 reflects the contribution made by small, weakly silicified forms from freshwater habitats with a low preservation potential.

Applications of diatoms

Few groups rival diatoms for the breadth of their potential applications. Their value as zonal indices in Cretaceous and Tertiary successions are outlined

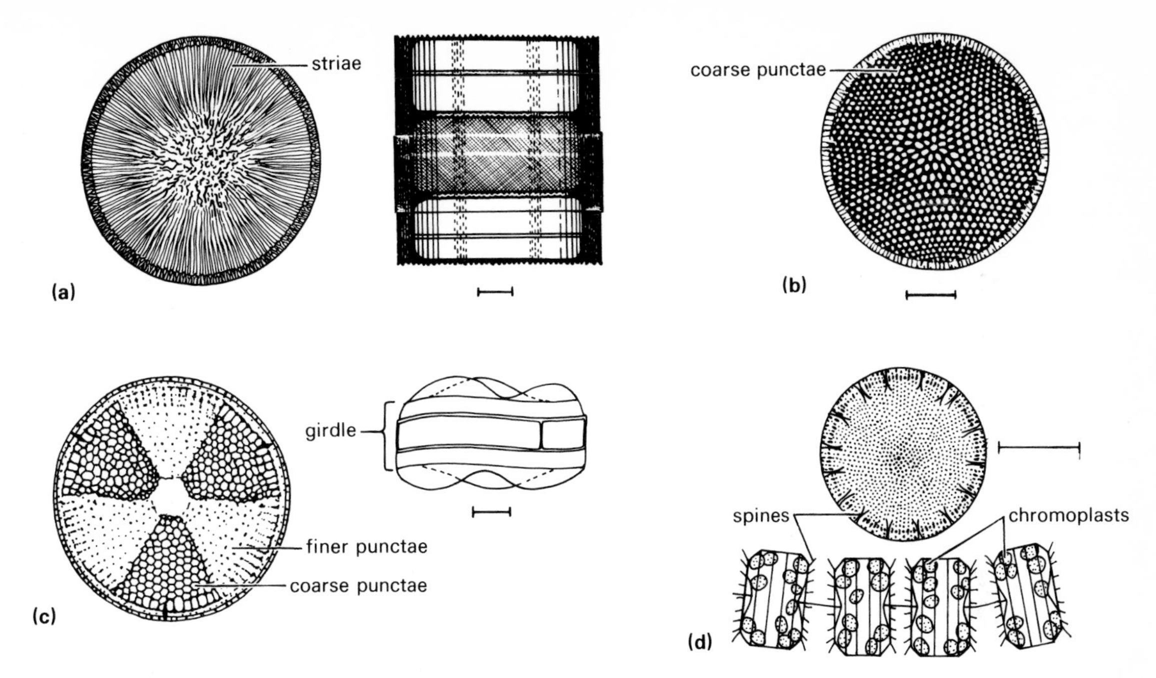

Figure 7.4 Centric diatoms. (a) *Melosira*, valve view (left) and girdle view of colony (right, about ×342); (b) *Coscinodiscus*, valve view, about ×535; (c) *Actinoptychus*, valve view (left, about ×277) and girdle view (right, about ×340); (d) *Thalassiosira*, valve view (above) and girdle view of colony (below, both ×670). Scale bar = 10 μm. (After van der Werff & Huls 1957–63)

for terrestrial sequences by Wornardt (pp. 690–714 *in* Brönnimann & Renz 1969, vol. 2) and Barron (1976), and for oceanic sequences by Gombos (1975, 1977a) and Schrader and Fenner (1976). Schrader (1973) also gives some examples of diatom evolution. Diatoms are of most value in diatom-rich, cool-water deposits, but the taxa can have a sufficiently wide geographic range to make them valuable for correlating warm and cool region biozones and strata (e.g. Kanaya, pp. 545–65 *in* Funnell & Riedel 1971). The palaeoecological value of diatoms is well demonstrated by Koizumi (1975, 1975a) in Pliocene and Pleistocene marine sediments, particularly for evidence of climatic cooling and changing sedimentation rates. Gardner and Burckle (1975) have shown that the *Ethmodiscus* oozes of the equatorial Atlantic were deposited during glacial maxima. The ratios between the oxygen isotopes ^{18}O and ^{16}O in the silica of fossil diatom frustules can also be used to indicate absolute temperatures in Quaternary deposits (Mikkelsen *et al.* 1978). Freshwater diatoms have been used to study the developmental history of lakes since the last glaciation, revealing the effects of changing acidity and climate (Round 1961) and the effects of human pollution (Bradbury 1975).

Mention should be made here of the economic value of diatomites. Those in California are marine, ranging in age from late Cretaceous to late Pliocene. In the Miocene they occur in units up to 1000 m thick with over six million frustules per cubic centimetre. Although freshwater diatomites rarely exceed 1 m thick, they are still of economic interest. The silica is graded and used for filtering, sugar refining, toothpaste, insulation, abrasive polish, paint and lightweight bricks.

Further reading

The biology and ecology of living diatoms are examined in Werner (1977) but a better introduction to these aspects and to taxonomy is Hendey (1964). Sieburth (1975) provides many beautiful photographs of living forms in their natural habitat. The distribution and significance of diatoms in oceanic sediments are reviewed in Funnell and Riedel (1971), whilst an article by Wornardt (pp.

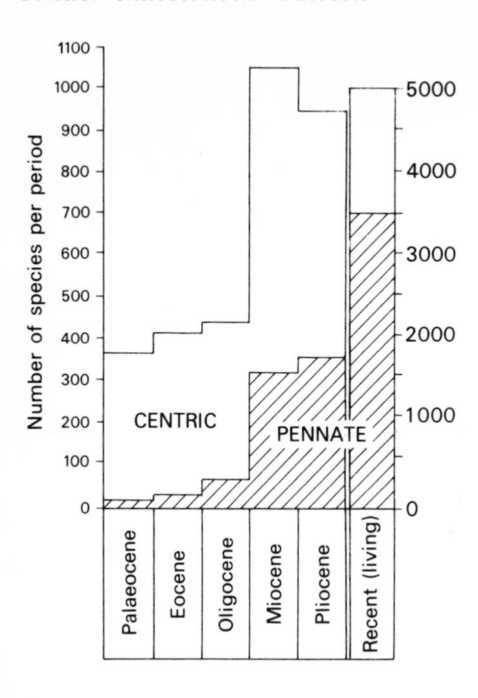

Figure 7.5 Changes in the species diversity of diatoms through the Cainozoic Era. (Based on Tappan & Loeblich 1973)

690–714 *in* Brönnimann & Renz 1969, vol. 2) clearly sets forth the geological value of diatoms and contains a useful bibliography. Recent and fossil genera and species may be identified with the aid of the catalogue by van Landingham (1967 to date).

Hints for collection and study

Recent diatoms are easily collected by scraping up the green scum from the floors of ponds or from the surfaces of mud, pebbles, shells and vegetation in shallow marine waters. Temporary mounts in distilled water can be prepared on glass slides and viewed with well-condensed transmitted light at about 400 × magnification or higher.

Fossil diatoms are readily studied in freshwater or marine diatomites. Any reputable aquarium shop or pet shop will sell the 'diatom powder' used for aquarium filters. These diatomites usually require little in the way of disaggregation or concentration, but diatoms in shales and limestones will need to be treated with methods B, C, D or E to release them and J to concentrate them (see Appendix). Treat the sample with formic (or even concentrated hydrochloric) acid if calcareous shells are not wanted (see method F). Temporary mounts can be prepared with distilled water and strewn on glass slides. For permanent mounts dry the residue on a glass slide, add a blob of Hyrax, Styrax or Canada Balsam to the cover slip and place this over the residue. When dry, examine with transmitted light. Some more-sophisticated techniques are given by Setty (1966).

8 Division Chrysophyta – Coccolithophores

Coccolithophores are unicellular planktonic protists with chrysophyte-like photosynthetic pigments, but they differ from most other Chrysophyta in having two flagella of equal length and a third whip-like organ called a **haptonema**. The group is an important component of the oceanic phytoplankton providing a major source of food for herbivorous plankton. Tiny calcareous scales called **coccoliths** (3–15 μm in diameter) form around these cells a protective armour which eventually falls apart on to the ocean floor to build deep-sea oozes and fossil chalks. Being both abundant and relatively easy to recover from marine sediments, coccoliths have been used increasingly for biostratigraphic correlation of the post-Triassic rocks in which they occur.

The living coccolithophore

A coccolithophore is generally a spherical or oval unicell < 20 μm in diameter, equipped with two golden-brown pigment spots with a prominent nucleus between, two flagella of equal length and a whip-like haptonema. The small, calcite coccoliths are formed in **vesicles** within the cell under the stimulus of light. These eventually move to the outside of the cell where they replace others, which are then shed. Reproduction is mostly asexual, by a simple division of the mother cell into two or more daughter cells. In some living genera there is also an alternation between a **motile** and a **non-motile** planktonic or benthic stage. The motile stage has a flexible skeleton with coccoliths embedded in a pliable cell membrane, but in the non-motile cysts, calcification of the membrane can take place, thereby forming a rigid preservable shell called a **coccosphere**.

Coccoliths

Coccolith morphology is the basis for classification of both Recent and fossil members of the group. Two basic modes of construction are known from electron microscope studies: **holococcoliths** are built entirely of submicroscopic calcite crystals, mostly rhombohedra, arranged in regular order; **heterococcoliths** are usually larger and built of different submicroscopic elements such as plates, rods and grains, combined together into a relatively rigid structure. As holococcoliths invariably disintegrate after they are shed, it is the heterococcoliths which provide the bulk of the microfossil record.

Heterococcoliths vary considerably in form and construction. The majority comprise discs of elliptical or circular outline (**shields**) constructed of radially arranged **plates**, enclosing a central area which may be empty, crossed by **bars**, filled with a **lattice** or produced into a long **spine**. The outward facing (**distal**) side of the shield is often more convex with a prominent sculpture and may be provided with a spine, whilst the other **proximal** face is flat or concave and may have a separate architecture (see Fig. 8.3).

Ecology of coccolithophores

Coccolithophores are predominantly autotrophic **nannoplankton** (i.e. between 5 and 60 μm in size), utilising the energy from sunlight to photosynthesise organic materials from inorganic ones. Living cells are therefore largely restricted to the photic zone of the water column (0–200 m depth) with the lighter, smaller cells living near the surface and heavier cells living lower down. They thrive either in zones of oceanic upwelling or of pronounced vertical mixing, for it is here that vital minerals are most readily available. In tropical areas, where they are most abundant, their numbers may reach as many as 100000 cells per litre of sea water.

Although a few species are adapted either to fresh or brackish waters, the majority are marine. Some of these, such as *Braarudosphaera* prefer to live in inshore waters, but most species are found in the open ocean. Here, surface water temperatures are important controls of distribution, giving rise to different assemblages adapted to subglacial, temperate, subtropical and tropical latitudes. In warmer waters there also appear to be assemblages related to different depths (see Honjo & Okada 1974, Honjo, pp. 951–72 *in* Ramsay 1977).

Coccoliths and sedimentology

After death coccolithophores sink through the water column at about 0·15 m per day and the coccoliths fall away. With increasing depth these scales tend to dissolve or disaggregate into finely dispersed carbonate matter (Fig. 8.1), this process operating first on holococcoliths or delicate heterococcoliths. For this reason, coccolith assemblages from sediments deeper than 1000 m are not truly representative of the original nannoflora. At depths of over 3000–4000 m, few coccoliths remain because most of the $CaCO_3$ has gone into solution so that coccolith and *Globigerina* oozes are replaced by the less-soluble diatom or radiolarian oozes, or by red clays. The boundary between calcareous and siliceous pelagic deposition approximates to the calcium carbonate compensation depth (or CCCD), this being the depth below which the rate of $CaCO_3$ solution exceeds the rate of supply. Many factors may cause this dissolution, including high hydrostatic pressures, high CO_2, low O_2, low pH, low temperatures, low $CaCO_3$ precipitation by organisms, or sluggish recycling of $CaCO_3$ from the land.

Figure 8.1 (*below*) Vertical distribution of coccoliths and coccolith-derived carbonates in the Pacific Ocean. (After Lisitzin Fig. 11.4 *in* Funnell & Riedel 1971)

Figure 8.2 (*right*) Coccolith concentrations in surface sediments of the Atlantic Ocean plotted in percentage by weight. (Modified from McIntyre & McIntyre Fig. 16.1 *in* Funnell & Riedel 1971)

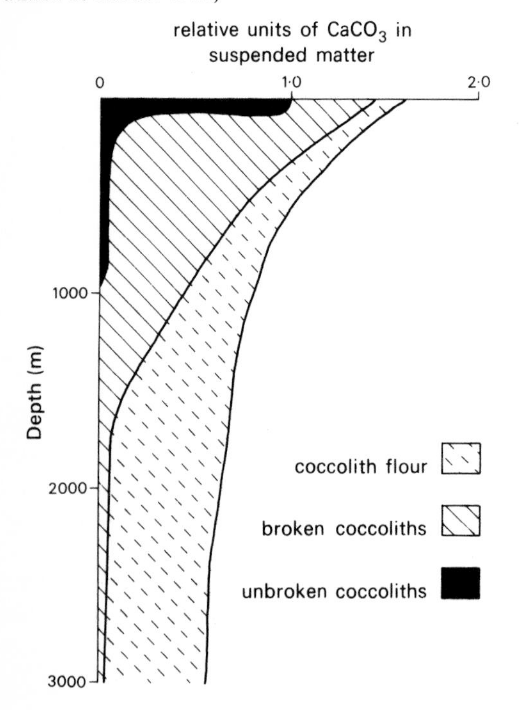

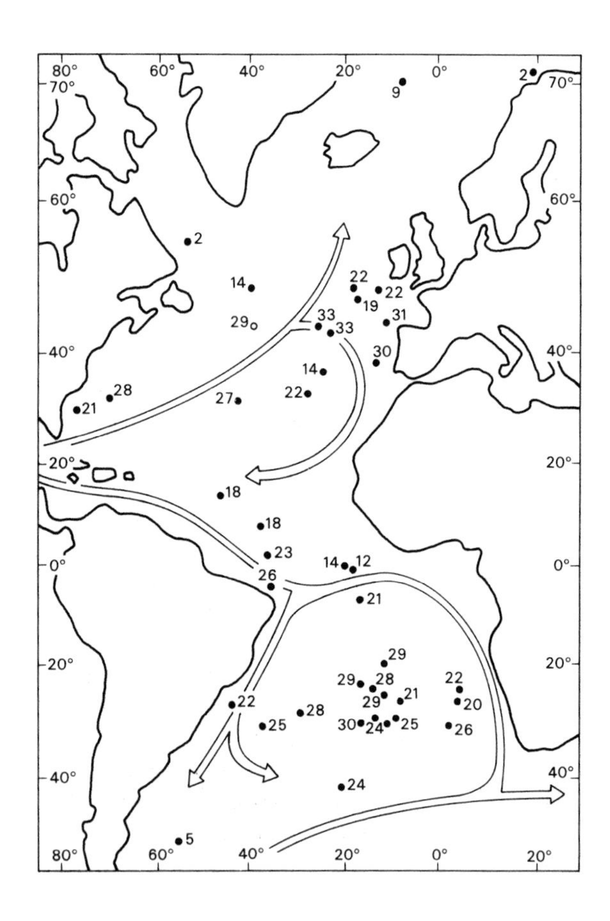

Honjo (1976) has shown, however, that coccoliths (and even whole coccospheres) can reach such depths intact by settling rapidly within the faecal pellets of copepod crustaceans.

The proportion of coccolithic material in Recent oceanic carbonates is greatest in subtropical and tropical regions underlying waters with high organic productivity. Here they may average 26% by weight of the sediment (Fig. 8.2). Coccoliths are likewise an important constituent of Cretaceous chalks (< 21%) and Tertiary chalks (< 90%). They are fewest in sediments from subglacial waters (about 1%) where both productivity and preservation conditions are unfavourable.

Unfortunately, there is a tendency for calcite overgrowths or recrystallisation to occur in coccoliths, obscuring their morphology (see Wise 1973). Solution of elements critical to the identification of fossil coccoliths may also present problems. Yet another disadvantage to the stratigrapher is the ease with which coccoliths are reworked into younger sediments without showing outward signs of wear. The rôles of coccolithophores in sedimentation are reviewed by Bramlette (1958), by various authors *in* Funnell and Riedel (1971), by Honjo (1976) and Schneidermann (pp. 1009–53 *in* Ramsay 1977).

Classification

Kingdom PROTISTA
Division CHRYSOPHYTA
Class COCCOLITHOPHYCEAE

Neither botanists nor palaeontologists can agree on how to classify the coccolithophores and their relatives. A relatively conservative botanist might regard them as belonging to the Division Chrysophyta and the Class Haptophyceae because they are unicellular, golden-brown algae with two equal flagella and a coat of scales. The palaeontological classification of Hay (pp. 1104–58 *in* Ramsay 1977), however, retains the coccolith-bearers in the older class Coccolithophyceae. Beyond this there is only a small measure of agreement between the classifications of Black (pp. 611–24 *in* Funnell & Riedel 1971, and 1972) and Hay (*ibid.*). These recent schemes are based on the ultrastructure of coccoliths and their arrangement about the cell, little of which can be seen without the aid of an electron microscope. The following genera exemplify some of the main types of heterococcolith.

Cyclococcolithina (Olig.–Rec., Fig. 8.3a, b) has a disc comprising two circular or elliptical rings (termed proximal and distal shields) built of overlapping radial plates arranged around a central, tubular pillar. Such an arrangement, with two shields connected by a central tube, is called a **placolith**. In *Pseudoemiliania* (U. Plioc.–L. Pleist., Fig. 8.3c), the radial plates of the two shields do not overlap and are arranged around a central space. The radial plates of *Helicopontosphaera* (Eoc.–Rec., Fig. 8.3e) are distinctively arranged into a single elliptical central shield surrounded by a spiral flange, also of radial elements.

The coccolith of *Zygodiscus* (U. Cret.–Eoc., Fig. 8.3f) comprises an elliptical ring built of steeply inclined and overlapping staves spanned by a cross bar. An open ring built of sixteen quadrangular grains spanned by cross bars is characteristic of *Prediscosphaera* (formerly called *Deflandrius*, M.–U. Cret., Fig. 8.3g). This genus contributed greatly to the deposition of the Cretaceous chalk. The solid spine of *Rhabdosphaera* (Plioc.–Rec., Fig. 8.3i) arises from a basal disc of fine and complex construction. Such **rhabdoliths** probably serve to reduce sinking of the cell below the photic zone.

Simpler in plan are the stellate coccoliths of the discoasters. *Discoaster* (U. Mioc.–Plioc., Fig. 8.3j) had a star-like disc up to 35 μm in diameter, built from 4–30 radiating arms of variable shape. The upper and lower surfaces also differ slightly in appearance. *Braarudosphaera* (Cret.–Rec., Fig. 8.3h) has five plates arranged with pentaradial symmetry. Discoasters are mostly found in fossil deep-sea carbonates, especially from warmer latitudes, and play an important rôle in Cainozoic biostratigraphy.

General history of coccolithophores

Being both a primary source of food in the oceans and a significant producer of atmospheric oxygen, the history of coccolithophores has a bearing on the overall history of life (see Tappan & Loeblich 1973). Precambrian and Palaeozoic records are few and dubious. The coccolith-like scales reported by Jost (1968) from the upper Precambrian of Michigan are

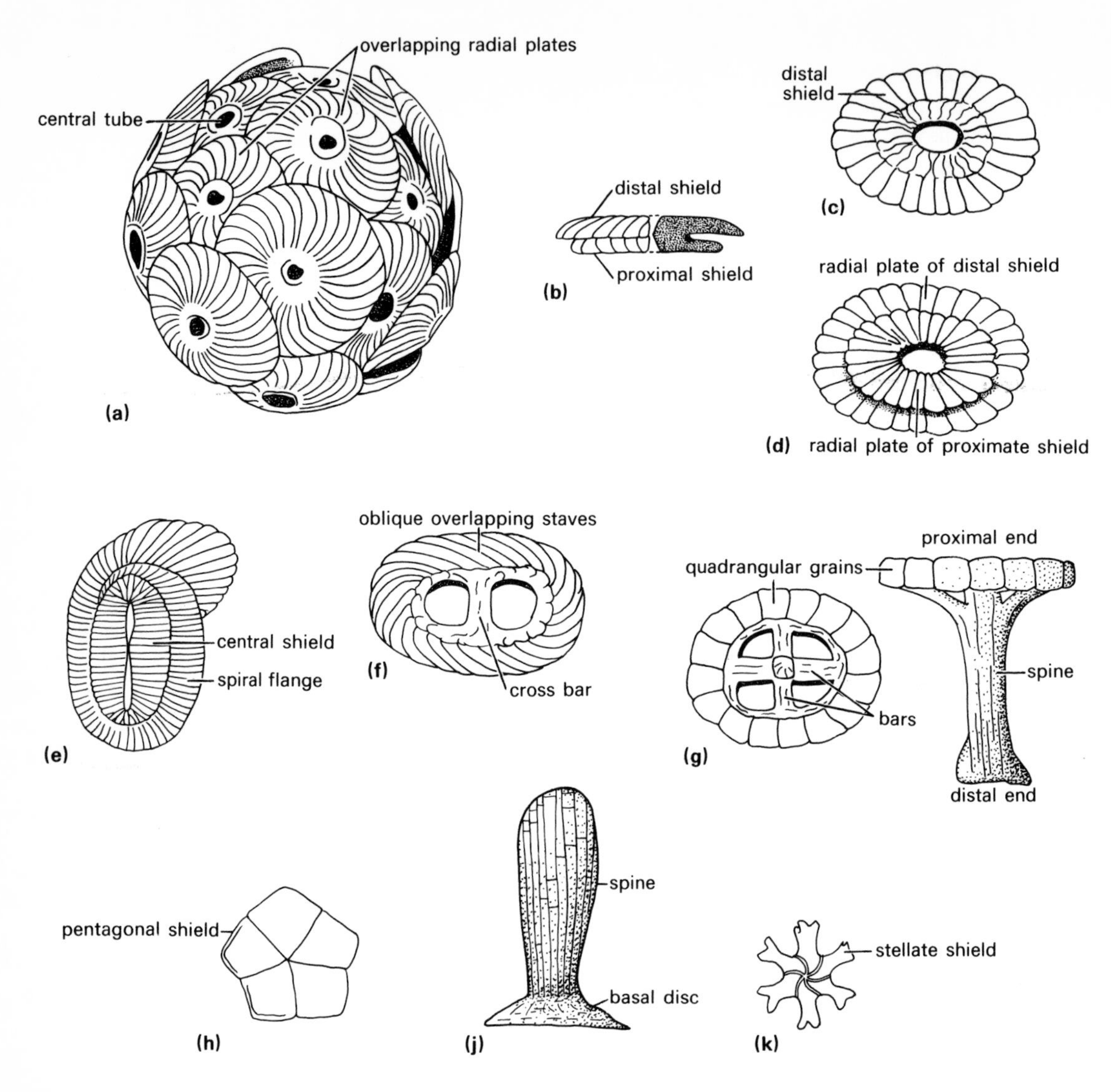

Figure 8.3 Coccoliths. (a) Recent coccolithophore *Cyclococcolithina* with coccolith shields ×2870; (b) side view of *Cyclococcolithina* coccolith, with cross section; (c) *Pseudoemiliania* distal view ×3600; (d) same from proximal side; (e) *Helicopontosphaera* ×2930; (f) *Zygodiscus* ×5340; (g) *Prediscosphaera* proximal and side view, ×4000; (h) *Braarudosphaera* ×2140; (j) *Rhabdosphaera*, side view ×4000; (k) *Discoaster* ×1000.

siliceous and more comparable in form with those of the freshwater coccolithophore *Chrysochromulina*, so it is possible that the ancestral stock of the calcareous coccolithophores has had a long history. The first generally accepted fossil coccoliths are rare and reported from upper Triassic rocks. Their diversification in the early Jurassic was a remarkable event that parallels the radiation of the peridinialean dinoflagellate cysts and both may be related to a series of transgressions connected with the opening of the Atlantic at this time.

Their numbers and taxonomic diversity increased steadily until the late Cretaceous period when there was a major marine transgression and a further, explosive radiation of many planktonic groups (Fig. 8.4). These conditions led to the deposition of chalk over vast areas of the continental platforms.

Regression at the end of the Cretaceous brought this to an end, exterminating many coccolithophore genera at the same time. There was another resurgence of forms in the Eocene, including the discoasters, many of them rosette-shaped with

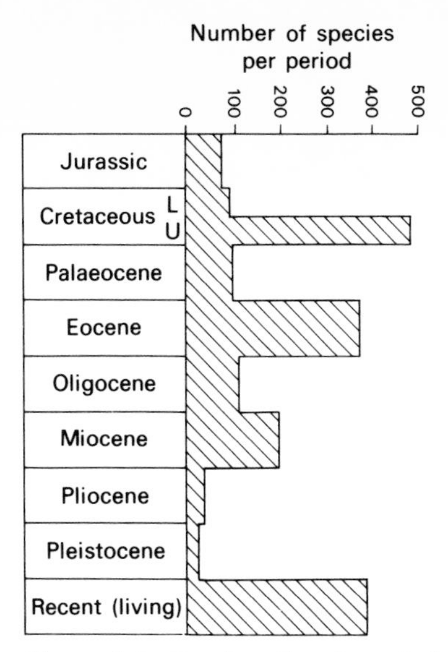

Figure 8.4 Species diversity of described coccoliths through time. (Based on Tappan & Loeblich 1973)

numerous rays. The latter kind died out at the end of the Eocene after which time there was a general dwindling in the diversity of coccoliths and discoasters, leading to extinctions at the end of the Pliocene. These may have been due to climatic cooling and regression. Certain of the placolith-bearing coccolithophores, however, thrived in the cooler waters of the Quaternary Era.

Applications of coccoliths

The biostratigraphic value of coccoliths and discoasters, leading to extinctions at the end of the (1954), since when electron microscopy has led to an information explosion. Each volume of the *Initial Reports of the Deep Sea Drilling Project* carries articles on the application of coccoliths to Mesozoic and Cainozoic stratigraphy, many with extensive illustrations (see Stradner 1973, Roth 1973, Wise & Wind 1977). The stratigraphic distribution of coccolith genera and the coccolith zones of the Mesozoic and Cainozoic are summarised by Hay (pp. 1069–104 *in* Ramsay 1977). Such zones now form the standard biostratigraphic scheme against which other groups are compared. Some interesting studies of distribution across the Mesozoic/Cainozoic boundary are discussed by Bramlette and Martini (1964) and Edwards (1973). Examples of coccolith evolution are given by Prins (pp. 547–59 *in* Brönnimann & Renz 1969, vol. 2) and Gartner (1970) and of discoaster evolution, by Bukry (1971).

Precise coccolith stratigraphy through long cores has allowed estimates of the changing rate of sedimentation through time (Hekel 1973) and Shafik (1975) uses similar data to demonstrate the presence of deep sea hiatuses. The ratio between coccoliths of warm and cool water type (e.g. *Discoaster*: *Chiasmolithus*) is a useful tool for indicating the changing palaeotemperature through late Cainozoic time (see Bukry 1973, 1975) but becomes decreasingly reliable for more remote periods. Haq and Lohmann (1976, 1977) have plotted the apparent migrations of 'warm' and 'cold' coccolith assemblages through the Cainozoic and estimated from this the changes in palaeotemperature. Worsley (1973) discusses similar palaeoclimatic aspects and the determination of depositional depth in coccolith-bearing sediments.

Further reading

For further information on collection, examination and identification to generic level, the most helpful single reference is Hay (pp. 1055–1200 *in* Ramsay 1977). Identification of genera and species may also be assisted by reference to Farinacci (1969 to date). Some aspects of their classification, ecology, distribution and evolution are brought together in a chapter by Haq *in* Haq and Boersma (1978, 79–107).

Hints for collection and study

Fossil coccoliths are abundant in Mesozoic and Cainozoic chalks and marls and are not uncommon in fossiliferous shales and mudstones. To extract them for study is relatively simple. Pulverise about 5–50 g of fresh sample (as in method A, see Appendix) and pour the liquid into a glass container to a depth of about 20 mm. After vigorous shaking allow the liquid to separate out for about two minutes and then pipette some of the supernatant liquid on to a glass slide. For a temporary mount, add a cover slip and examine the strew at 800 × magnification (or higher) with highly-condensed transmitted light under a petrographic microscope. The light should be polarised with crossed nicols so that rotation of the stage (or the slide) brings out the position of the small wheel-like coccoliths with black cross optical figures. Once located these coccoliths can be examined under normal transmitted light for evidence of their gross morphology. Permanent mounts can be prepared from strews dried on glass slides: add a drop of Caedax or Canada Balsam to the cover slip and place this over the strew.

9 Divisions Chlorophyta and Rhodophyta – Green algae and red algae

The green algae (Division Chlorophyta) and red algae (Division Rhodophyta) comprise two major groups of the Kingdom Plantae whose colour differences arise from their different photosynthetic pigments. The chlorophytes live predominantly on land or in fresh water today but there are some important marine forms, whilst the rhodophytes are dominantly marine.

Major subdivisions of these algae are based on soft-part morphology and aspects of life history, neither of which are usually revealed to the palaeontologist. For example, the green algae have reproductive cells with two or four whip-like flagella of equal length while the red algae lack these motile cells altogether. Both divisions of algae range from unicellular to complex multicellular species. Many of the latter are calcareous and although they could hardly be called microfossils in their natural state, a tendency to break apart after death soon renders them as such. Their abundance in limestones, plus their general ecological value, make these calcareous algae a significant part of micropalaeontological inquiry.

DIVISION CHLOROPHYTA

The chlorophytes owe their green colour to the photosynthetic pigments chlorophyll *a* and *b*, but also contain ß-carotene and various xanthophylls. Living and fossil green algae vary considerably in morphology, biology and ecology, hence these aspects are best commented upon separately. There are three classes of interest here: the Prasinophyceae (which may include some very ancient plankton), the Chlorophyceae (with oil-forming and limestone-building forms) and the Charophyceae, which produce the curious gyrogonites.

Class Prasinophyceae. These simple marine algae possess one or two layers of plate-like scales on the cell walls and the flagella. Both benthic and planktonic types are known, but it is with the latter type that certain ancient organic-walled microfossils have been compared (Wall 1962). The living examples have single uninucleate, spherical cells with a perforate or imperforate outer membrane. Their fossil counterparts resemble them closely. *Tasmanites* (Camb.–Rec., Fig. 9.1a) for example, had a perforate spherical cell wall about 100 to 700 μm in diameter. It thrived in Palaeozoic lagoons and sometimes accumulated to form coals. Similar vesicles of less certain affinities, such as *Leiosphaeridia* and *Chuaria* are generally classed with the Acritarcha (*q.v.*).

Class Chlorophyceae. Recent members of this class are recognised by having cells with a large central vacuole surrounded by protoplasm. Their morphology ranges from simple unicellular to complex multicellular, including types more akin to the vascular land plants, to which they are thought to have given rise.

The Order Volvocales is 'primitive' because the cells are motile, each one equipped with a pair of flagella. This order may have been represented in middle Precambrian times by *Eosphaera* (Part **c** of the figure on p. 22) and in Devonian times by *Eovolvox* (Fig. 9.1c). Small, hollow **calcispheres** called *Calcisphaera* (Fig. 9.1b) are found in lagoonal limestones from the Devonian onwards and these may also be formed by permineralisa-

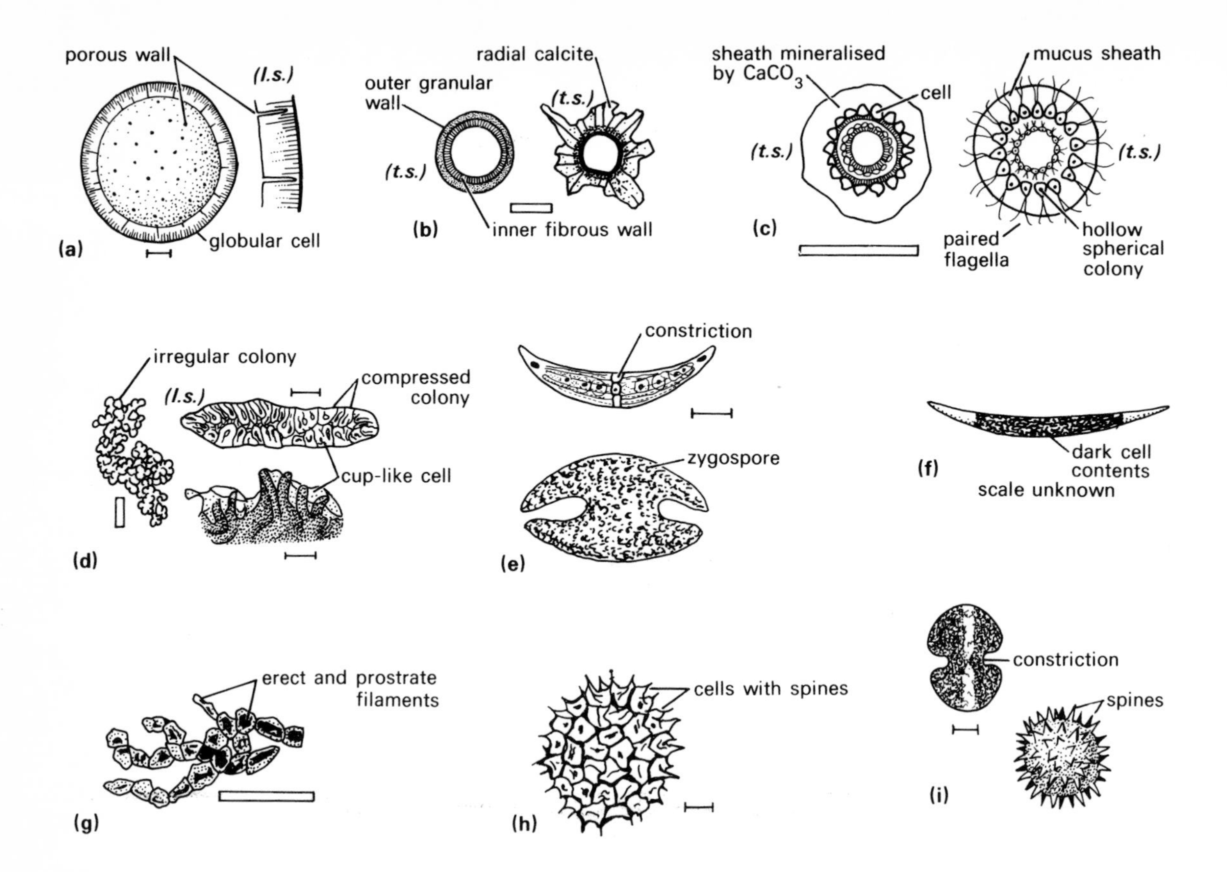

Figure 9.1 Green algae. (a) *Tasmanites*; (b) *Calcisphaera*: left = non-radiosphaerid, right = radiosphaerid; (c) *Eovolvox*: left, as preserved in $CaCO_3$; right, reconstruction with inner daughter colony; (d) *Botryococcus*: left, colony; right, sections through thallus; (e) Recent *Closterium* above, with zygospore below; (f) fossil *Closterium*-like desmid (scale unknown); (g) endolithic chaetophoralean-like alga; (h) *Pediastrum*; (i) Recent *Cosmarium* with zygospore to right. Scale: single bar = 10 μm; double bar = 100 μm; l.s. and t.s. = longitudinal and transverse sections. ((a) after Wall 1962; (b) & (c) after Kazmierczak 1976a; (d) partly after Harris 1938; (f) based on Baschnagel 1966 and (g) on Kazmierczak & Golubić 1976)

tion of *Eovolvox*-like colonies (Kazmierczak 1976, 1976a). **Radiosphaerid** calcispheres have prominent external spines and a radial calcite microstructure whilst the **non-radiosphaerid** kind have a smooth surface and microgranular structure.

The Recent colonial alga *Botryococcus* (Order Chlorococcales) flourishes today in freshwater lakes, the dead cells sometimes accumulating to form an oily deposit. Almost identical fossils, once called *Pila* or *Reinschia*, are known from at least Ordovician times onwards and likewise are associated with the formation of oils and boghead coals (Fig. 9.1d, see Traverse 1955). *Pediastrum* is another freshwater chlorococcalean but it is planktonic. It comprises from 2 to 128 cells united in a flat, discoidal colony one cell thick, the whole embedded in mucilage (Fig. 9.1h). Fossil *Pediastrum* sometimes turn up in palynological preparations and can be useful indicators of fresh to brackish water conditions of deposition (see Evitt 1963a).

Septate filamentous green algae belong mainly to the Orders Ulotrichales, Oedogoniales and Zygnematales, the last two being freshwater orders. All these may participate in the building of algal mats. The Family Desmidaceae (Order Zygnematales) have a highly ornamented cell with the contents divided into two sections by a constriction. They are known from Devonian times onwards. The Recent desmid *Cosmarium* (Fig. 9.1i) has a dumb-bell-shaped cell with a median constriction. The cyst formed after sexual reproduction (called a **zygospore**) is spiny and somewhat resembles an acritarch. The middle Devonian desmid from freshwater cherts (Fig. 9.1f) closely resembles Recent *Closterium* (Fig. 9.1e) in its lunate shape. Fossil desmids have not been widely reported, but they can be used to indicate the nature of ancient freshwater environments, such as those in the ditches around prehistoric settlements (van Geel 1976).

The Order Chaetophorales comprises forms with prostrate and erect filaments. It may have been this group which gave rise to vascular land plants in the Silurian, so it is interesting to note their possible presence in the fossil record at this time (Fig. 9.1g; see Kazmierczak and Golubić 1976) although these specimens had a boring, endolithic habit.

CALCAREOUS GREEN ALGAE. There are two orders of calcareous marine Chlorophyceae: the Dasycladales and Siphonales. Both are known as fossils since Cambrian times and both are important contributors to carbonate sedimentation in tropical shallow seas.

The Dasycladales are essentially cylindrical, spherical or club-shaped macroscopic algae attached to the sea floor by a basal holdfast (**rhizoid**, Fig. 9.2b). Regular whorls of branches (**rays**) arise from the central axis (**stipe**) and may subdivide further into secondary or tertiary branches (Figs 9.2a, b). An extracellular crust of micrite-sized ($< 5\ \mu m$) aragonite needles forms either around the stem and grows outwards or around the branches and grows inwards. After organic decay a skeleton is left, with rays represented by hollow canals or infilled with secondary $CaCO_3$ or sediment. Ultimately, the aragonite skeleton may either recrystallise to the more stable calcite crystal form or break down into lime mud. In *Oligoporella* (U. Carb.–Trias, Fig. 9.2a) for example, the rays are not branched as they are in *Petrascula* (U. Jur., Fig. 9.2b). The evolution of dasyclads is discussed by Herak *et al.* (pp. 143–53, *in* Flügel 1977) and their classification is

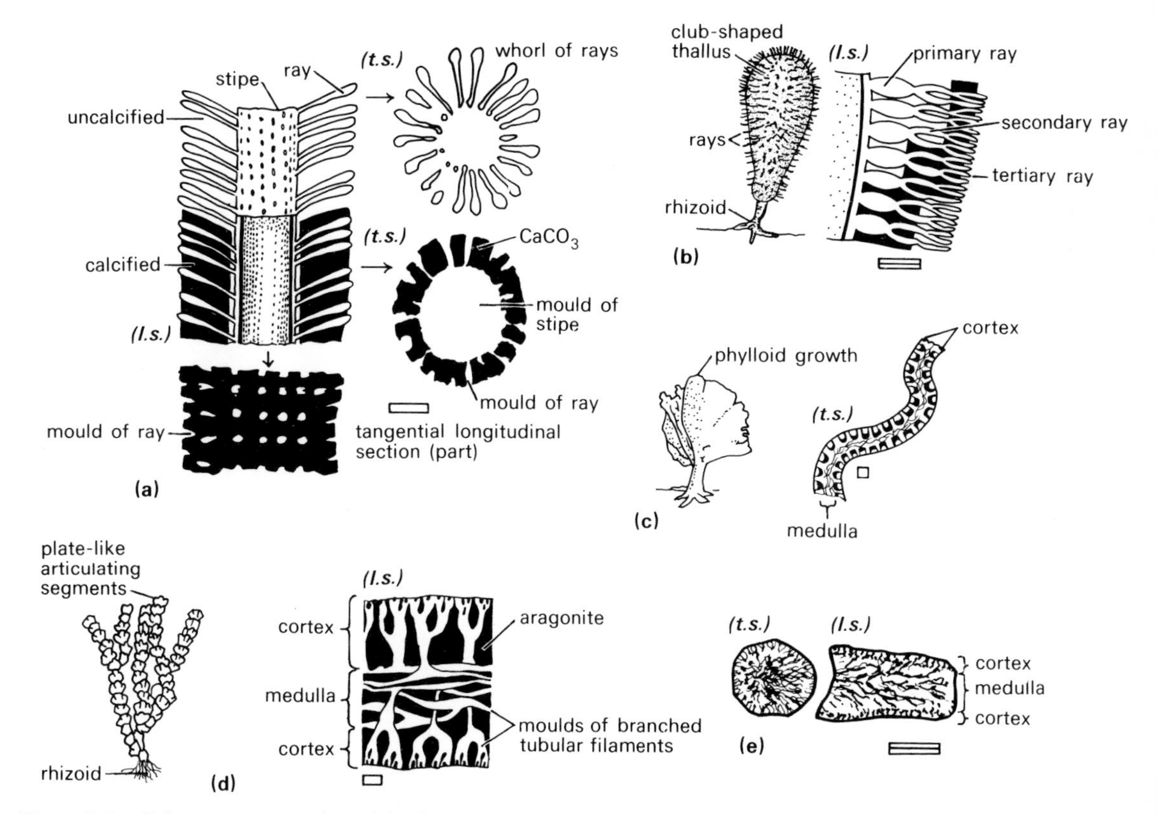

Figure 9.2 Calcareous green algae. (a) *Oligoporella*; (b) *Petrascula*; (c) *Eugonophyllum*; (d) *Halimeda*; (e) *Boueina*. Scale: double bar = 100 μm; treble bar = 1 mm; l.s. and t.s. = longitudinal and transverse sections.

considered by Bassoullet *et al.* (pp. 154–66, *in* Flügel 1977).

The Siphonales differ primarily in having few septa, if any, dividing the cell contents. Sparsely septate and branched filaments of this type are known from late Precambrian times (*c.* 1300 Ma BP, Part **d** of the figure on p. 22) but calcified forms did not appear until the late Cambrian period. These calcareous algae belong mostly to the Family Codiaceae and generally consist of tubular, freely branching filaments packed or interwoven together and bound by an aragonitic cement into sheets (e.g. *Eugonophyllum*, U. Carb.–Perm., Fig. 9.2c), articulating rods (e.g. *Boueina*, Jur.–L. Cret., Fig. 9.2e) or plates (e.g. *Halimeda*, Cainozoic, Fig. 9.2d) of virtually constant form. An outer **cortex** and an inner **medulla** can be distinguished in sections through these calcified units. Cellular decay after death usually results in their disarticulation. These porous elements can be distinguished from similar dasycladalean fragments by the absence of both radial symmetry and a central stipe in the codiaceans.

Class Charophyceae. These rather specialised algae are found mainly in fresh or brackish waters to depths of 12 m. Their fossil remains are known in calcareous shales of late Silurian and younger age. The stems, which can reach 2 m in length, consist of long **cortical cells** wound around another long **central cell** (e.g. Recent *Chara*, Figs 9.3a, b). Whorls of similar tubular, branches arise at regular intervals (**nodes**) along the stem. Reproduction is entirely sexual with distinct male and female sex organs, called **antheridia** and **oogonia** respectively. The oogonia are ovoid, consisting of a single **egg cell** spirally enclosed by long **tubular cells** (e.g. *Stellatochara*, Trias.–L. Cret., Fig. 9.3c). Calcification of stems, branches and oogonia, and especially the latter (which are sometimes called **gyrogonites**) is not uncommon. This group, with their tough, desiccation-resistant oogonia, have adapted to conditions on land in a manner comparable with that of vascular plants (see Ch. 10). Marine algae have little need for resistant fruits (zygospores) like the charophycean oogonia but they are produced by some other non-marine forms (e.g. *Cosmarium*, Fig. 9.1i) and these may be preserved as fossils.

Charophyte oogonia are sometimes of great stratigraphic value, especially in Jurassic to Oligocene freshwater deposits (see Currey 1966, Fiest-Castel 1977).

DIVISION RHODOPHYTA

The Rhodophyta or red algae are a largely marine group that vary in form from unicellular to complex multicellular, but the basic growth mode is a branched filamentous one. Their red and blue photosynthetic pigments allow them the use of the blue light of deeper (< 250 m) waters, hence they often occur at greater depths than the green algae.

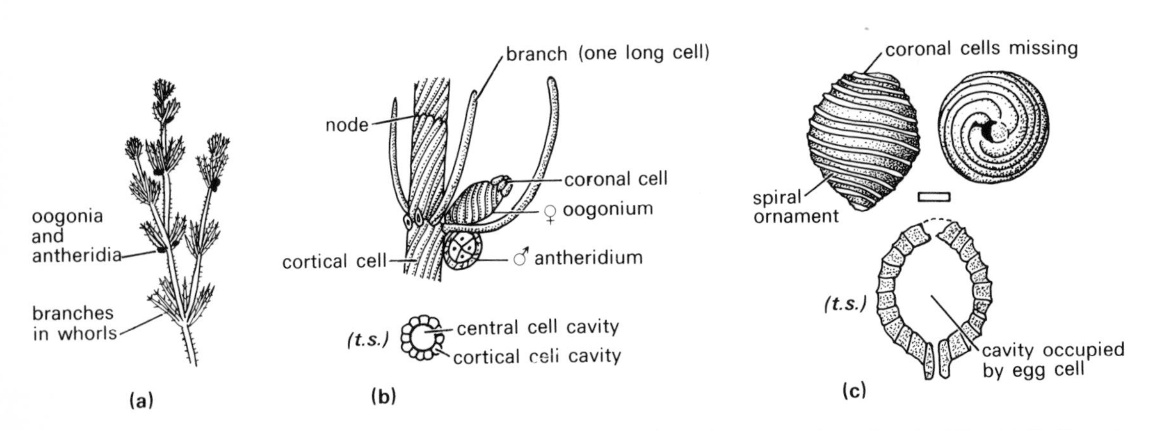

Figure 9.3 Charophytes. (a) *Chara* plant × 1; (b) Detail of stem, branches, oogonia and antheridia; (c) *Stellatochara* gyrogonite (double bar = 100 μm); t.s. = transverse section. ((a) & (b) modified from Andrews 1947; (c) based on Peck 1957)

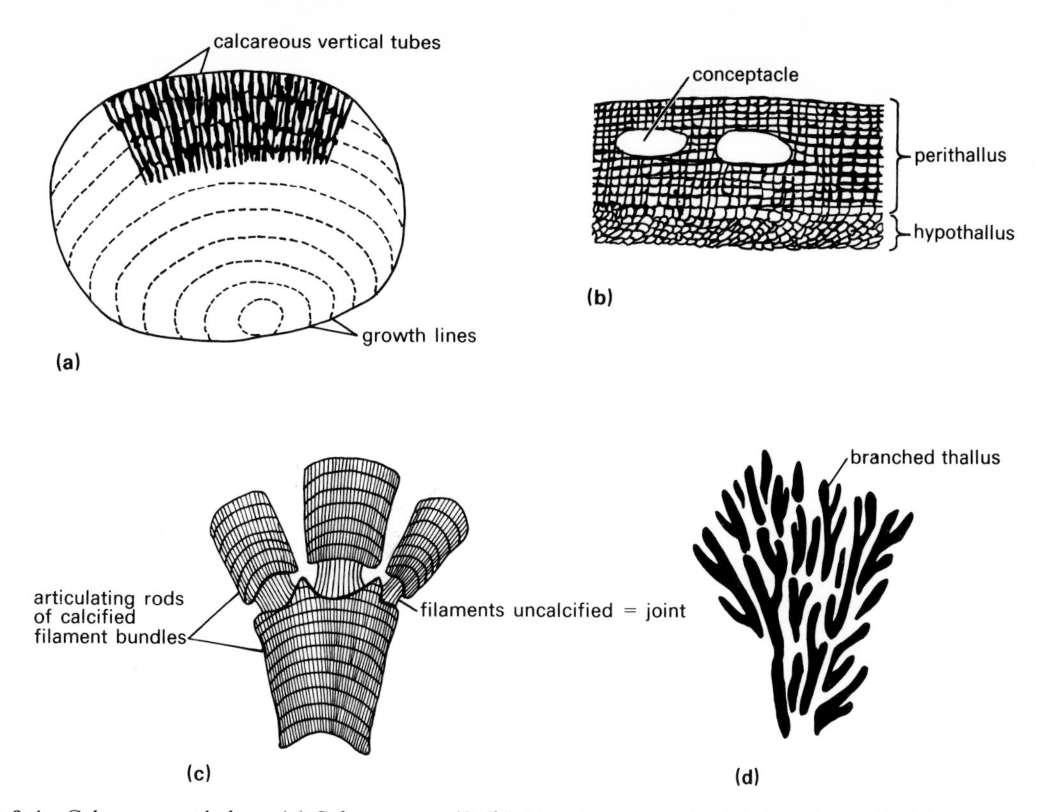

Figure 9.4 Calcareous red algae. (a) *Solenopora* ×40; (b) *Lithothamnion* ×50; (c) *Corallina* ×50; (d) *Epiphyton* ×13.

The geological history and significance of fossil red algae is a large topic that belongs more properly to palaeobotany and carbonate petrology (see Johnson 1971, Ginsburg *et al.* 1971) but microscopy is nonetheless essential for their study.

The mound-forming alga *Epiphyton* (Camb.–Dev., Fig. 9.4d) may have been a rhodophyte. The thallus was filamentous and branched, trapping or precipitating lime mud on the sticky filaments and in the spaces between.

Coralline red algae have calcified, calcitic thalli that have contributed prolifically to reef building and carbonate sediments both past and present (Wray 1971). There are two families of geological importance: the Solenoporaceae and the Corallinaceae.

The Solenoporaceae (Camb.–Mioc.) formed nodular, crustose masses of calcite comprising closely packed vertical tubes that once housed rows of cells with polygonal cross sections. Reproductive structures such as sporangia (for asexual reproduction) and conceptacles (for sexual reproduction) are never preserved and were probably external and uncalcified. In *Solenopora* (L. Camb.–L. Cret., Fig. 9.4a) the cross partitions across the cell rows are thin, widely spaced (or absent) and rarely at similar levels in adjoining tubes.

The Corallinaceae (Carb.–Rec.) form closely packed calcified threads of cells that have rectangular cross sections and usually occur in distinct layers (**perithallus** and **hypothallus**). Reproductive structures also develop within the tissues and become calcified. The important Subfamily Melobesieae contains crustose and nodular forms (e.g. *Lithothamnion*, Cret.–Rec., Fig. 9.4b) whereas the Corallineae contains filamentous forms with non-calcified articulating joints between the calcareous tubes or plates (e.g. *Corallina*, Eoc.–Rec., Fig. 9.4c).

Applications of calcareous algae

Although calcareous algae may assist in the correlation of carbonate rocks (see Poignant 1974) their

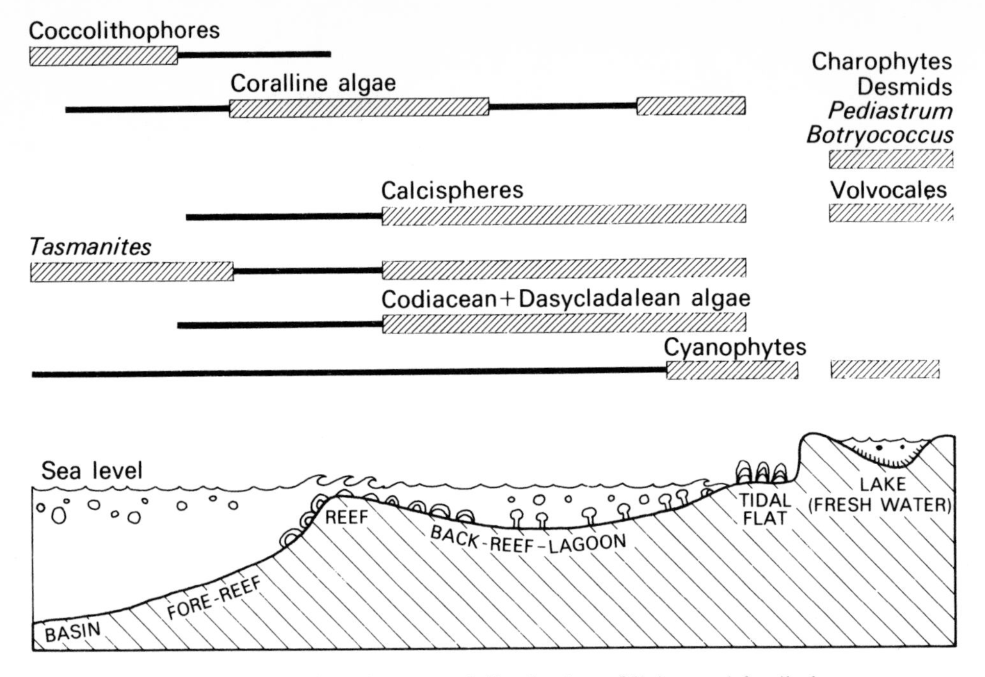

Figure 9.5 General environmental distribution of living and fossil algae.

major value is palaeoenvironmental. Wolf (1965), Patriquin (1972) and Perkins *et al.* (1972) for example, outline various aspects of their importance to carbonate production and Alexanderson (pp. 261–9 *in* Flügel 1977) discusses their rôle in carbonate cementation. The growth forms of many coralline algae tend to vary with turbulence and depth (see Bosence 1976) and may therefore prove useful as palaeoenvironmental indicators. Even transported algae can be used as indicators of different marine environments (Elliott 1975). Their general value in outlining various shallow water biofacies in carbonate sediments is illustrated in Figure 9.5 and further developed in chapters within Flügel 1977 (especially pp. 279–359).

Further reading

The main groups of marine and non marine calcareous algae are discussed by Wray (1977) in a resumé of their classification, geological distribution and palaeoenvironmental value. The latter aspect is also well illustrated by Ginsburg *et al.* (1971) and Flügel (1977).

Hints for collection and study

Recent microscopic green algae can be collected from freshwater ponds and lakes, either by scraping up the green scum from the floor or by trawling the water with a very fine nylon or muslin plankton net. Temporary and permanent mounts can be made on glass slides and studied with transmitted light, as with cyanophytes and diatoms (*q.v.*). Calcareous green and red algae may be gathered from marine rock pools and carbonate environments. They are best studied in petrographic thin sections or in acetate peels (see method N in Appendix). Specimens of larger coralline algae can be cut, polished, etched and peeled directly but smaller fragments should be embedded first in polyester resin: place about 5 cc of calcareous algal material in the cup of a polystyrene egg box or a plastic ice cube cup. Pour in the polyester resin slowly and stir to remove bubbles. Allow to dry and then section, polish and peel as in method N or prepare thin sections.

Fossil non-calcareous algae may be found in non marine argillaceous and calcareous rocks using the methods quoted for spores, pollen, dinoflagellates and acritarchs (*q.v.*). Calcispheres, codiaceans, dasycladaleans and corallines are best studied in thin sections or peels, especially from reef-associated carbonates. Charophyte gyrogonites and stems may be released from calcareous rocks by methods B, C, and D (or F if silicified) and examined by reflected light alongside ostracods and molluscs.

10 Division Tracheophyta – Spores and pollen

The term **spore** refers to any single-celled or few-celled body, produced as a means of propagating a new individual. Spores formed by bacteria, fungi, algae and protists are unfortunately rarely preserved but those of vascular land plants are very common as fossils. These terrestrial spores possess a wall that is remarkably resistant to microbial attack and to the effects of temperature and pressure after burial. Produced in vast numbers, the tiny spores can travel widely and rapidly in wind or water, eventually settling on the floors of ponds, lakes, rivers, seas and oceans. Such features make them valuable to biostratigraphy, particularly when correlating continental and nearshore marine deposits of Devonian or younger age. Where the ecology of the parent plant is known, pollen and spores can be utilised for palaeoecological and palaeoenvironmental studies, especially in Quaternary sediments.

The study of spores and pollen grains, called **palynology**, is an extensive interdisciplinary field with applications to petroleum and coal exploration, to Quaternary history, archaeology, botany and medicine. Spores and pollen are also fundamental clues in the story of land plant evolution.

Life cycles in land plants

Vascular land plants differ from their algal forebears not only in their choice of habitat, but also in their development of special conducting (**vascular**) tissues. Nonetheless, they have inherited from the algae an **alternation of generations**, which is a life cycle alternating between a spore-producing **sporophyte** generation (reproducing asexually with spores) and a gamete-producing **gametophyte** generation (reproducing sexually with male and female **gametes**). In algae it is generally the gametophyte which is larger, but in the vascular plants the sporophyte generation predominates. A simplified kind of life cycle is shown in Figure 10.1.

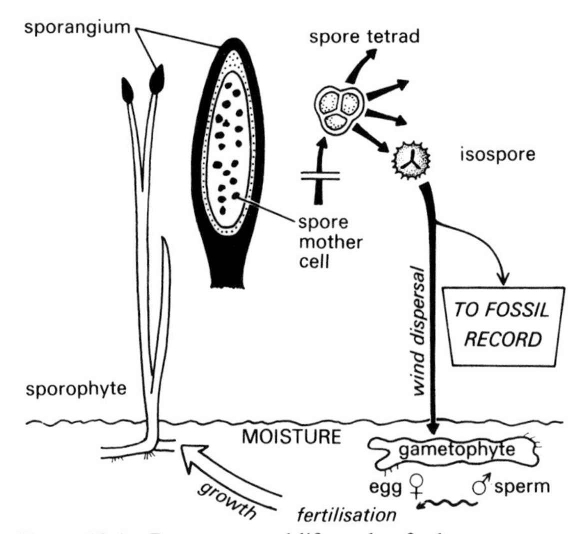

Figure 10.1 Reconstructed life cycle of a homosporous plant, the Devonian psilopsid *Rhynia*.

The sporophyte may be a small herb, a shrub or a tree, but all of these produce, at certain times of the year, spore-containing capsules called **sporangia**. At first the spore-mother-cells within these sporangia each contain the same number of chromosomes as the sporophyte plant itself (i.e. a **diploid** condition). Soon, however, each spore-mother-cell divides its contents into four new cells, collectively called a **spore tetrad**. Each cell now has half the chromosome complement of the spore-mother-cell (i.e. a **haploid** condition). This reduction division, called meiosis, initiates the gametophyte generation.

When the spores are ripe, the sporangium splits open and they are scattered to the winds. If a spore settles on a suitable, moist substrate it germinates and grows into a small gametophyte plant called a

prothallus. The mature prothallus bears male sperm cells and female egg cells, the former swimming over to the latter to fertilise them. Fertilisation, of course, re-doubles the chromosome complement of the egg cell, which now begins to divide and grow into a new sporophyte plant which rapidly dwarfs its antecedent. The whole cycle is then ready to revolve once more.

In order to understand further the nature and function of spores and pollen we must examine the major modifications of this life cycle, especially homospory, heterospory and seed-bearing.

Homospory. This is found in the life cycles of the more lowly vascular plants such as those of the Psilopsida (e.g. living *Psilotum*), primitive club-mosses (e.g. living *Lycopodium*), small horsetails (e.g. living *Equisetum*) and most ferns. As outlined above, it comprises the production of numerous identical spores, called **isospores**, that give rise to a free-living gametophyte generation (Fig. 10.1). The prothallus bears both male and female cells. Such homosporous plants probably dominated land vegetation in Silurian and early Devonian times (Fig. 10.18).

Heterospory. This is characteristic of the larger and more advanced free-sporing vascular plants such as the giant clubmosses (e.g. fossil *Lepidodendron*), giant horsetails (e.g. fossil *Calamites*) and water ferns (e.g. fossil and living *Azolla*) and appeared in separate stocks during middle and late Devonian times. In *Lepidodendron*, for example, small male **microspores** 20–50 μm in diameter and large female **megaspores** 200–400 μm in diameter were produced within special **cones** (Fig. 10.2). The microsporangia, which contained over 200 microspores, occurred near the tip of these cones, and the megasporangia, with from one to 16 megaspores, occurred near the base of the cone. After liberation and wind dispersal, the microspores of such plants developed into small male gametophytes and the megaspores developed into independent female gametophytes (Fig. 10.2).

Seed bearing. In the foregoing kinds of life cycle, fertilisation still requires water to bear the sperm on its uncertain travels. Pressure to occupy drier ground during the late Palaeozoic, however, encouraged some ferns to short-circuit the life cycle slightly. This was achieved by imprisoning the megaspore within the megasporangium and letting

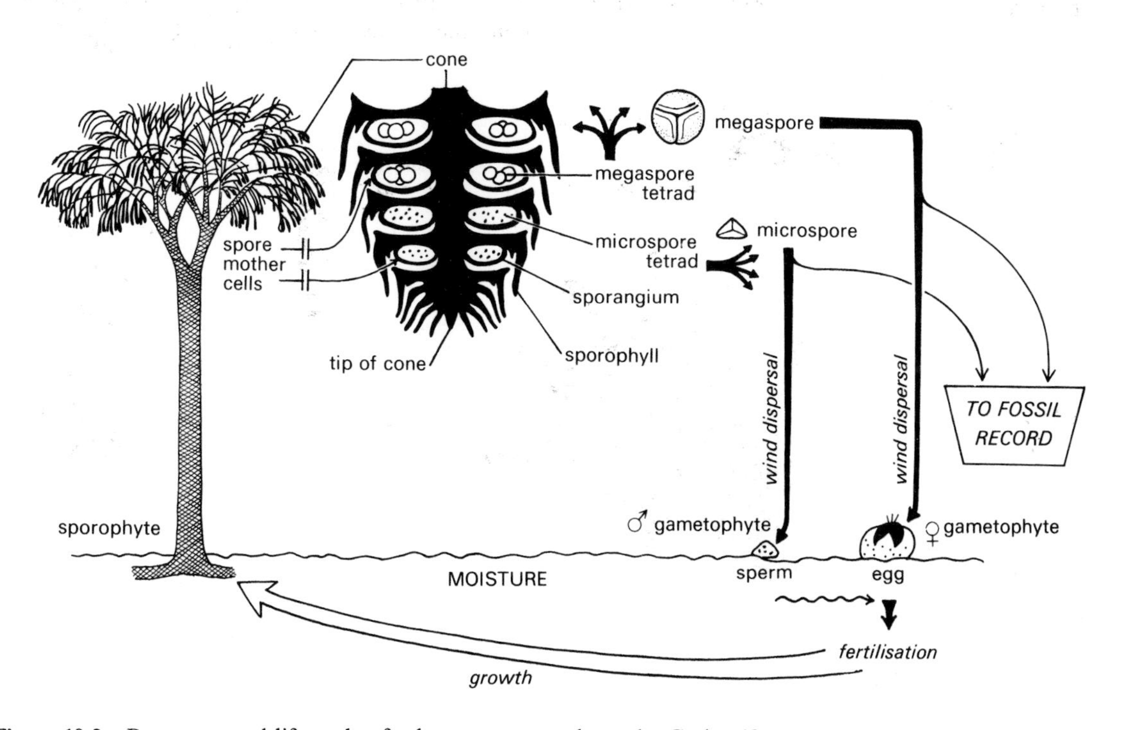

Figure 10.2 Reconstructed life cycle of a heterosporous plant, the Carboniferous lycopsid *Lepidodendron*.

it undergo its development there in the confinement of an **ovule**. Development of the spore was also telescoped, so that it divided into a tiny, few-celled prothallus before liberation. This multicellular spore is the **pollen grain**.

The function of a pollen grain, like a spore, is one of dispersal but its aim is also to reach the female cone directly and to proceed with fertilisation. The mode of fertilisation varies between the two main groups of seed bearers, the gymnosperms and the angiosperms.

GYMNOSPERMS. The **gymnosperms** (i.e. conifers and their allies) bear more or less exposed megasporangia in cones or in flower-like structures (see Fig. 10.3). The pollen grain alights on the ovule and grows a long **pollen tube** to connect with the egg cells via a slit called the **micropyle**. Sperm cells then travel down this tube and fertilisation takes place leading to the growth of a young sporophyte (i.e. a **seed**) within the cone. This seed is provided with a protective and nutritious covering and is now the main mode of propagation and dispersal. The first seed-bearing gymnospermous plants (seed ferns) appear to have evolved by mid-Carboniferous times, although seed-like structures are known in upper Devonian rocks.

ANGIOSPERMS. The **angiosperms** (i.e. flowering plants) have gone a stage further to protect the female ovule, by enclosing it within an outer jacket called a **carpel** (Fig. 10.4). This passes upwards into a prominent projection called the **style**, with a terminal knob-like **stigma** to catch the passing pollen. The microsporangia (**anthers**) are also borne on stalks, called **stamens**, and liberate their pollen to the wind, to water, or to visiting insects. Pollination involves the growth of a pollen tube down the style, through the micropyle to the ovule. After fertilisation, the seed is able to develop, protected from the climate and from ravaging insects. Such flowering plants are known first in lower Cretaceous rocks and by late Cretaceous times were a principal part of the land flora.

Spore morphology

The morphology of spores and pollen grains can be described according to their shape, apertures, wall-structure, wall-sculpture and size.

Basic shape. The shape of a spore owes much to the nature of the meiotic divisions of the spore-mother cell. In simultaneous meiosis, the mother cell splits simultaneously into a **tetrad** consisting of four

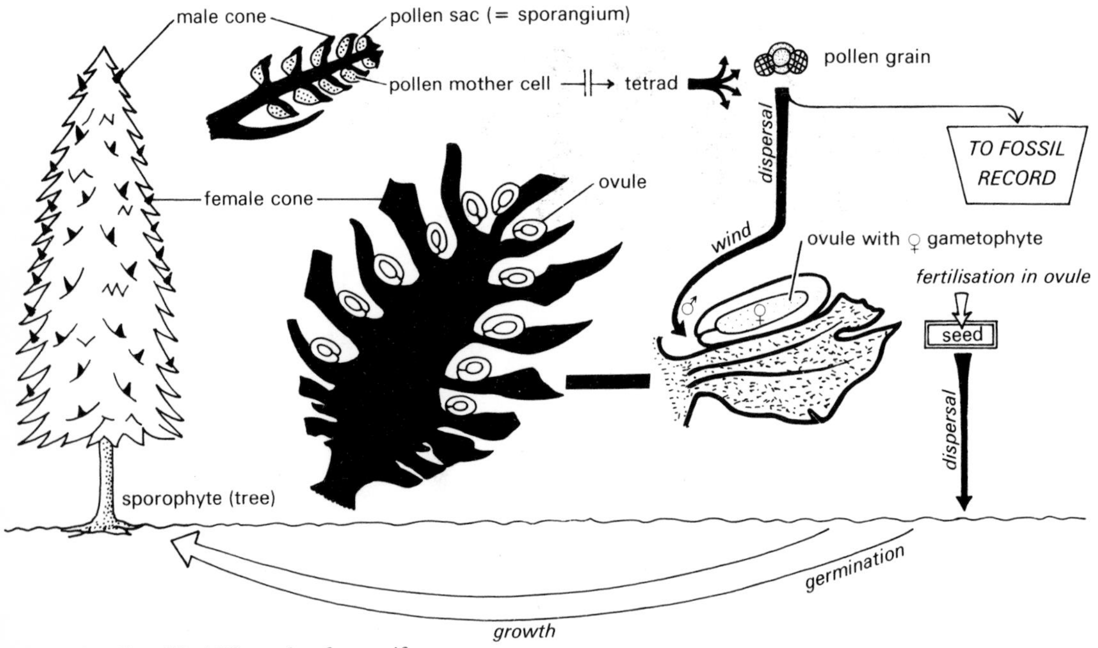

Figure 10.3 Simplified life cycle of a coniferous gymnosperm.

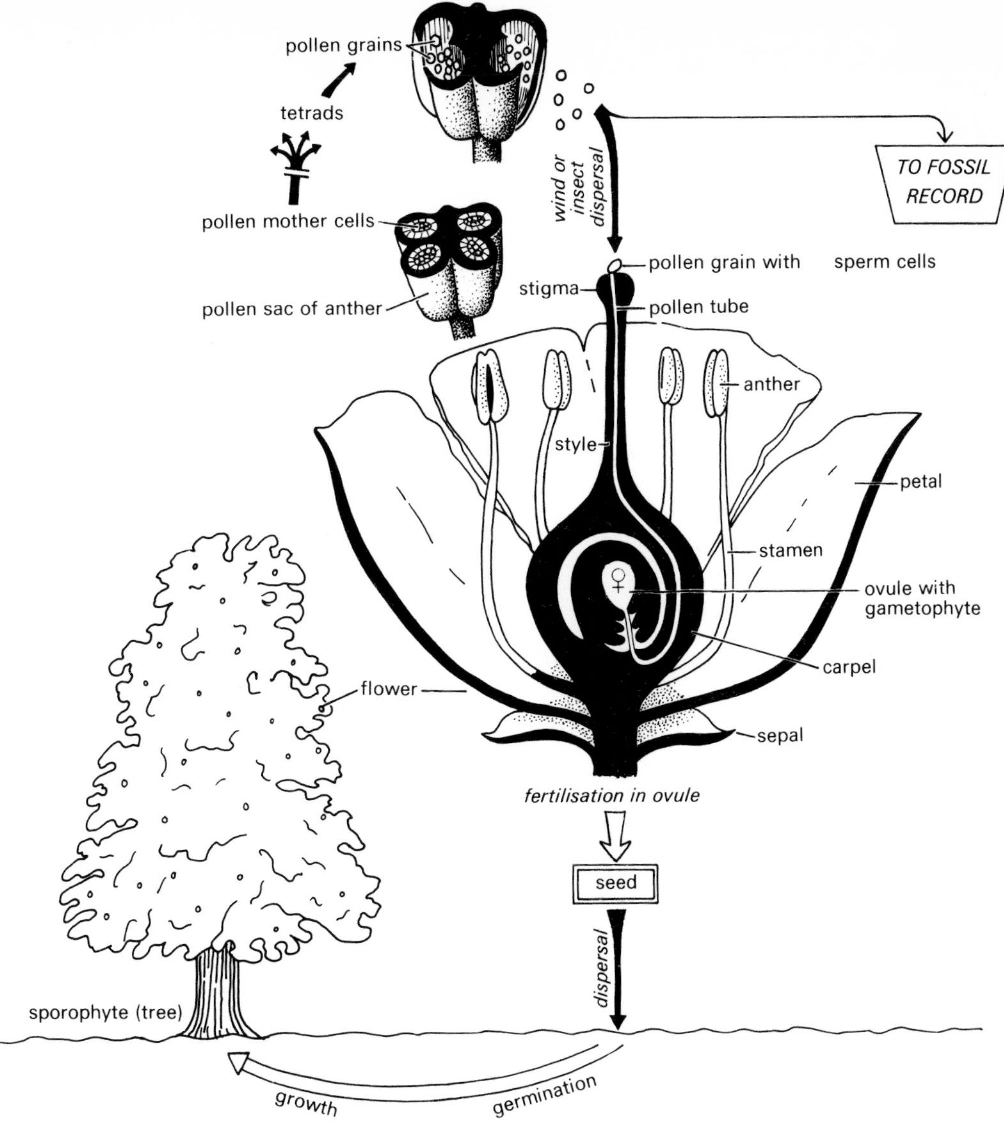

Figure 10.4 Simplified life cycle of an angiosperm.

smaller cells, each arranged as if at the corner of a tetrahedron. Such tetrahedral tetrads have ample room for all-round growth and are often subspherical in shape (Fig. 10.5). In successive meiosis, the mother cell divides at first into two cells which then subdivide further along a single plane at right angles to the first division, or along two planes at right angles (Fig. 10.5). The tetrads here are tetragonal and may more resemble the segments of an orange in shape.

That surface of a spore turned towards the centre of a tetrad is called the **proximal polar face** (Figs 10.6, 10.7). The opposite surface of the spore, forming part of the tetrad exterior, is the **distal polar face**, the two being connected by an imaginary polar axis and separated by an imaginary equator. Where the polar axis exceeds the equatorial axis in length, the spore is called **prolate**. The converse condition is known as **oblate** and a more or less equidimensional condition is termed **spheroidal**.

Description of spores is assisted by a proximal view (thereby revealing the equatorial contour or

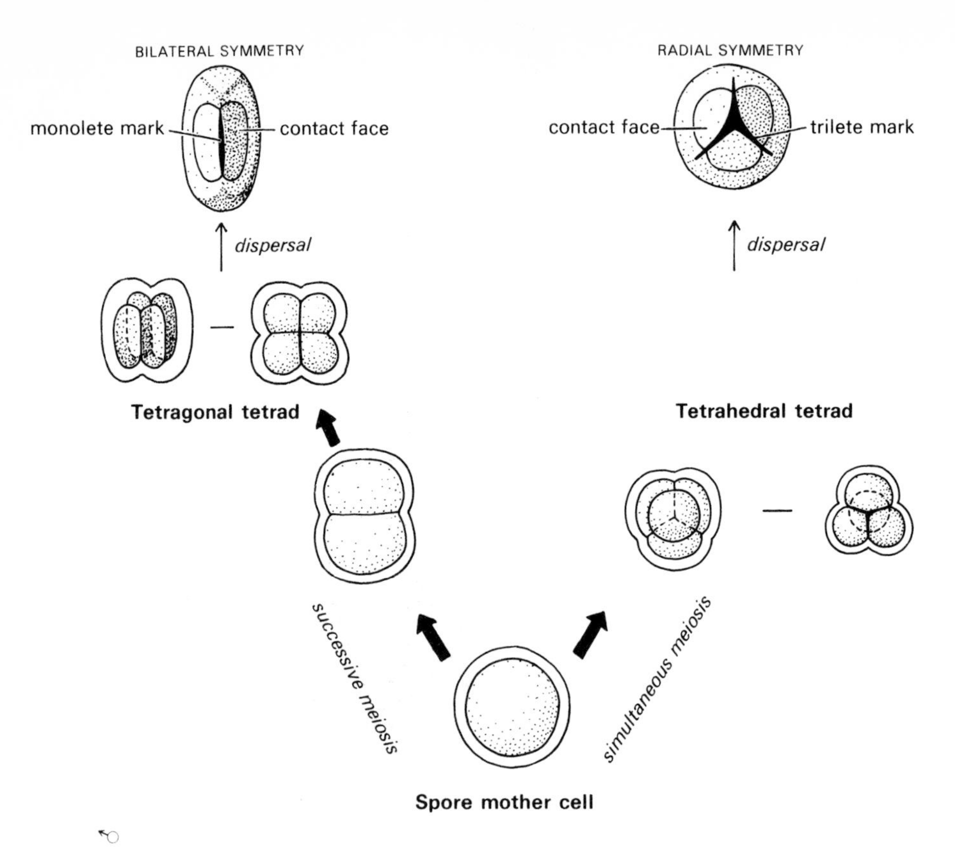

Figure 10.5 Meiosis and the production of bilateral or radially symmetrical spores.

amb, and the aperture) and by an equatorial view. In pollen grains a distal rather than a proximal view is more useful. The recognition of these features is aided by a study of the apertures.

Apertures. Spores of vascular plants are characterised by well-formed and consistently placed **germinal apertures**. These serve to allow ready germination of the prothallus and to accommodate size changes caused by humidity fluctuations. The form and position of these apertures are of great value in describing and classifying fossil spores and pollen. There are six basic kinds, with many modifications, some of which are outlined below.

TRILETE SPORES. **Trilete** spores are typical of the free-sporing vascular plants. In these the spores usually arise from simultaneous meiosis of the spore-mother cell, leading to tetrahedral tetrads (Fig. 10.5). Each spore is generally spheroidal or oblate, with three flattened **contact surfaces** around the proximal pole, marking the sites of contact with its three neighbours (Fig. 10.6). Dividing these contact surfaces are three sutures called **laesurae**, radiating from the proximal pole at 120° apart. Together these comprise the **trilete mark**. Trilete apertures serve as the site for germination of the male or female prothallus.

The symmetry of trilete spores is therefore **radial**, but **heteropolar**, i.e. with differently formed polar faces.

MONOLETE SPORES. **Monolete** spores are also produced by free-sporing vascular plants but tend to be less common. In this case the meiosis was successive, leading to oblate spores with only two flattened contact surfaces (Fig. 10.5). Hence only one laesura separates these two faces, being proximal in pos-

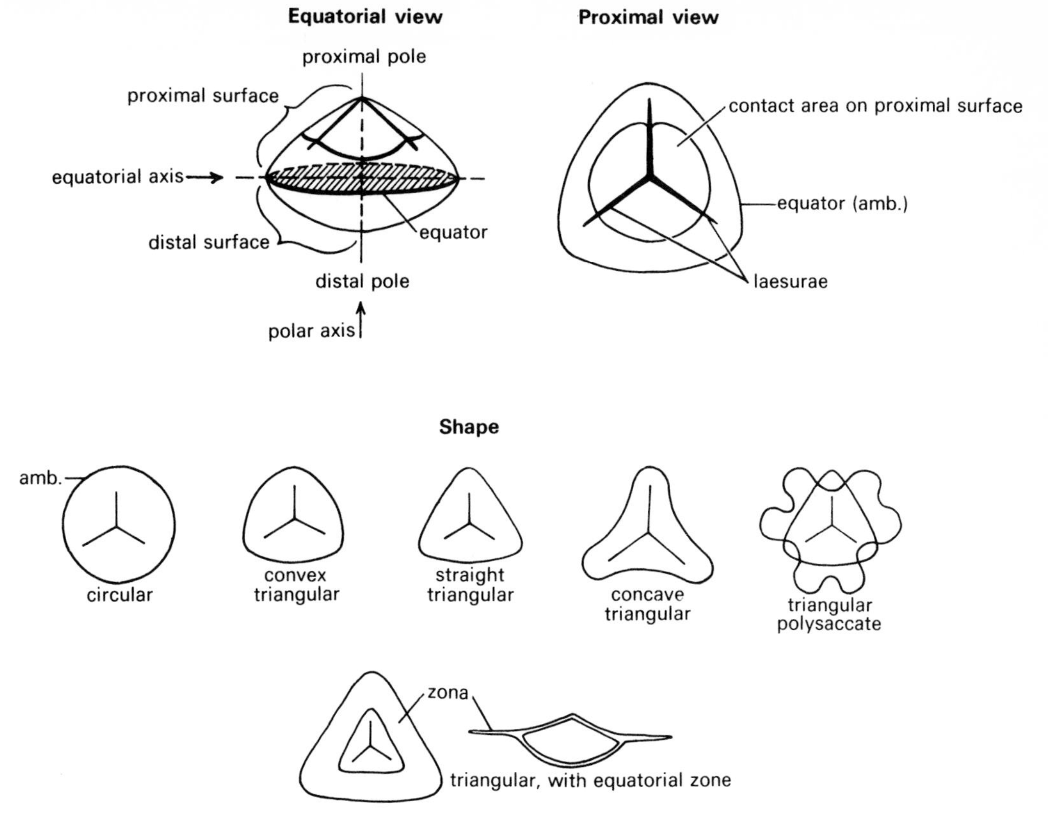

Figure 10.6 Morphology and terminology of trilete spores.

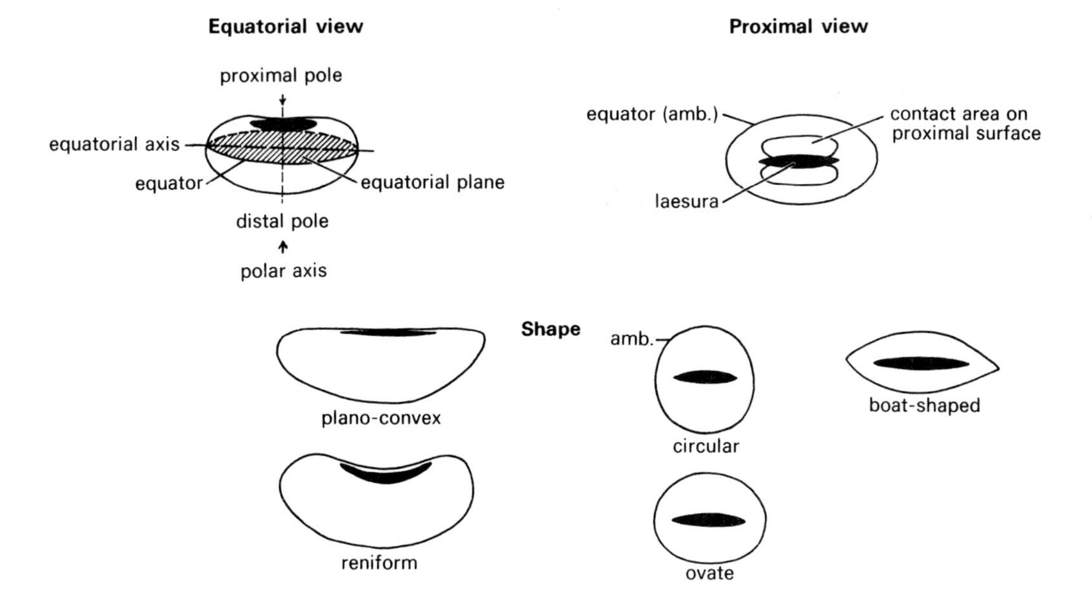

Figure 10.7 Morphology and terminology of monolete spores.

ition and comprising the **monolete mark** (Fig. 10.7). Germination takes place through this proximal aperture.

The symmetry of monolete spores is therefore **bilateral** and heteropolar.

ALETE SPORES. A number of spores have no obvious trilete or monolete mark (e.g. Fig. 10.12a) and are therefore called **alete**. These may have been produced singly rather than in tetrads.

MONOSULCATE POLLEN. In the pollen grains of seed-bearing plants, the germinal apertures are equatorial or distal in position rather than proximal as in spores. Distal furrows are called **sulci** (singular = sulcus) whilst furrows perpendicular to the equator are called **colpi** (singular = colpus; Figs 10.8, 10.9).

Monosulcate pollen bears a single sulcus along the distal face of the grain, which is usually oblate in shape (Fig. 10.8). Such pollen is typical of the gymnosperms and monocotyledonous angiosperms and arises from successive meiosis.

Like monolete spores, monosulcate pollen is bilaterally symmetrical and heteropolar.

TRICOLPATE AND MULTICOLPATE POLLEN. **Tricolpate** pollen grains are typical of the dicotyledonous angiosperms. They arise from simultaneous meiosis of the pollen-mother-cell and are typically spheroidal or prolate. There are three furrow-like germinal apertures (colpi) arranged 120° apart (Fig. 10.9).

Many variants are known on the tricolpate theme. Reduction of the furrows to small pores results in **triporate** apertures. Intermediate between these is **tricolporate** pollen, with pores situated within the colpi (Fig. 10.9). A trend towards apertural elaboration leads to the development of **hexacolpate** apertures or to a complex meshwork of colpi termed **pantocolpate** or of pores called **pantoporate** (Fig. 10.7f). Grains with more than four furrows are commonly referred to as **stephanocolpate** (or **stephanoporate** or **stephanocolporate** respectively).

The symmetry of such pollen is usually radial but **isopolar**, i.e. with similarly developed polar faces.

ASULCATE OR ACOLPATE POLLEN. Pollen grains which have no obvious germinal apertures are termed **asulcate** or **acolpate** (e.g. Figs 10.14c, 10.15b, 10.17e).

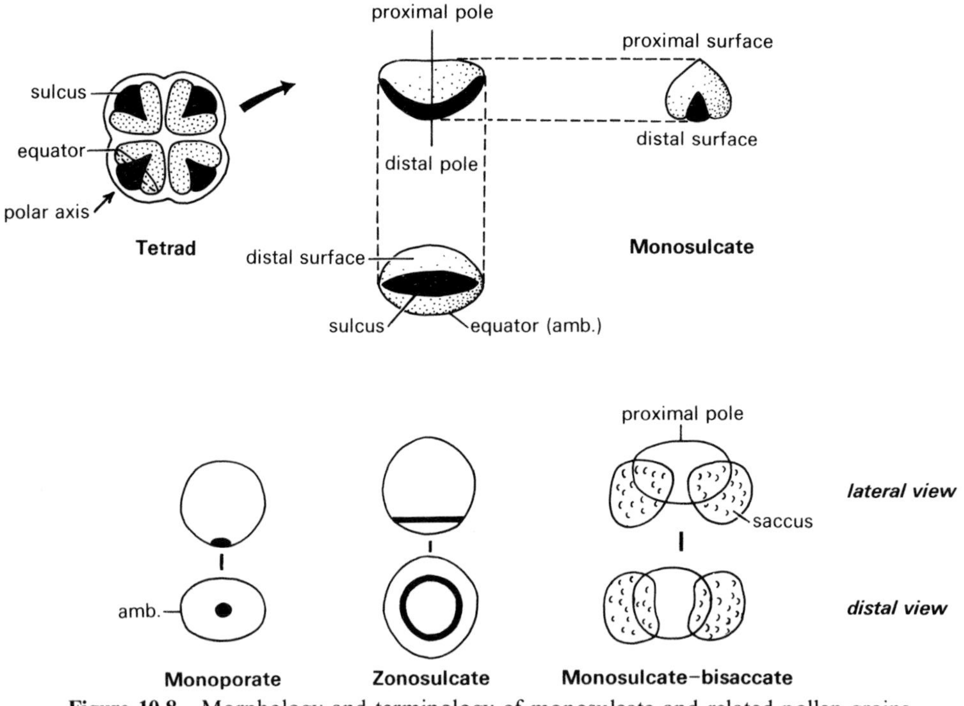

Figure 10.8 Morphology and terminology of monosulcate and related pollen grains.

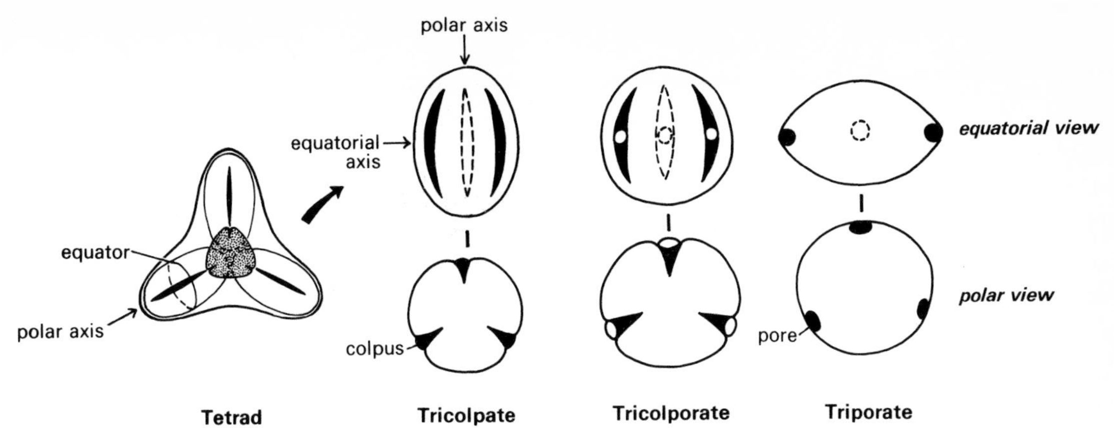

Figure 10.9 Morphology and terminology of tricolpate and related pollen grains.

Wall composition and structure. The spore wall serves to protect the young gametophyte from desiccation and microbial attack, but must also be amenable to rapid germination at the required moment. It therefore has a double wall comprising an inner **intine** and an outer **exine**.

The intine is an ordinary cell wall made of cellulose. It is this layer which expands rapidly to form the prothallus or pollen tube. The exine layer, however, is no ordinary structure. It is a highly resistant waxy coat of a material called **sporopollenin**, the 'simple' formula of which varies from $C_{90}H_{134}O_{20}$ to $C_{90}H_{142}O_{35}$. The structure of this exine provides a valuable means of classifying spores and pollen grains, defining two main groups: atectate and tectate.

Atectate spores are typical of the free-sporing vascular plants and atectate pollen grains are typical of most gymnosperms and a few primitive angiosperms. In this condition the exine may be more or less a homogenous layer or it may be stratified. For example, in **cavate** spores and **saccate** pollen, two layers of the exine coat are separated by an **air sac** (e.g. Figs 10.13f, 10.14b & c, 10.15a & c).

Tectate exine is typical of most angiosperm pollen grains. Here the exine comprises an inner, non-sculptured **nexine** and an outer sculptured **sexine** layer, the latter divided into rod-like **columellae** supporting the roof-like **tectum** (Fig. 10.10).

Sculpture. The superficial sculpture of the exine is of considerable importance in the description and classification of spores and pollen grains (Fig. 10.10). In atectate spores and pollen the surface may be smooth (**psilate** or **laevigate**), covered with small grains (**granulate**), pitted (**foveolate**), grooved (**fossulate**), with mesh-like sculpture (**reticulate**), with fine parallel grooves (**striate**), warty (**verrucate**), with rod-like projections (**baculate**), with pointed projections (**echinate**) or with club-shaped projections (**clavate**). In many spores the equatorial region bears a thickened exine (a **cingulum**) or a thin exine flange (a **zona**) extending beyond the edge of the spore (see Figs 10.11b, d).

In pollen grains the clavate condition, by expansion of the tops, may give rise to a perforate tectum supported by columellae (i.e. tectate). The tectum surface may be smooth or sculptured in much the same way as outlined above.

Size. Ideally, radially symmetrical grains should be measured in two directions – length of polar axis and length of equatorial axis; bilateral grains should be measured from the length of the axis parallel to the laesura or sulcus and the length of the maximum transverse direction. However, fossil spores and pollen grains are invariably so flattened that measurements can only be taken in one plane for a given specimen.

Isospores, microspores and pollen grains (known collectively as **miospores**) generally range in size from 5–200 μm, and megaspores tend to exceed 200 μm in maximum diameter. Size variation within a species is quite limited, but measurements should take into account whether the grains have become swollen with water or by treatment with alkali during maceration, as this is their natural tendency.

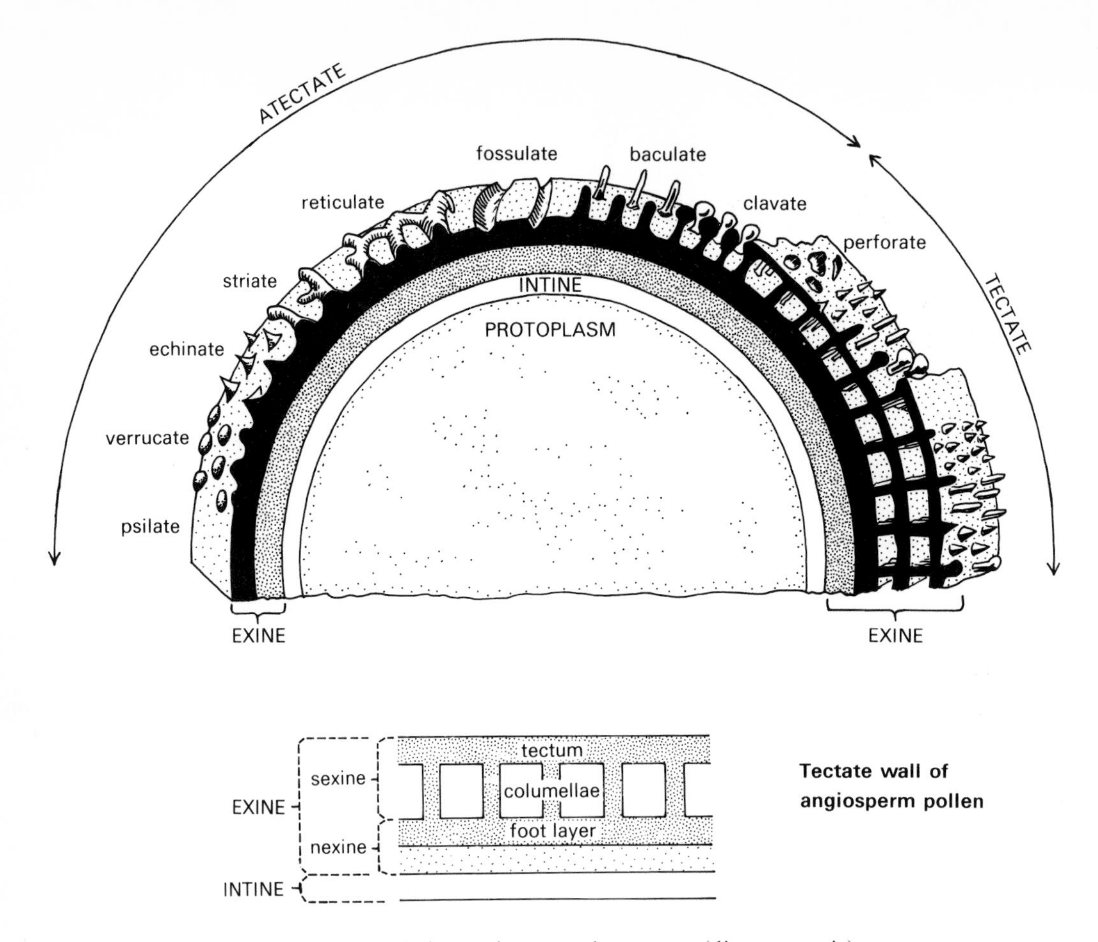

Figure 10.10 Exine sculpture and structure (diagrammatic).

Functional morphology of the exine. The exine coat has four main functions: protection, accommodation of size changes, dispersal and germination.

PROTECTION. Sporopollenin is an extremely stable material that resists biological enzymes and acidic decay. It also serves to protect the inner protoplasm from desiccation and mechanical abrasion. Among animals, only the goat is so far known to digest the exine with enzymes; in other creatures, spores and pollen tend to pass through the gut with mechanical damage alone or are unharmed. This resistance to acid makes possible, of course, the treatment of sediments with a variety of acids for the release of their contained spores and pollen.

SIZE CHANGES. Increases in atmospheric humidity cause the protoplasm of the spore to expand and vice versa. These changes in volume require the exine to be elastic, although much may be taken up by expansion or contraction of the germinal apertures. Air sacs and tectate walls may also help to accommodate these volume changes.

DISPERSAL. Spores are usually dispersed in the wind, but pollen grains may also be dispersed in water or by insects. Aerial dispersal is favoured by small size, low specific gravity and a high surface-area-to-volume ratio. Most floating pollen grains are therefore small (17–58 μm in diameter) and smooth walled with a thin exine. The air sacs of gymnosperm pollen grains may also assist floating. Insect-carried pollen grains tend to have a thicker, sculptured exine with prominent spines and a coat of sticky material. They are either relatively large (59–200 μm) or small (5–16 μm). Water pollination occurs in sea grasses, but here the exine coat has been lost, presumably because the risk of desiccation at sea is negligible.

GERMINATION. If conditions are sufficiently moist, the germinal apertures in the spore exine expand and the prothallus is able to push through. Likewise, the monosulcate aperture of gymnosperm pollen allows the pollen tube to find the exposed micropyle and begin fertilisation. In angiosperms, however, the pollen grain must be even more efficient. Not only must it penetrate the stigma, style and carpel before reaching the micropyle, it must also achieve fertilisation quickly. This is because many flowering plants live in ecological niches that favour brief flowering and growth seasons or short one- or two-year life cycles. The cavities in their tectate pollen walls therefore contain chemicals that react rapidly to contact with the correct kind of stigma and initiate germination. The multiple apertures of dicotyledonous pollen may also ensure that no time is lost in sending out the pollen tube.

Distribution and ecology

The ecology of spores and pollen reflects, of course, the ecology of their parent plants. Because of size sorting in sediments, however, the leaves, wood, seeds and spores of a plant are rarely preserved together. The habitat and ecology of the spore-producing plants can, none the less, be inferred, but this must be preceded by an understanding of dispersal and sedimentation.

Dispersal and sedimentation. Wind-dispersed spores and pollen are produced in far greater numbers than those dispersed by insects, mostly because of the uncertainty of the former process. Many of those carried by insects are also larger and therefore may not travel so far in wind and water. The distance travelled by air-borne pollen depends greatly on their size, weight, surface sculpture and on atmospheric conditions. Miospores are most frequently found about 350–650 m above the land surface during the day, but many sink to the surface at night or are brought down by rainfall. Under favourable conditions pollen grains have been known to drift for at least 1750 km, although about 99% tend to settle within one km of the source. Very little ever reaches the oceans by aerial dispersal.

Once the pollen grains or spores have settled, they stand a chance of entering the fossil record, either by falling directly into bogs, swamps or lakes, or by being washed into them and into rivers, estuaries and seas. By this stage the pollen record has already been filtered by differential dispersal in the air and may now undergo a similar filtering in the water. For example, large miospores and megaspores will tend to settle out in rivers, estuaries, deltas or shallow shelf areas, whereas small miospores may settle out in outer shelf and oceanic conditions. Those which are not covered by water will tend either to germinate or become oxidised.

Being both light-weight and resistant, spores and pollen may suffer several cycles of reworking and redeposition, leading to some confusion in the fossil record. Experienced palynologists detect these reworked forms by differences in preservation (e.g. colour, corrosion, wear), ecological or stratigraphical inconsistencies and associated evidence for reworking.

Pollen analysis. Pollen analysis involves the quantitative examination of spores and pollen at successive horizons through a core, particularly in bog, marsh, lake or delta sediments. This method yields remarkable information on regional changes in vegetation through time, especially in Quaternary sediments where the parent plants are best known, but it has also been used with success in older deposits such as Carboniferous coals.

The relative frequencies of different spore and pollen types are calculated for each of a number of closely spaced sample horizons through the core. The tree pollen (e.g. oak, elm) is often summed together, whilst the non-tree or non-arboreal pollen (NAP, e.g. willow, hazel) may be summed separately, although expressed as a percentage in relation to the tree pollen. Bog, heath and lake-vegetation spores and pollen (e.g. those of sedges, grasses and heather) may also be expressed independently, but again in relation to tree pollen. The **pollen spectra** of each species are then arranged alongside to give a **pollen diagram** of palynological changes through the core (Fig. 10.20).

Such diagrams will invariably give a biased impression of the flora. Apart from the adverse effects of dispersal discussed above, one should bear in mind the frequency of flowering or dehiscence, the number of sporangia, cones or flowers, their position relative to the dispersal agencies and the preservation potential of various spores and pollen.

Reviews of pollen analysis are given by Erdtman (1943) and West (1971).

Classification

Kingdom PLANTAE
Division TRACHEOPHYTA

Most fossil spores and pollen grains are placed in genera and species based entirely on their exine morphology (i.e. form taxa). The ultimate aim, however, must be to relate these morphotypes to the 'natural' biological classification. As comparative morphology with Recent grains is not always possible or even reliable, it is therefore necessary to examine fossil sporangia, cones and flowers for *in situ* associations.

Many pre-Quaternary dispersed grains, however, are of unknown botanical affinity. In Palaeozoic and Mesozoic microfloras this problem is solved by the use of form genera and species which, under the International Code of Botanical Nomenclature (ICBN), cannot be allocated to the usual supra-generic taxa. The form taxa are therefore grouped into categories based on the word **turma**. Spores are included in the Anteturma Sporites and pollen in the Anteturma Pollenites (see Potonié & Kremp 1955, 1956). Below this rank there is a descending hierarchy of taxa: Turma, Subturma, Infraturma and so forth, each based on rigid morphological criteria. The problem for Tertiary palynologists is more acute because of the similarities between Tertiary and extant pollen grains. This has led some workers to use the name of the extant genus. Others argue that this could be misleading because of different evolutionary rates between the reproductive and vegetative parts of plants; they prefer to use a form genus for the grains, or to add a suffix to the root of the generic name. Hence oak-like pollen grains of Tertiary age may variously be referred to as *Quercus, Quercoides, Quercoipollenites* or *Tricolpites*.

There are other approaches to this problem. Oil company palynologists save time by giving each morphotype a computerised serial number. Boulter and Wilkinson (1977) have used a rapid grid system for naming Tertiary pollen, whilst Hughes and Croxton (1973) devised a binomial nomenclature outside the statutes of the ICBN.

The suprageneric classification outlined below follows the established botanical scheme of Fuller and Tippo (1949). Because the roots, stems, leaves, cones, flowers, seeds, pollen and spores of fossil plants tend to be preserved separately, they have usually received independent binomial names (i.e. parataxa) which, none the less, remain valid under the ICBN. These fossil spores and pollen grains are generally described and classified on the basis of their exine structure, germinal apertures, shape and symmetry, sculpture and size range.

Vascular land plants belong to the Kingdom Plantae, the Division Tracheophyta and to one of the four subdivisions: Psilopsida, Lycopsida, Sphenopsida and Pteropsida.

Subdivision Psilopsida. The Psilopsida (U. Sil.–Rec.) are small plants without leaves that bear the exposed sporangia directly on the stem (Fig. 10.1).

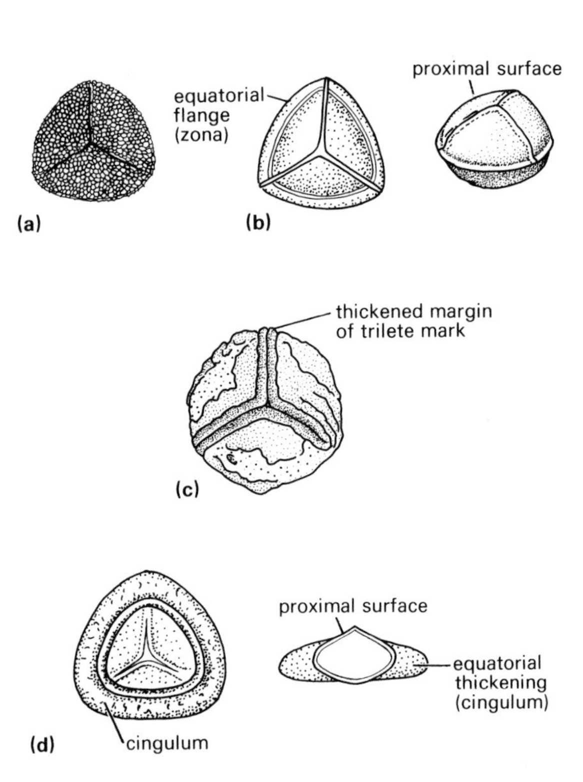

Figure 10.11 Psilopsid and lycopsid spores. (a) *Rhynia* isospore, proximal view ×267; (b) *Lycospora*, proximal and proximolateral view ×333; (c) *Triletes*, proximal view ×27; (d) *Densosporites*, proximal view and transverse section, about ×534. ((a) after Scagel *et al.* (1965) from Kidston & Lang; (b) & (c) modified from Potonié 1962 and (d) from Tschudy *in* Tschudy & Scott 1969)

The life cycle is homosporous with trilete or monolete isospores. Those of the fossil *Rhynia* (Dev., Fig. 10.11a) for example, are 35–65 μm in diameter, with a trilete mark, subtriangular amb and a coarsely granular to baculate sculpture.

Subdivision Lycopsida. The Lycopsida (U. Sil.–Rec.) or club mosses, have ranged in size from the large Carboniferous *Lepidodendron* some 30 m tall, to the small mossy *Selaginella* and *Lycopodium* of Recent times. Small leaves (**microphylls**) are developed around the stem whilst the sporangia are borne on special leaves (**sporophylls**) that are usually collected together in cones. Lycopsid spores are usually trilete (rarely monolete) and often heterosporous.

Lepidodendron is known primarily from its stems and leaves, but its cones (*Lepidostrobus*) are also known, being up to 300 μm long and containing microsporangia near the tip and megasporangia near the base (Fig. 10.2). The microspores (*Lycospora*, Dev.–Perm., Fig. 10.11b) were some 20–50 μm across, trilete, circular to subtriangular in amb, with a smooth wall extending into a narrow flange around the equator. The megaspores (*Triletes*, Dev.–Rec., Fig. 10.11c) were 200–400 μm in diameter, with a prominent trilete mark and a subtriangular amb. *Triletes* is often smooth on the proximal surface and roughened on the distal surface.

Densosporites (Dev.–Carb., Fig. 10.11d) is a dispersed microspore genus some 28–48 μm across, with a faint trilete mark, a smooth-to-spinose exine and a thickened equatorial zone. It is locally abundant in coals formed in marginal swamp conditions and was produced by herbaceous lycopods such as *Selaginellites*.

Subdivision Sphenopsida. The Sphenopsida or horsetails (L. Dev.–Rec.) are mostly herbaceous plants although the Carboniferous *Calamites* grew to heights of at least 10 m. Their stems are distinctly jointed, each joint bearing whorls of smaller branches or small leaves. Sporangia occur at the tips of stems or branches and may be homosporous or heterosporous.

Spores of the living horsetail *Equisetum* (Fig. 10.12a) are alete and provided with photosynthetic

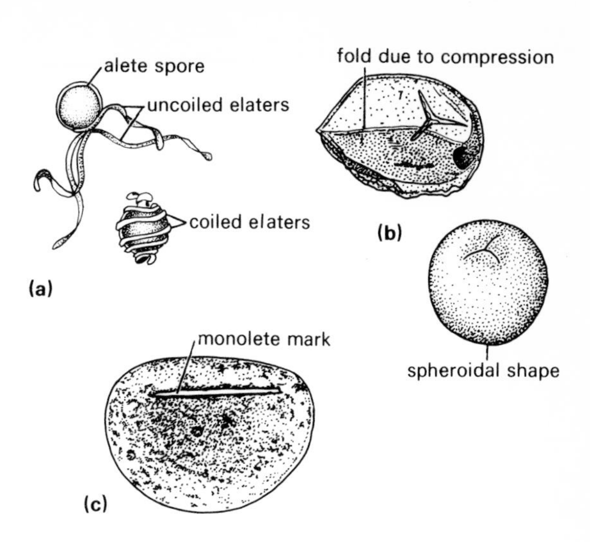

Figure 10.12 Sphenopsid spores. (a) *Equisetum* with coiled and uncoiled elaters, about ×120; (b) *Calamospora*, proximal view of compressed spore (above) and appearance before compression (below), about ×500; (c) *Laevigatosporites*, proximolateral view ×534. ((b) partly after Potonié & Kremp 1955)

pigments, hence they have a very thin exine. They bear four flexible **elaters** that can coil or uncoil with changes in humidity. Apparently their function is to split open the sporangium and to ensure dispersal of the spores in clusters. Similar spores, called *Elaterites*, have been found in Carboniferous cones of calamitean type. Most such cones, however, have yielded the spores called *Calamospora* (Fig. 10.12b). These range from 30–1000 μm in diameter and are spheroidal with thin smooth walls and short trilete marks. The size range is thought to indicate that *Calamites* was incipiently heterosporous.

Laevigatosporites (U. Carb., Fig. 10.12c) is a dispersed spore that may have had sphenopsid origins. It was monolete, bilateral and occasionally very large (20–500 μm).

Subdivision Pteropsida. The Pteropsida comprise the ferns and all the seed-bearing plants. These have developed larger leaves called **macrophylls** and their sporangia are borne either on such macrophylls or on modifications of them. Three classes are recognised: the Filicineae, Gymnospermae and Angiospermae.

CLASS FILICINEAE. The Filicineae or ferns (M. Dev.–Rec.) are the most primitive pteropsids because they have retained an independent gametophyte in their life cycle. Terrestrial ferns are homosporous with trilete or monolete spores, usually from 15–90 μm across and of very varied sculpture.

Spores of the Carboniferous fern *Oligocarpia* are about 30 μm across and subtriangular with smooth walls and a trilete mark (Fig. 10.13a). Spores of the Jurassic tree fern *Coniopteris* (Fig. 10.13b) are about 55 μm across, triangular in amb, with thick walls, a finely granulate to verrucate sculpture and a trilete suture. These may be referred to the dispersed-spore genus *Deltoidospora*. Recent *Polypodium* ferns produce bilateral, monolete spores about 40 μm long often with a verrucate or baculate sculpture (Fig. 10.13c). Dispersed-spore genera such as this may be called *Polypodiisporites*.

The delicate water ferns are rare as fossils but their spores are known in Jurassic and younger rocks. These are usually spheroidal and trilete and comprise both microspores (50–75 μm across) and megaspores (200–800 μm across).

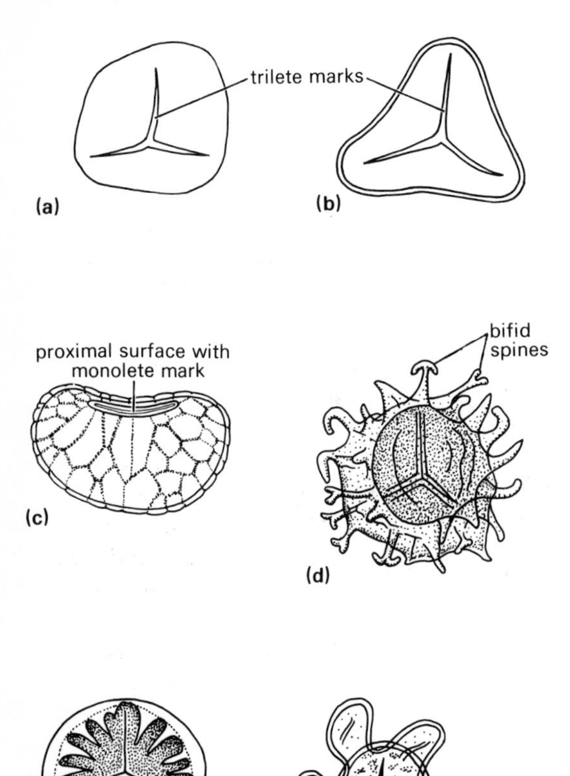

Figure 10.13 Fern spores and dispersed spores of uncertain affinity. (a) *Oligocarpia* spore, proximal view ×666; (b) *Deltoidospora*, proximal view ×333; (c) *Polypodium* spore, proximolateral view ×666; (d) *Ancyrospora*, proximal view, about ×250; (e) *Emphanisporites*, proximal view, about ×387; (f) *Alatisporites*, proximal view ×333. ((a) after Abott 1954; (c) after Scagel *et al.* 1965)

DISPERSED SPORES. A great many dispersed spores are of uncertain biological origin but of considerable stratigraphic value. For example, *Ancyrospora* (M. Dev.–L. Carb., Fig. 10.13d) was typical of its time, occurring in lacustrine sediments. This spore was spheroidal with a trilete mark and distinctive grapnel-ended spines, perhaps to assist its dispersal by arthropods. *Emphanisporites* (L. Dev.–L. Carb., Fig. 10.13e) was also a trilete miospore with a subcircular to subtriangular amb but with radial ribs at the proximal end and a smooth or sculptured distal surface. *Alatisporites* (Carb., Fig. 10.13f) possessed air sacs like younger coniferous pollen, but its sacs were more numerous and the aperture was trilete and proximal.

CLASS GYMNOSPERMAE. This class includes the seed ferns, conifers, maidenhair trees and cycads. In these the female gametophyte is not free-living but enclosed within an ovule. The microspore develops into a few-celled pollen grain before dispersal from the microsporangium and is commonly monosulcate and bilateral. Fertilisation results in a naked seed, not enclosed in a fruit.

The Pteridospermales or seed ferns are an extinct order known from Carboniferous to Cretaceous rocks. Although their foliage was fern-like, they reproduced with seed and pollen-producing organs as in other gymnosperms. The pollen grains are very variable in form, however. Usually they are trilete or monolete like those of ferns. In the Carboniferous cone genus *Whittleseya*, for example (which had *Neuropteris*-like foliage), *Monoletes*-type grains were produced (Fig. 10.14a). These were monolete and bilateral with two distal grooves which may have been the forerunners of a sulcus. The pollen found in the Jurassic pollen cone *Caytonanthus* resembles the dispersed grain *Vitreisporites*

(Fig. 10.14b) in being bisaccate and less than 40 μm in diameter.

The cycads (Orders Bennettitales and Cycadales) probably arose from seed ferns in Triassic times and flourished throughout the remainder of the Mesozoic Era but only the Cycadales have survived from then until the present. Both orders were palm-like trees and shrubs with flower-like pollen cones and seed cones. Typically the pollen is monosulcate, bilateral and boat-shaped and could have been carried by insects. The Jurassic bennettitalean flower *Weltrichia* contains pollen grains of the *Cycadoptites* type (Fig. 10.14d) a dispersed-grain genus that is very widespread in Mesozoic rocks. Similar grains are found in the Recent cycadalean *Zamia* (Fig. 10.14f).

The Cordaitales are an extinct order of tall, palm-like trees that flourished during the Carboniferous period but ranged from late Devonian to Triassic times. Cordaitaleans produced pollen grains about 100 μm long, each provided with a broad monosulcate furrow and a single air sac. Those found in cones of the tree *Cordaites* resemble the asulcate, monosaccate dispersed-grain called *Florinites* (Carb.–Perm., Fig. 10.14c).

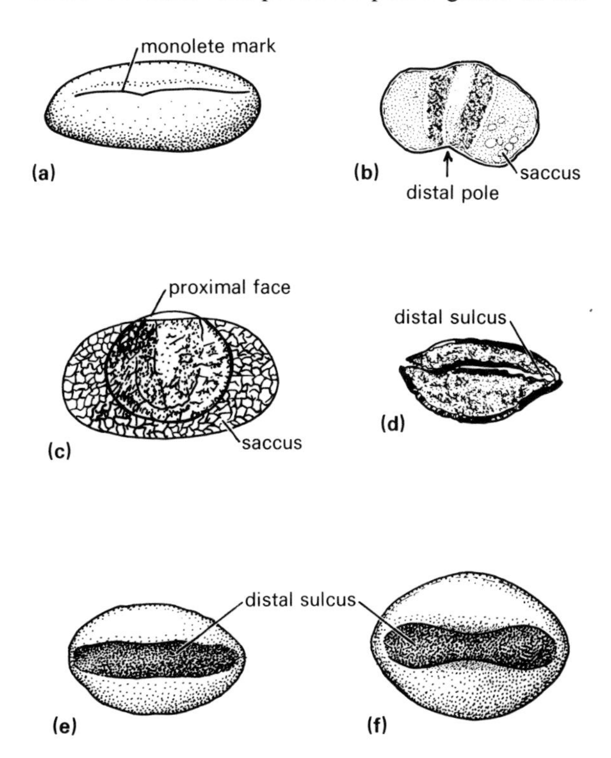

Figure 10.14 Gymnosperm pollen. (a) *Monoletes*, proximal view, about ×666; (b) *Vitreisporites*, lateral view ×666; (c) *Florinites*, lateral view, about ×426; (d) *Cycadoptites*, distal view ×333; (e) *Ginkgo*, distal view ×870; (f) *Zamia*, distal view ×666. ((b) after Harris 1964; (d) after van Konijnenburg 1971; (e) after Wodehouse 1935 and (f) after Scagel *et al.* 1965)

The maidenhair trees (Order Ginkgoales, Perm.–Rec.) were once important members of the Mesozoic land flora. Their leaves are distinctively spatulate or fan-shaped but their pollen grains closely resemble those of the cycads. The *Ginkgo* tree (Jur.–Rec.) for example, produces subspherical to ovate, monosulcate grains some 23–32 μm long of the *Cycadoptites* type (Fig. 10.14e), although the sulcus is less sinuous than those of *Zamia* and the grain is more elongate.

The conifers (Order Coniferales, Perm.–Rec.) were also much more important in Triassic and Jurassic land floras than they are today. Frequently their pollen grains are bisaccate, monosulcate, smooth and thin-walled with few distinctive features. Those of the pine tree, *Pinus* (Fig. 10.15a) are from 50–70 μm in diameter with two sculptured sacci on either side of the distal sulcus. Dispersed pollen grains of this kind, called *Pityosporites*, are known since late Carboniferous times. Pollen grains of the monkey puzzle tree, *Araucaria* (Fig. 10.15b) are some 60–95 μm across and spheroidal with a thin, finely granulate exine. A weakly defined area of very thin exine outlines the eventual site of germination, i.e. they are asulcate. Dispersed pollen grains of this type, called *Araucariacites*, are known at least since early Jurassic times. *Classopollis* (U. Trias.–Cret., Fig. 10.15d) is an unusual but common grain that had coniferous origins. Slightly prolate to spheroidal in shape, it possessed a trilete mark on the proximal surface and a single pore on the distal surface. Between the pore and the equator, however, is a circular groove of thin exine. This **zonosulcus** may have served as the germinal aperture.

Dispersed gymnosperm pollen grains of uncertain affinity are also very common in Permian and Mesozoic strata. *Taeniaesporites* (Perm.–Jur., Fig. 10.15c) for example, is bisaccate with from 3 to 5 longitudinal striae across the central body. Such striate, bisaccate miospores are typical of Permian microfloras and may have been produced by conifers or by *Glossopteris*-like plants.

CLASS ANGIOSPERMAE. The Angiospermae or flower-

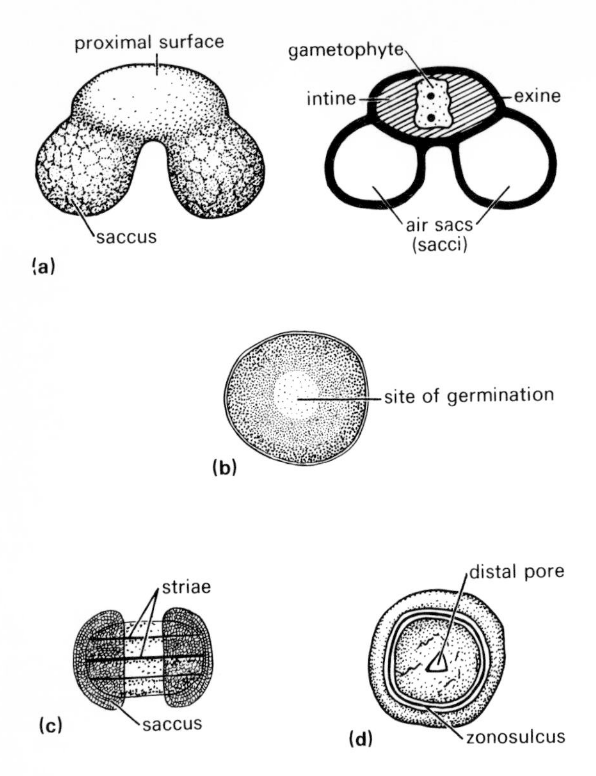

Figure 10.15 Gymnosperm pollen. (a) *Pinus* pollen, lateral view of external surface (left) and in section (right) ×666; (b) *Araucaria* pollen, distal view ×333; (c) *Taeniasporites*, distal view, about ×200; (d) *Classopollis*, distal view, about ×500. ((a) after Wodehouse 1935; (b) after Scagel *et al.* 1965; (c) based on Hart Fig. 13.3f *in* Tschudy & Scott 1969)

ing plants are known with certainty in early Cretaceous rocks and appear to have flourished ever since. The female gametophyte is here enclosed in an ovule, protected by a carpel, with a pollen-receiving style and stigma. The male gametophyte is borne in anthers and shed as pollen grains. These reproductive structures are protected and enhanced by sterile structures called petals and sepals, the whole comprising a flower (Fig. 10.4). Fertilisation then gives rise to a seed enclosed within a fruit.

Many flowering plants are insect pollinated; others have abandoned this method for wind pollination (e.g. grasses) or water pollination (e.g. sea-grasses). Flowers are generally delicate and short-lived structures that are very rare as fossils, hence it can be difficult to relate fossil pollen grains to their parent plants. As many extant genera appear to have been present in the late Cretaceous, however, some reference can be made by comparison with modern pollen. Angiosperm pollen is typically tectate with distal or equatorial germinal apertures. There are two subclasses: the Monocotyledonae and Dicotyledonae.

The Monocotyledonae (or monocots) produce a single seedling leaf or **cotyledon** after germination of the seed. Their leaves have unbranched, parallel veins and the flower parts are arranged in units or multiples of three. Familiar examples include grasses, lilies, onions, tulips and palm trees. Pollen grains of this group confirm that they are the more primitive stock for they often resemble cycad pollen in being monosulcate, bilateral and boat-shaped. For example, those of the date palm *Phoenix* (Fig. 10.16a) are of this shape, from 12–30 μm long with a thin, almost smooth tectate exine and a single deep sulcus. Dispersed pollen grains of palm type may be called *Palmaepollenites* (U. Cret.–Rec.). The grasses are a younger stock, having dominated much of the post-Oligocene land vegetation. Their pollen is typically spheroidal and **monoporate** with a thin smooth exine. That of the

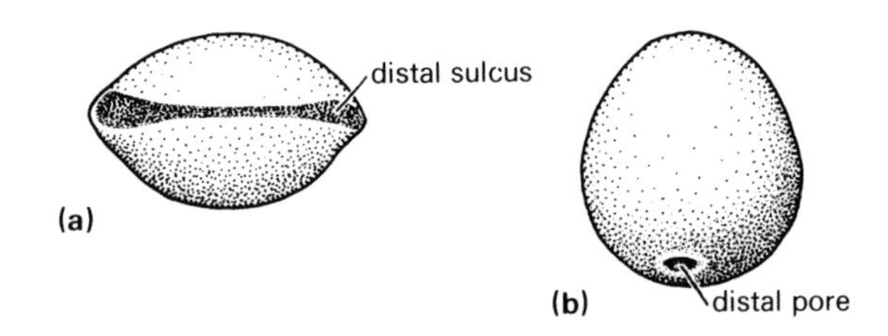

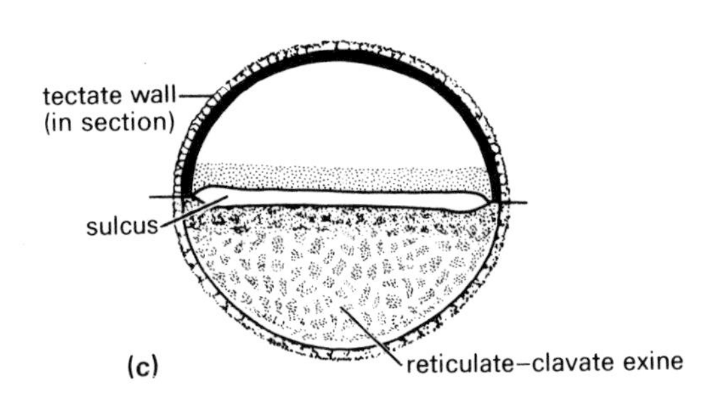

Figure 10.16 Monocotyledonous angiosperm pollen. (a) *Phoenix* pollen, distal view ×1065; (b) *Festuca* pollen, distolateral view ×666; (c) *Clavatipollenites*: bottom, external view of distal surface; top, optical section, ×1333. ((a) & (b) after Wodehouse 1935; (c) after Kemp 1968)

fescue grass, *Festuca* (Fig. 10.16b) for example, is of this kind, and resembles the dispersed fossil pollen genus *Monoporites* (Palaeoc.–Rec.).

Of great palaeontological interest is the miospore *Clavatipollenites* from the lower Cretaceous (Barremian) rocks. This, the oldest of angiosperm pollen grains, is monosulcate and ovate with a clavate exine supporting a thin and discontinuous tectum (Fig. 10.16c).

The Dicotyledonae (dicots) produce two seedling leaves after germination of the seed. They also have net-like leaf venation and flower parts arranged in units or multiples of four or five. Typically their pollen grains are multicolpate or multiporate. Pollen grains of the oak tree, *Quercus*, for example, are spheroidal, 30–36 μm across, with tricolpate apertures and a triangular amb (i.e. *Quercoipollenites*, U. Cret.–Rec., Fig. 10.17a). Those of the birch tree, *Betula*, are oblate, 20–40 μm across with triporate or sometimes 4- to 7-pored apertures (i.e. *Betulaepollenites*, U. Cret.–Rec., Fig. 10.17b). The ash, *Fraxinus* (U. Cret.–Rec.) has grains which are usually oblate with a quadrate amb, tetracolpate apertures and a reticulate sculpture (i.e. *Fraxinoipollenites*, Fig. 10.17c).

Pollen grains of the closely related willow, *Salix* (U. Cret.–Rec., fig. 10.17d) and poplar, *Populus* (U. Cret.–Rec., Fig. 10.17e) are quite different in appearance. In the willow, which is insect pollinated, the grains are tricolpate with a thick reticulate exine. In the poplar, wind dispersal has led to a thinner exine with weak sculpture and no furrows (i.e. acolpate). Pollen grains of the plantain, *Plantago* (Fig. 10.17f) are spheroidal with a thin granular exine and from 4 to 14 scattered pores (i.e. pantoporate).

The earliest dicotyledonous pollen is *Tricolpites* from the Albian stage of the lower Cretaceous (Fig. 10.17g).

General history of spores and pollen

Acritarchs with trilete-like marks have been reported from upper Precambrian and lower Palaeozoic marine rocks (see Schopf, pp. 171–3 *in* Tschudy & Scott 1969) but these have not received general acceptance. Non-marine and deltaic rocks remain rare in the Palaeozoic column until the upper Silurian at which point the vascular plant

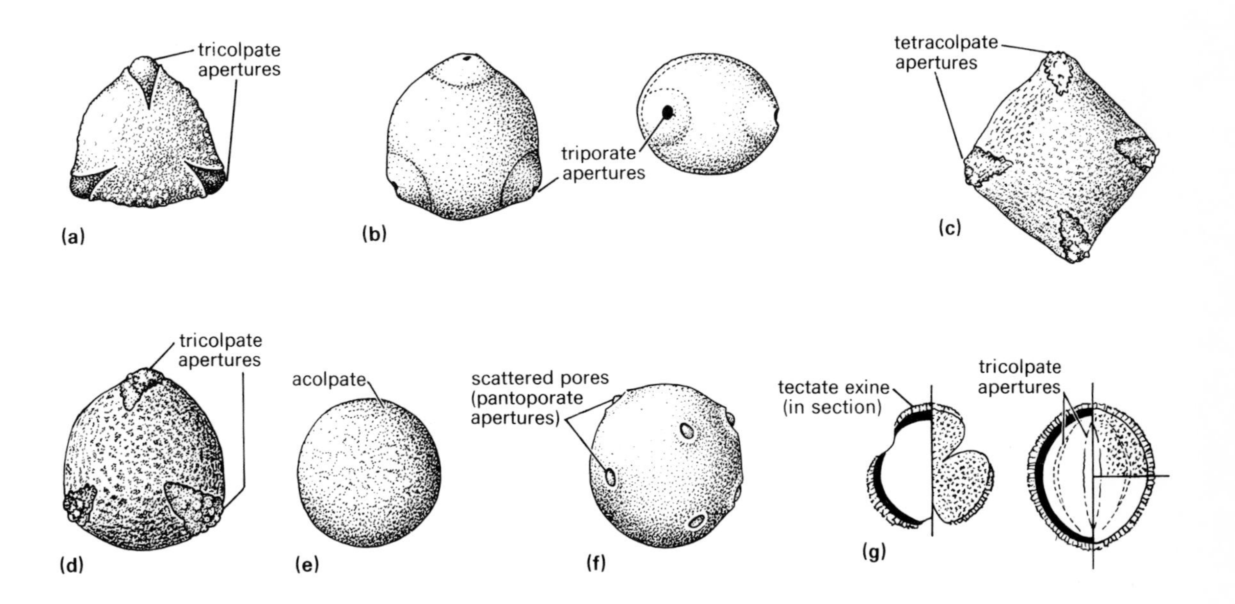

Figure 10.17 Dicotyledonous angiosperm pollen. (a) *Quercus* pollen, polar view ×666; (b) *Betula* pollen, polar and equatorial views ×1165; (c) *Fraxinus* pollen, polar view, about ×1000; (d) *Salix* pollen, polar view, about ×870; (e) *Populus* pollen, polar view, about ×770; (f) *Plantago* pollen, polar view, about ×600; (g) *Tricolpites*, polar and equatorial views of external surface and optical sections ×1333. ((a)–(f) after Wodehouse 1935; (g) after Kemp 1968)

record begins in earnest. Note should be made, though, of the occurrence of a few early and middle Silurian trilete spores, for these represent evidence of an older land flora.

During the late Silurian and early Devonian, widespread continental conditions encouraged an explosive radiation of land plants, especially small herbaceous and homosporous lycopsids and psilopsids (Fig. 10.18). Their spores were small, moderately sculptured and less than 60μm in diameter, but by mid-early Devonian times larger, probably lycopsid megaspores had appeared and sculpturing also became more prominent. Spores with grapnel-ended spines are typical of middle Devonian to lower Carboniferous lacustrine and fluviatile sediments (e.g. *Ancyrospora*, Fig. 10.13d). Fossil fern fronds are known from the middle Devonian and some of these ferns had developed heterospory along with the lycopsids by late Devonian times. The first seed ferns appeared for certain in mid Carboniferous times.

The Carboniferous flora is, of course, well known from the abundant coal deposits it left behind. It included a wealth of arborescent, heterosporous lycopsids, no doubt liberating clouds of *Triletes* and *Lycospora* into the air. The horsetails (with *Calamospora* and *Laevigatosporites*), seed ferns (with spores and bisaccate pollen), and cordaitaleans (with *Florinites*) were also important elements.

During the drier Permian period, the lycopsid-sphenopsid swamp vegetation was replaced by seed-bearing floras that were less dependent on moisture. Striated bisaccate pollen and monosulcate pollen of gymnospermous origin became common at this time.

Ferns, seed ferns, cycads, maidenhair trees and conifers were all significant spore and pollen producers during Triassic and Jurassic times. Angiosperm fruits and monosulcate pollen grains (e.g. *Clavatipollenites*, Fig. 10.16c) made their first appearance in rocks of early Cretaceous (Barremian) age, whilst dicotyledonous pollen (e.g. *Tricolpites*, Fig. 10.17g) appeared later in the Aptian. Multicolpate pollen rapidly became more abundant and diverse as the flowering plants displaced gymnosperms on land in the late Cretaceous. This remarkable takeover may have been encouraged by the increasingly seasonal climate of this period and the ensuing Cainozoic, calling for new reproductive strategies. The late Cretaceous flora was relatively

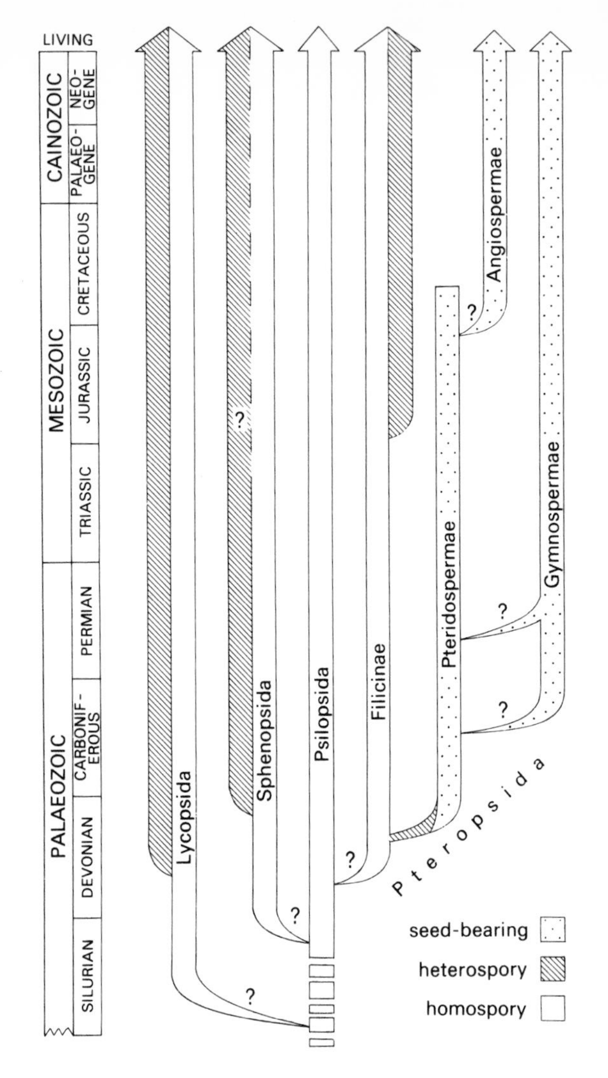

Figure 10.18 History of homospory, heterospory and seed-bearing in vascular land plants.

modern in aspect, with broad-leaved woodlands, but pollen evidence suggests that grasslands did not develop until the early Cainozoic Era, nor become significant until the late Cainozoic. Miocene through to Recent microfloras provide the most remarkable record of vegetational change in response to climatic and geographic fluctuations (see examples below).

The story of the evolution of spores, pollen and land plants is well summarised by Chaloner (1967 and pp. 1–14 *in* Ferguson & Muller, 1976) and Kuprianova (1969). Evolutionary developments in spores are also discussed by Kremp (1967), in gymnosperm pollen by Millay and Taylor (1976) and in angiosperm pollen by Nair (1968) and Laing (pp. 15–26 *in* Ferguson & Muller 1976).

Applications of spores and pollen

Spores and pollen grains are valuable stratigraphic markers in terrestrial, lacustrine, fluviatile and deltaic sediments and helpful in correlating non-marine and marine strata. Their utility is well demonstrated in a general review by Hopping (1967) and for each period from the Devonian to the Quaternary by various authors in Tschudy and Scott (1969). It is difficult to select from the many examples of their use but we may cite some papers on the Devonian (Richardson 1967), Carboniferous (Smith & Butterworth 1967), Triassic (Warrington 1970) and Jurassic (Pocock 1972). Some of the problems of freshwater-to-marine correlation are outlined in a paper by Hughes and Pacltová (1972).

A distinctive mode of palynological correlation has long been employed for coal seams. The problem here is that little evolution is discernible between successive seams and long-ranging spores tend to dominate. Raistrick (1934–5) found that the average miospore assemblage of a coal seam was virtually unchanged when traced laterally but quite different from those of older or younger seams. Hence coal seams can be correlated locally by comparison of the relative proportions of different miospores, as demonstrated by Smith & Butterworth (1967) in Great Britain and Navale and Tiwari (1968) in India (Fig. 10.19).

A similar approach can be used to correlate Quaternary interglacial deposits on a local scale. For example, the familiar divisions of the British Holocene from pollen zone IV (Pre-Boreal with birch woodland, about 10000 yrs BP) to pollen zone VIII (Sub-Atlantic alder-oak woodland, modern) were based on the changing pollen spectra (see West 1968, pp. 279–83; 292–325). Most Quaternary interglacial deposits in temperate latitudes record, in their microfloras, a change from glacial to cool birch forest with abundant small herbs and shrubs, through pine forest, to a climatic optimum with elm, oak, lime, alder and hazel, followed by a climatic deterioration with pine, birch and then renewed glacial conditions (e.g. Fig. 10.20). Deep-sea sediments may also record these changes, with the added advantage of preserving glacial microfloras as well (Musich 1973). Even more remarkable is the story of uplifting of the Andes in late Cainozoic times, enshrined in microfloras that range from tropical lowland to temperate high mountain and ultimately to glacial (van der Hammen *et al.* 1973). Mio–Pliocene microfloras in

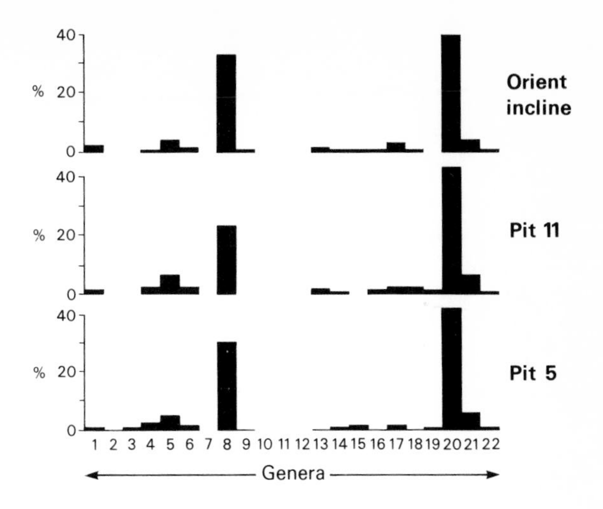

Figure 10.19 Correlation of three Gondwana coal seams on the similar relative frequencies of 22 miospore genera. (Modified from Navale & Tiwari 1968)

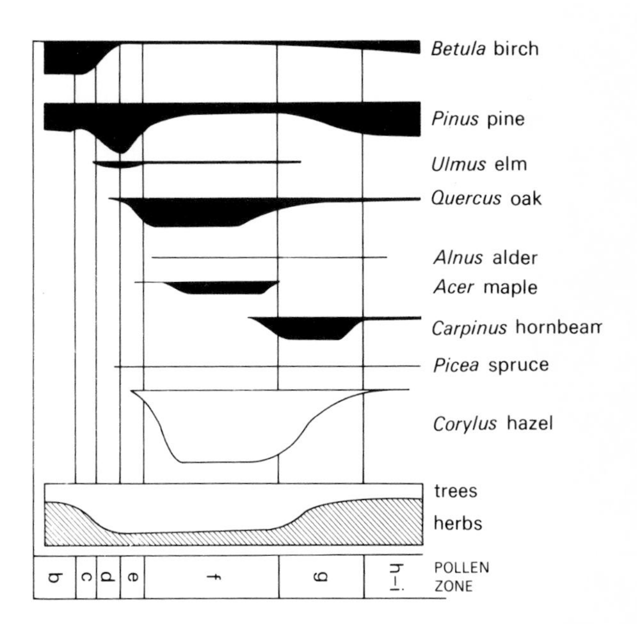

Figure 10.20 Generalised pollen diagram from the Ipswichian (or 'Eemian') interglacial deposits. (Modified from Leopold, Figs 17–19, *in* Tschudy & Scott 1969 after West and Pearson)

the Netherlands, however, suggest climatically controlled vegetational cycles at this time, like those of the Quaternary (Zagwijn 1967). At more remote periods, the cause of microfloral changes are less certain but ecological successions and biofacies can be recognised in the Eocene (see Martin 1976), Palaeocene (Kedves 1960), Jurassic and Cretaceous (Hughes & Moody-Stuart 1967) and in the Carboniferous (Smith 1962, Marshall & Smith 1964, Habib 1968).

Pollen analysis is also of great assistance to archaeologists, not only because it provides a framework for Quaternary stratigraphy but because of the view it gives of man's early environment and his effect upon it. There was, for example, a curious sudden decline in the tree pollen at the horizon of the late-middle Acheulian (palaeolithic) hand axe culture in the Hoxnian interglacial (West, p. 421 *in* Tschudy & Scott 1969) that might have been due to forest clearance. A change from elm–oak forest to maple–lime bigwoods in Minnesota some 500 years ago could, conversely, have been due to depleted Indian populations and fewer fires (McAndrews 1968). According to van Zeist (1967) the palynology of burial mounds can reflect the changing patterns of early land use by man. Godwin (1967) has even outlined the remarkable evidence for cultivation of *Cannabis* in England by Saxons, Normans and Tudors. A short review of archaeological palynology is given by Dimbleby (1969). Oldfield (1970) examines some of the problems of these pollen analytical methods.

Pollen and spores can help sedimentologists to discover the provenance of fine-grained sediments, as demonstrated by Rossignol-Strick (1973). As with other fossils, they can also be used to estimate the rate of sedimentation (see Davis 1967) or the degree and source of sediment reworking (Muir 1967, Phillips 1974). Although reworking may be a natural hazard for palynologists, Stanley (1967) has shown how horizons rich in reworked miospores can be used as correlation markers in deep-sea sediments, in this case corresponding with glacial maxima and periods of greatly lowered sea level.

Further reading

An invaluable and well-balanced introduction is Tschudy & Scott (1969). Evolutionary topics are tackled from various points of view in Ferguson and Muller (1976), and West (1971) provides a succinct account of methods and results in Quaternary pollen analysis. A fuller treatment of analytical techniques and a key to pollen identification are given by Faegri and Iversen (1975). Although rather specialised, the texts by Wodehouse (1935) and Erdtman (1943) may also assist in studies of Quaternary pollen grains. Fossil spores and pollen can be diagnosed with the aid of the *Catalog of Fossil Spores and Pollen* (Traverse *et al.*, 1957 to date). An illustrated dictionary of palynological terms with a valuable bibliography has been prepared by Kremp (1965).

Hints for collection and study

To understand and recognise fossil spores and pollen, it is particularly worthwhile to look at Recent material. A collection of common spore types from trees, shrubs and ferns can readily be made by removing the flowers, cones or sporangia when just on the point of opening. If not examined directly they should be stored in alcohol. Strew slides can be made by removing the anthers, pollen sacs or sporangia with a scalpel and placing these on a glass slide with a drop of distilled water. While looking down a microscope, bruise the anthers, etc., with a seeker or the blunt edge of a scalpel and spread the released miospores over part of the slide. To make the structures more distinct, allow the strew to dry and then add a drop of Gray's spore stain (0.5% malachite green and 0.05% basic fuschin in distilled water; the slide should then be warmed over a bunsen for one minute), or basic fuschin stain (0.5% basic fuschin in distilled water) or safranin stain (1 g of safranin '0' in 50 ml of 95% alcohol plus 50 ml of distilled water). After ten minutes rinse the slide with a little distilled water and dry at a low temperature. Mount the cover slip with water or glycerine (30% aqueous solution) for temporary preparations, or in Canada Balsam or glycerine jelly for permanent ones. View with well-condensed transmitted light at 400 × magnification or higher.

Fossil miospores are most readily prepared from plant-bearing muds or shales and from peats, lignites and coals. They can also be very abundant in dark marine shales and mudstones. Palynological laboratories invariably remove the siliceous material with hydrofluoric acid, (a dangerous chemical without proper facilities), the calcareous material with hydrochloric acid and the vegetative plant tissues with a variety of strong acids, alkalis and oxidants. Miospores can be prepared for study without these sophisticated techniques but the results are inevitably diluted with mineral and vegetable matter. Disaggregation should follow methods A to F (see Appendix), wash as in method G and concentrate as in method H or K. If the organic material is dark and opaque, treat with method E. Temporary mounts in water or glycerine and permanent mounts in glycerine jelly or Canada Balsam can be prepared on glass slides.

Examine the strewn slide with well condensed transmitted light, using oil immersion objectives for the higher magnifications if possible. Miospores can usually be distinguished from other vegetable matter by their shape, their sharper outlines and often by their amber colour. Much information on palynological techniques can be found in Gray (pp. 470–706 *in* Kummel & Raup 1965).

11 Phylum Ciliophora – Tintinnids and calpionellids

The Class Ciliophora contains protists that are covered with numerous tiny flagella called **cilia**. These cilia are arranged in neat rows embedded within an outer skin called the **pellicle** and they serve both for locomotion and food gathering by beating together in waves. Many of the 'ciliates' have two kinds of nuclei: a large macronucleus for normal cell functions and a tiny micronucleus for reproductive purposes. A distinct **cell mouth** and a **buccal cavity** are further distinctive features of these active protists (Fig. 11.1).

Although the Ciliophora are an abundant and widely distributed group, only the Suborder Tintinnina are of much geological interest. The latter, which comprise about 40% of all known ciliates, dwell for the most part as motile plankton in marine waters around the world. Unfortunately, however, few of these become fossilised and the fossil record (Ord.–Rec.) is very patchy but calcareous forms, known as **calpionellids**, are sufficiently abundant in some Mesozoic facies to provide useful biostratigraphic zones.

The living tintinnid

The tintinnid cell is generally cylindrical, conical or bell-shaped with the posterior end drawn out into a stalk or **pedicel** for attachment to the external shell, called a **lorica** (Fig. 11.1). The anterior end of the cell is broader and fringed by a feathery crown of tentacle-like **membranelles**, which are actually bundles of fused cilia. Beneath these membranelles occur the buccal cavity and the cell mouth, whilst the remainder of the pellicle is traversed by spiralling rows of cilia.

The cell attaches by the pedicel low down inside the lorica but is otherwise not in contact with it (Fig. 11.1). From the **aperture** of the lorica projects the anterior crown of the cell, the beating membranelles serving to propel the whole structure backwards with a spiral motion. Captured food such as bacteria, green algae, coccolithophores, dinoflagellates and diatoms, are passed by the cilia to the mouth for ingestion, leading to internal digestion within food vacuoles. This diet imparts a green colour to the cell.

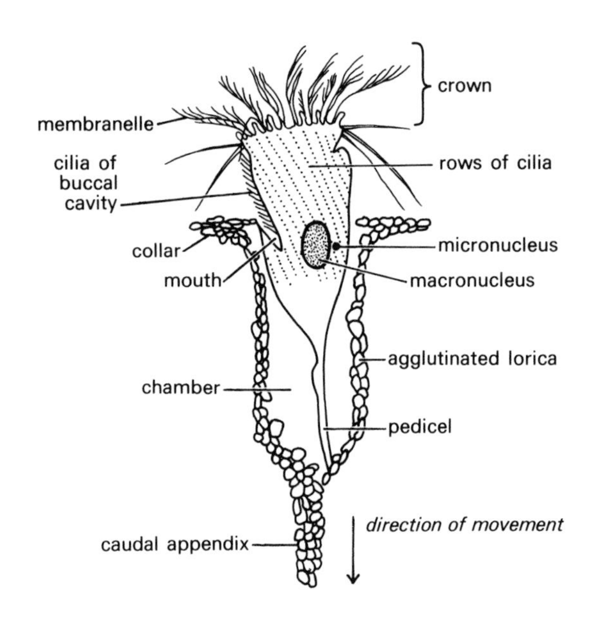

Figure 11.1 Recent tintinnid *Tintinnopsis*, about ×400. (Modified from Colom 1948 after Fremiet)

The lorica

The lorica may be from 20–1000 μm long although most are 120–200 μm in length. In outline, the loricae vary from globular, through conical, bell-

and bottle-shaped to bullet- or nail-shaped. All have an aperture at the **oral** end and most have a closed **aboral** region of rounded or pointed form at the opposite end (Fig. 11.2). Within is a very spacious, single **chamber** that can be up to ten times as voluminous as the cell itself.

Sculpture takes the form of spines, costae, fins, transverse grooves, longitudinal grooves or spiral grooves, reticulate patterns or fenestrate (i.e. window-like) structures. Minute cavities within the wall, called **alveoli**, are filled with low gravity fluids that no doubt help to keep the lorica more buoyant. The relatively great surface area and the development of **collars** (Figs 11.1, 11.2a) may help to retard sinking in some forms, but stream-lining for efficient motility appears to be the major function of the tintinnid lorica.

The tintinnid wall is principally a delicate organic structure of chitin or xanthoprotein but it may be strengthened by the agglutination of tiny particles like coccoliths and diatom frustules. Calcareous walls are unknown in living tintinnids; fossil calpionellids, however, are thought to have secreted a primary calcite shell, which, Remane (*in* Haq & Boersma 1978, p. 163) argues, is against a ciliate affinity.

Distribution and ecology of tintinnids

Whilst tintinnids feed on the nannoplankton, they are themselves on the menu of silicoflagellates, larger zooplankton and fish. Tintinnina occur in the photic zone in all seas and oceans, but are rarely abundant except in the Antarctic where they are exceeded in number only by the diatoms on which they feed. Sensitivity to temperature and salinity gives rise to Recent assemblages typical of arctic, antarctic, temperate, subtropical and tropical seas, and of estuaries, lakes and peat bogs. However, freshwater forms comprise only about ten of the 840 species and these are mostly isolated relics left by retreating seas of the Tertiary, such as in the Caspian Sea and Lake Baikal.

Fossil calpionellids are best known from pelagic fine-grained limestones deposited in the Mesozoic Tethys ocean, occurring in densities of up to $21/mm^3$ of rock alongside coccoliths (including *Nannoconus*), planktonic foraminifera and radiolarians. The agglutinated loricae of tintinnids may be found in neritic limestones and glauconitic clays or in estuarine, lacustrine and peat bog deposits. Like other microfossils they are easily reworked but more serious is the fragility of the organic and agglutinated forms, the former rarely surviving and the latter requiring careful preparation if they are to be extracted from the rocks intact.

Classification

Kingdom PROTISTA
Phylum CILIOPHORA
Class CILIATA
Order SPIROTRICHIDA
Suborder TINTINNINA

Taxonomy is based mainly on the shape, composition, wall structure and sculpture of the lorica. Caution is necessary, though, because the size and shape of the lorica can vary with ecological conditions, often being larger at lower temperatures. Furthermore, there has been disagreement as to the significance of wall composition. Campbell (1954) united the calcareous calpionellids with the organic or agglutinated tintinnids of similar shape, whilst Tappan and Loeblich (1968) have preferred to distinguish many of the organic walled forms. Remane (pp. 574–87 *in* Brönnimann & Renz 1969, vol. 2) is amongst those who argue that the calpionellids may not even have been tintinnids or ciliates at all.

As most fossil assemblages occur in limestones, they are usually studied from randomly orientated thin sections, a process requiring considerable practice. *Tintinnopsis* (Rec.; Figs 11.1, 11.2a) has an organic and agglutinated wall with a slightly constricted aperture surrounded by a flaring collar. The fossil calpionellid *Tintinnopsella* (U. Jur.–L. Cret., Fig. 11.2b) is very similar but has a fine-grained calcareous wall. In *Calpionella* (U. Jur.–L. Cret., Fig. 11.2c) the wall is also calcareous and the collar is short and erect. *Tytthocorys* (U. Eoc., Fig. 11.2d) has a three-layered calcareous lorica, the aperture constricted by a shelf just below the short, flaring collar. *Salpingella* (Rec., Fig. 11.2e) has a wholly organic, nail-shaped lorica with a flaring collar and longitudinal fins. The superficially similar calpionellid *Salpingellina* (Fig. 11.2f) occurs in lower Cretaceous rocks and has a calcareous lorica.

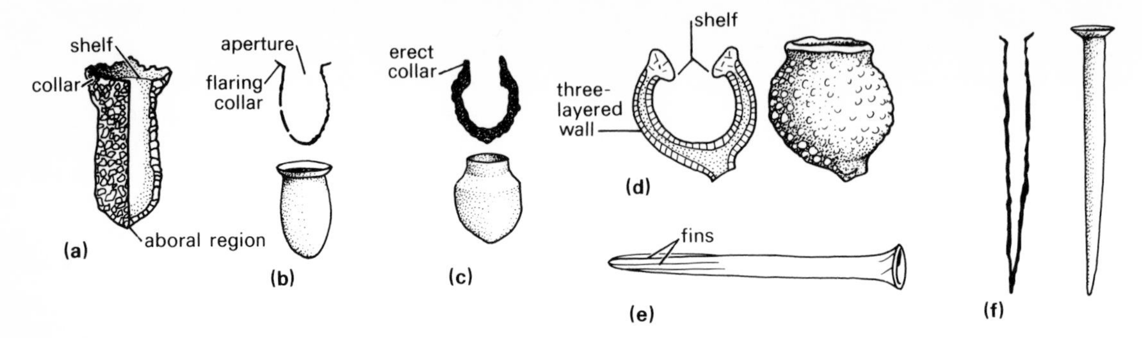

Figure 11.2 Tintinnid and calpionellid loricae. (a) *Tintinnopsis*, external surface and longitudinal section ×133; (b) *Tintinnopsella*, longitudinal section and reconstruction ×133; (c) *Calpionella* longitudinal section and reconstruction ×333; (d) *Tytthocorys* longitudinal section and external view ×150; (e) *Salpingella* ×133; (f) *Salpingellina* longitudinal section and reconstruction ×166. ((a) after Remane, Fig. 2.14 *in* Brönimann & Renz 1969; (b), (c) & (f) partly after Colom 1948; (d) after Tappan & Loeblich 1968; (e) after Kofoid & Campbell 1939)

General history of tintinnids

The fossil record provides a very patchy view of tintinnid history. They are rare in both Palaeozoic and Tertiary rocks and have not yet been reported from Cambrian, Carboniferous, Permian, upper Cretaceous, Palaeocene, Miocene and Pliocene sediments (Tappan & Loeblich 1968). Even the rare Pleistocene records do little justice to the 840 or so living species, so we conclude that the group has generally had a poor preservation potential. However, in the late Jurassic and early Cretaceous, the more readily preserved calpionellids bloomed from Mexico to the Caucasus in the Tethys ocean, building deep-sea limestones together with the coccoliths. Their dramatic decline at the end of the early Cretaceous is surprising and not paralleled in other plankton groups. It could, perhaps, have been caused by vigorous competition from the thriving planktonic foraminifera, radiolarians and dinoflagellates.

Applications of tintinnids

Tintinnids can be used to correlate those Tethyan limestones in which they abound (e.g. Remane, pp. 559–73 *in* Brönnimann & Renz 1969, vol. 2). Echols and Fowler (1973) have investigated the potential of brackish water species for palaeosalinity and shoreline studies in Pleistocene sediments of the North Pacific.

Further reading

An important introduction to the ecology and classification of tintinnids and calpionellids is the paper by Colom (1948). Stratigraphically useful species are described and illustrated in Borza (1969) and their classification and geological value are reviewed in a chapter by Remane *in* Haq and Boersma (1978, pp. 161–70).

Hints for collection and study

Calpionellids are best studied in thin sections or peels of Mesozoic Tethyan limestones of deep-water origin (see method N in Appendix). The morphology of the species must then be reconstructed from the various unorientated cross sections, an axial (longitudinal) section being the most helpful for identification.

12 Phylum Sarcodina – Radiolarians and heliozoans

Both the radiolarians and heliozoans are free-floating protists with roughly spherical cells and thread-like pseudopodia extending radially over a delicate endoskeleton. These two groups of the Actinopoda differ, however, in that modern radiolarians are entirely marine with representatives in Cambrian times, whereas the Heliozoa are an exclusively freshwater subclass with no fossil representatives before the Pleistocene. Radiolarians are most useful for biostratigraphy of Mesozoic and Cainozoic deep-sea sediments and have great potential as palaeoenvironmental indicators.

SUBCLASS RADIOLARIA AND SUBCLASS ACANTHARIA

The living radiolarian

The single-celled radiolarians average between 100–2000 μm in diameter, although colonial associations up to 250 mm long are reported. The protoplasm of each cell is divided into an outer **ectoplasm** and an inner **endoplasm**, separated by a perforate organic membrane called the **central capsule** (Fig. 12.1a). Radiating outwards from this central capsule are the pseudopodia, either as thread-like **filopodia** or as **axopodia**, which have a central rod of fibres for rigidity. Near the centre of the dense endoplasm occurs a large nucleus or several nuclei. The ectoplasm typically contains a zone of frothy, gelatinous bubbles, collectively termed the **calymma** and a swarm of yellow symbiotic algae called **zooxanthellae.**

A **skeleton** is usually present within the cell and comprises, in the simplest forms, either radial or tangential elements, or both (Fig. 12.1b). The radial elements consist of loose **spicules**, external **spines** or internal **bars**. They may be hollow or solid and serve mainly to support the axopodia. The tangential elements, where present, generally form a porous **lattice** shell of very variable morphology, such as spheres, spindles and cones. Often there is an arrangement of concentric or overlapping lattice shells. Skeletal composition differs within the

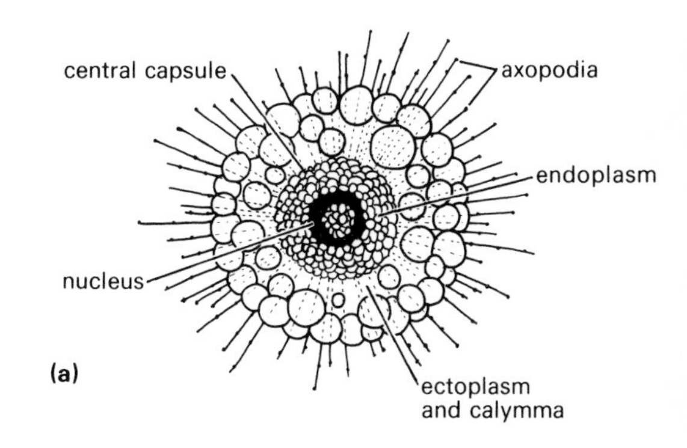

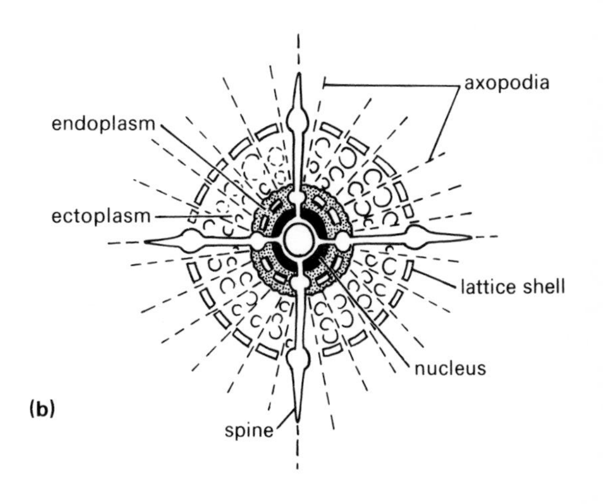

Figure 12.1 (a) Cross section through a naked radiolarian cell (*Thalassicola*); (b) cross section through a spumellarian showing the relationship of the nucleus, endoplasm and ectoplasm to three concentric lattice shells and radial spines. ((a) after Westphal 1976)

radiolarian group, being of strontium sulphate (i.e. celestine, $SrSO_4$) in the Subclass Acanthara, opaline silica in the Superorder Polycystina (Orders Spumellaria and Nassellaria) and organic with up to 20% opaline silica in the Order Phaeodaria.

As far as is known, reproduction is entirely asexual, by division of the cell into two daughter cells, with either one daughter keeping the old skeleton or with both evacuating it and secreting new ones. In the Family Collosphaeridae (Spumellaria), the cells remain attached to form colonies. Individual radiolarians are thought to live no longer than one month.

Radiolarians are marine zooplankton, using their sticky radiating axopodia to trap and paralyse passing organisms (e.g. phytoplankton and bacteria). Food particles are digested in vacuoles within the calymma and nutrients are passed through the perforate central capsule to the endoplasm. Those living in the photic zone may also contain zooxanthellae and can survive by symbiosis with these alone.

Buoyancy is maintained in several ways. The specific gravity is lowered by the accumulation of fat globules or gas-filled vacuoles. Frictional resistance is increased by the development of long rigid axopods borne on skeletal spines. Holes in the skeleton allow the protoplasm to pass through and also reduce weight. The spherical and discoidal skeletal shapes are further devices to reduce sinking, as in foraminifera, coccolithophores and diatoms. The turret and bell-like skeletons of the Nassellaria appear to be adaptions for areas of ascending water currents, the mouth being held

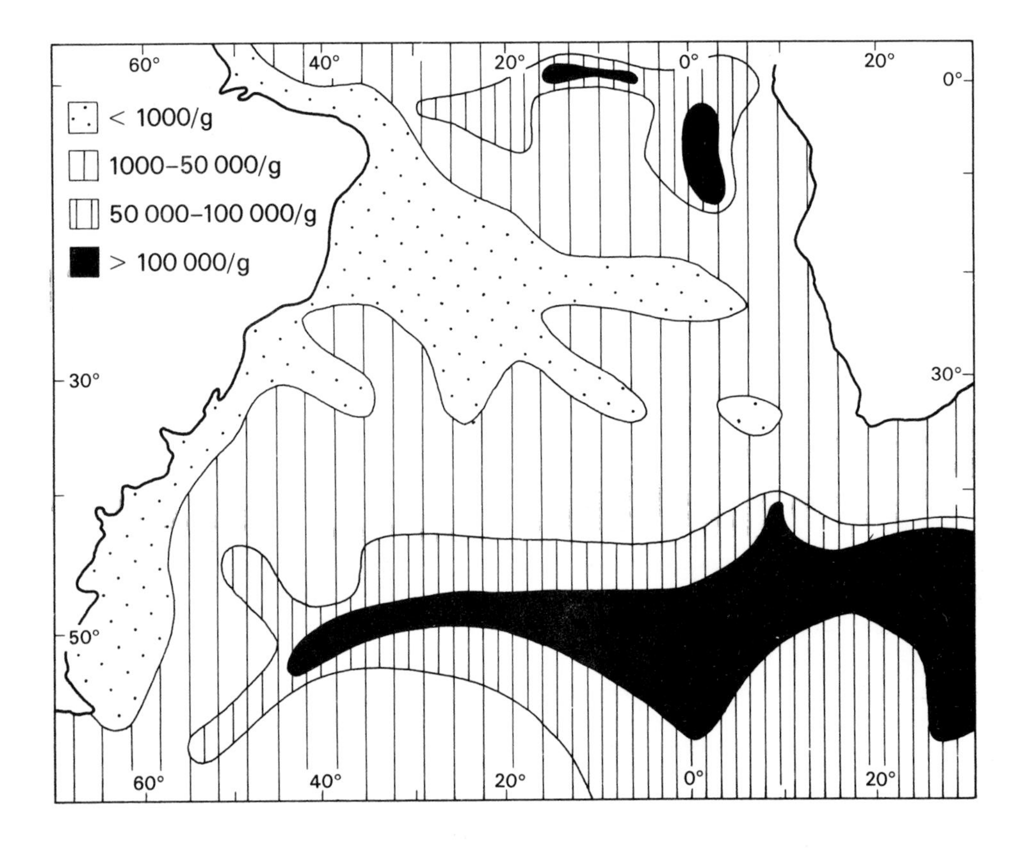

Figure 12.2 Abundance of Radiolaria in surface sediments of the South Atlantic. (Modified from Goll & Bjørklund 1974)

downwards and the axis held vertically, much as in silicoflagellates.

Radiolarian distribution and ecology

Radiolarians prefer oceanic conditions, especially just seaward of the continental slope, in regions where divergent surface currents bring up nutrients from the depths and planktonic food is plentiful. Although most diverse and abundant at equatorial latitudes, where they may reach numbers of up to 82000 per m^3 water, they also thrive with diatoms in the subpolar seas (Fig. 12.2). Radiolarians tend to bloom seasonally in response to changes in food and silica content, currents and water masses.

Different species may occur in assemblages at distinct water depths, each approximately corresponding to discrete water masses with certain physical and chemical characteristics (e.g. Fig. 12.3). Assemblage boundaries at 50, 200, 400, 1000 and 4000 m are reported, with Acantharia and Spumellaria generally dominating the photic zone (< 200 m) and Nassellaria and Phaeodaria dominating depths below 2000 m. Understandably, these Phaeodaria lack symbiotic algae. Some radiolarian species have very great depth ranges, though, with juveniles and small adults thriving at the shallower end of the range and the larger adults living in the deeper waters.

As with the Foraminiferida, some cold water species that live near the surface in subpolar waters occur at greater depths near the Equator (Fig. 12.3). Surface assemblages typical of polar, subpolar, subtropical and tropical latitudes are recognised in the present oceans, each more or less delimited by boundary currents. Such assemblages can even be used to plot the changing history of currents and water masses through the Cainozoic (see Casey, pp. 331–41 *in* Funnell & Riedel 1971). Recent ecology and distribution are reviewed by Casey (pp. 151–9 *in* Funnell & Riedel 1971, and pp. 809–45 *in* Ramsay 1977).

Radiolarians and sedimentology

Both the $SrSO_4$ skeletons of the Acantharia and the weakly silicified, tubular skeletons of the Phaeodaria are very prone to dissolution on the deep-sea

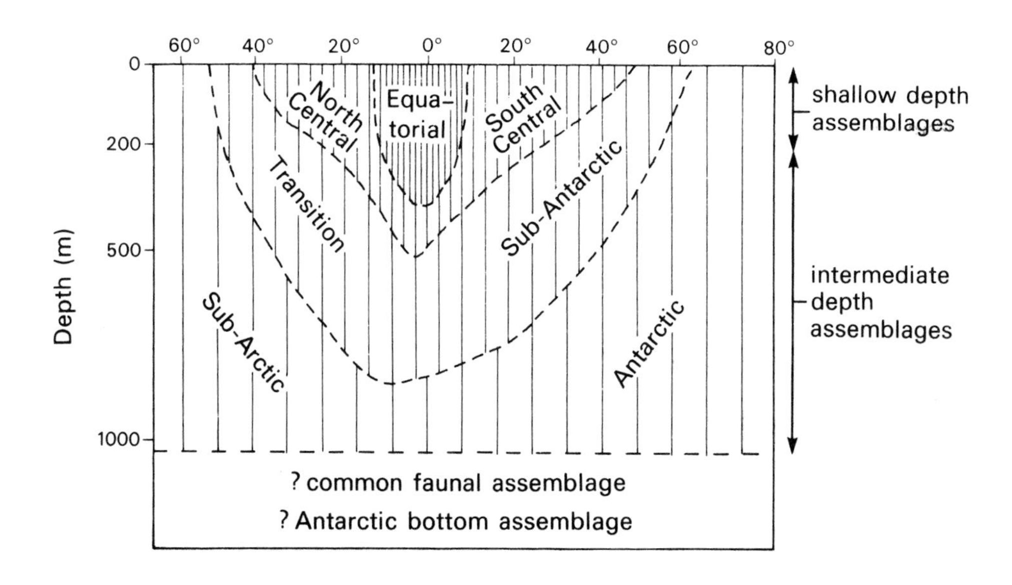

Figure 12.3 Latitudinal and vertical assemblages of polycystine radiolarians in the Pacific along a 170° W transect. (Modified from Casey, Fig. 7.1 *in* Funnell & Riedel 1971)

floor and are therefore rare as fossils (see Reshetnjak, pp. 343–9 *in* Funnell & Riedel 1971). Conversely, the solid opaline skeletons of the Spumellaria and Nassellaria tend to be even more resistant than those of silicoflagellates and diatoms, although all are susceptible to dissolution because seawater at any depth is very undersaturated relative to silica. Below the calcium carbonate compensation depth (usually 3000–5000 m) nearly all $CaCO_3$ enters into solution so that siliceous radiolarian or diatomaceous oozes tend to accumulate. Radiolarian oozes are mostly found in the equatorial Pacific below zones of high productivity at 3000–4000 m depth and can contain as many as 100 000 skeletons per gram of sediment, but they may also occur abundantly in marine diatomaceous oozes or in *Globigerina* and coccolith oozes.

With increasing depth and dissolution, the abundance of Radiolaria in deep-sea sediments decreases, through a progressive loss of the more delicate skeletons (Fig. 12.4). If the settling or sedimentation rate is slow, the chances for solution of skeletons will also increase, eventually lending a bias to the composition of fossil assemblages (see Holdsworth & Harker 1975). Consequently, red muds of the abyssal plains mainly consist of volcanic and meteoritic debris barren of all but the most resistant parts of radiolarian skeletons. At these great depths, the best preservation of radiolarians is encouraged by rapid sinking within the faecal pellets of copepod crustaceans (Casey, p. 542 *in* Swain 1977).

Fossil radiolarians are frequently found in chert horizons. Nodular cherts found interbedded with calcareous pelagic sediments of Mesozoic and Cainozoic age are probably deep-water deposits formed below belts of upwelling plankton-rich waters, as at the present day (see Ramsay 1973). Such cherts are thought to be organic in origin. The massive and ribbon-bedded cherts found in on-land sequences of Palaeozoic age may also contain Radiolaria but were less certainly formed by them. Many of these so-called radiolarian cherts are interbedded with black shales and basic volcanic rocks in settings which have been interpreted as ancient oceanic crust. All the good radiolarian assemblages of this age, however, come from continental shelf sea facies, such as goniatite–bivalve shales (Holdsworth, pp. 167–84 *in* Swain 1977).

Like other microfossils, Radiolaria are very prone to exhumation and reburial in younger sediments. These and other aspects of radiolarians in sediments are reviewed more fully in Funnell and Riedel (1971).

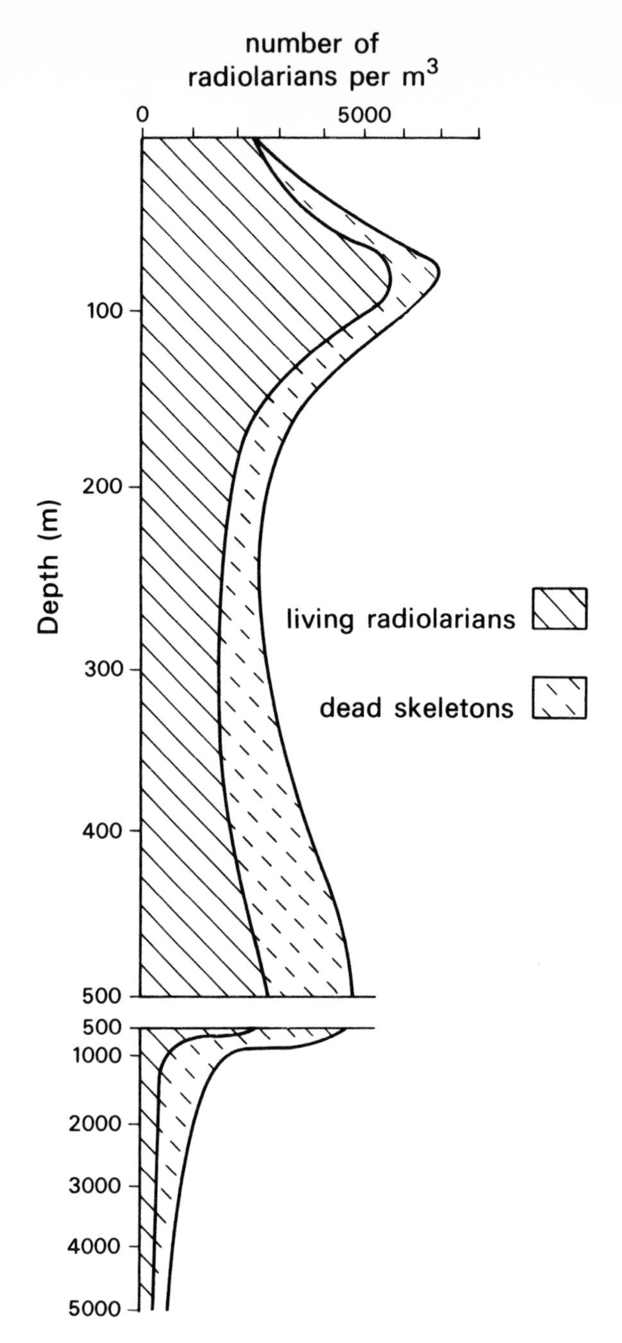

Figure 12.4 Vertical distribution of living and dead radiolarians through the water column at a station in the Central Pacific (After Petrushevskaya, Fig. 21.4 *in* Funnell & Riedel 1971)

Classification

Kingdom PROTISTA
Phylum SARCODINA
Class ACTINOPODA

The Subclass Radiolaria is primarily subdivided on morphology of the unmineralised (and therefore unfossilised) central capsule as well as on the composition and geometry of the skeleton. This use of soft-part criteria inevitably poses problems for palaeontologists working with skeletons of extinct Palaeozoic or early Mesozoic forms. The classification of Campbell (1954a), based on the works of Haeckel, has been overtaken by more recent studies of biology and phylogeny (see Riedel, pp. 649–61, *in* Funnell & Riedel 1971, Petrushevskaya 1971, and authors *in* Ramsay 1977) and the systematics is in a state of flux. The Spumellaria and Nassellaria with opaline skeletons are now usually united in the Superorder Polycystina; zoologists treat the old Suborder 'Acantharina' as a distinct Subclass Acantharia because of their slightly different cellular organisation.

Subclass Acantharia. The Subclass Acantharia have their skeletons generated at the cell centre rather than peripherally as is usual in the other groups. This skeleton generally comprises 20 spines of $SrSO_4$ joined at one end (in the endoplasm) and arranged like the four spokes of five wheels in different planes and of varying diameters (e.g. *Zygacantha*, ?Mioc.–Rec., Fig. 12.5a). *Acanthometra* (Rec., Fig. 12.5b) has thin radial spines embedded in protoplasm which invariably disarticulate after death. *Belonaspis* (Rec., Fig. 12.5c) has an ellipsoidal lattice formed by fused spine branches (**apophyses**) with 20 projecting radial spines.

Subclass Radiolaria. SUPERORDER POLYCYSTINA. The Order Spumellaria have skeletons of solid opaline silica, often in the form of a spherical or discoidal lattice, with several concentric shells bearing radial spines and internal supporting bars. In *Thalassicola* (Rec., Fig. 12.1a) a skeleton is either lacking or consists merely of isolated spicules. *Actinomma* (Rec., Fig. 12.6c) has three concentric, spherical lattice shells with large and small radial spines and bars. *Dictyastrum* (Jur.–Rec., Fig. 12.6d) has a flattened skeleton with three concentric chambers

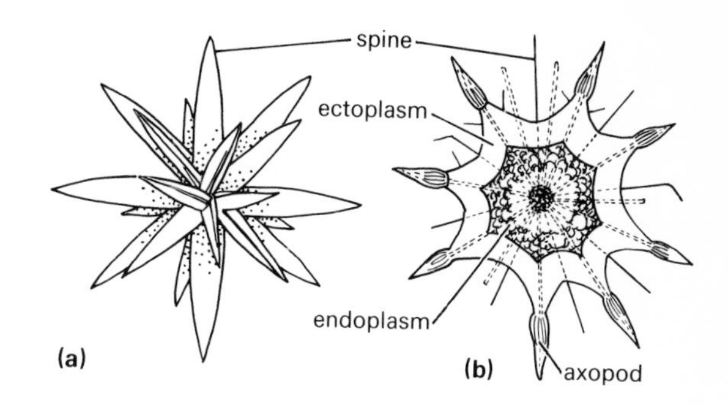

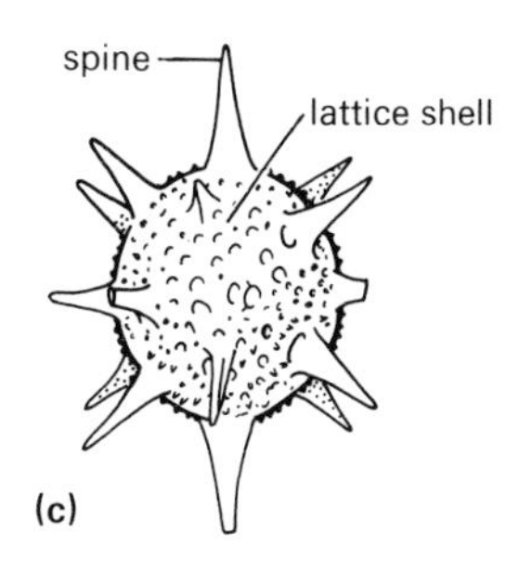

Figure 12.5 Acantharian radiolarians. (a) *Zygacantha* skeleton ×160; (b) *Acanthometra* cell with spicules ×71; (c) *Belonaspis* skeleton ×100. ((a) after Campbell 1954a from Popofsky; (b) after Westphal 1976; (c) after Campbell 1954a from Haeckel)

leading to three radiating chambered arms. Related genera also have radial beams to subdivide the chambers into chamberlets.

The Palaeozoic spherical Radiolaria may not be closely related to the younger Spumellaria and comprise several but as yet little-studied groups (see Holdsworth, pp. 167–84 *in* Swain 1977). *Entactinosphaera* (U. Dev.–Carb., Fig. 12.6a) for example, has a six-rayed internal spicule supporting two or more concentric lattice shells.

Albaillella (Carb., Fig. 12.6b) belongs to a group of radiolarians with bilaterally symmetrical, triangulate skeletons that flourished in Silurian to Carboniferous times. Their systematic position is uncertain, with Holdsworth (*op. cit.*, p. 168) placing them in a separate suborder, Albaillellaria; but they have also been compared with the later Nassellaria.

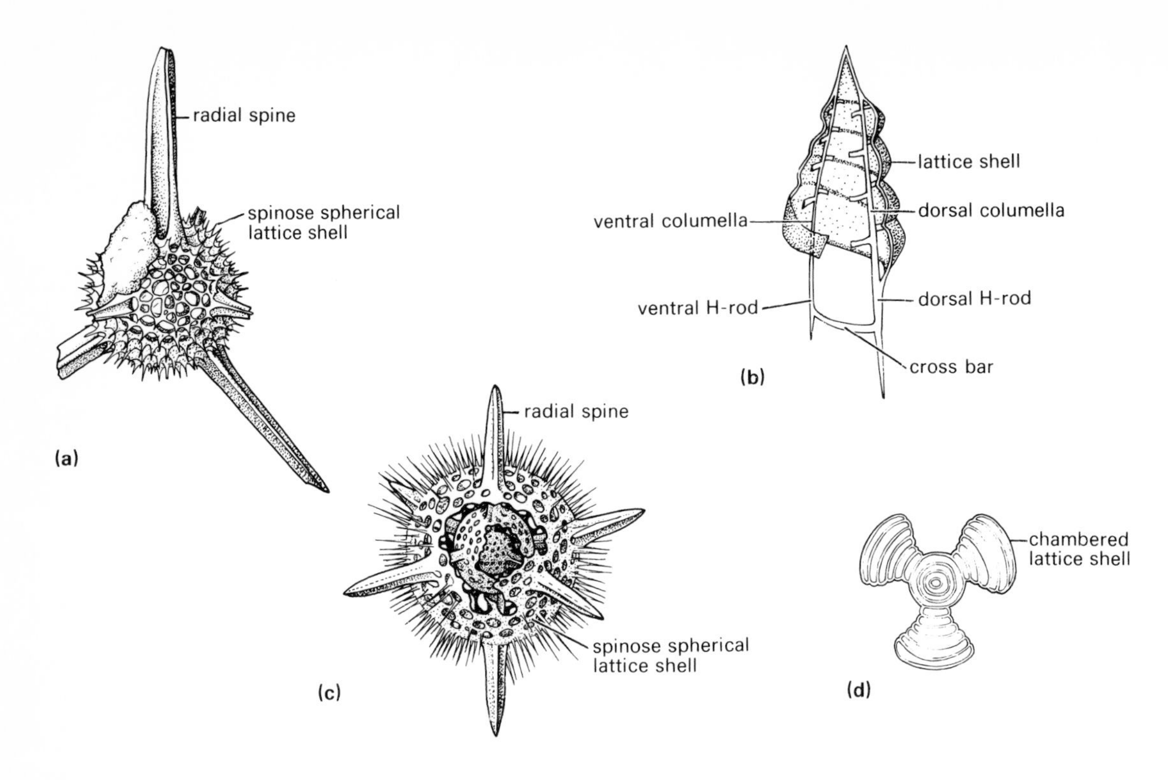

Figure 12.6 Polycystine Radiolaria. (a) *Entactinosphaera* × 195; (b) *Albaillella* (scale unknown); (c) *Actinomma* (scale unknown); (d) *Dictyastrum* ×66. ((a) after Foreman 1963; (b) after Holdsworth 1969; (c) after Campbell 1954a; (d) after Campbell 1954a from Haeckel)

The Order Nassellaria have solid opaline skeletons, usually comprising a primary spicule, a ring or a lattice shell. The **primary spicule** comprises three, four, six or more rays which may be simple, branched or anastomosing. In *Campylacantha* (Rec., Fig. 12.7a) for example, the skeleton comprises a three-rayed spicule, each ray bearing similar but smaller branches. Evolutionary modifications of these rays led in certain stocks to a **sagittal ring** which may bear spines, sometimes in the form of tripod-like basal feet. In *Acanthocircus* (Cret., Fig. 12.7b) the ring bears three simple spines, two of them projecting inwards. The phylogeny of taxa with more-elaborate lattice shells can be traced from the study of the form of the primary spicule or ring elements (see Campbell 1954a, Fig. 8). The lattice may be spherical, discoidal, ellipsoidal or fusiform and constructed of successive chambers (**segments**) that partially enclose earlier ones. The skeletons differ from those of Acantharia and Spumellaria in having a wide aperture (**basal shell mouth**) at the terminal pole. This may be open or closed by a lattice. The initial chamber (**cephalis**) is closed and contains the primary spicule elements referred to above. The cephalis may also bear diagnostic features such as an **apical horn**. The second chamber is called the **thorax** and the third the **abdomen** with, sometimes, many more **post-abdominal** segments, each separated by a 'joint' or **constriction**. *Bathropyramis* (Cret.–Rec., Fig. 12.7c) has a conical lattice with rectangular pores and about nine radial spines around the open basal shell mouth. *Podocyrtis* (Cret.–Rec., Fig. 12.7d) has a conical, segmented skeleton with an apical horn and a tripod of three radial spines around the open mouth. Successive chambers of the fusiform *Cyrtocapsa* (Jur.–Rec., Fig. 12.7e) form prominent segments and the mouth is closed by a lattice.

SUPERORDER TRIPYLEA. The Order Phaeodaria have skeletons which are about 95% organic and 5% opaline silica constructed in the form of a lattice of hollow or solid elements, often with complex dendritic spines called **styles**. The central capsule also

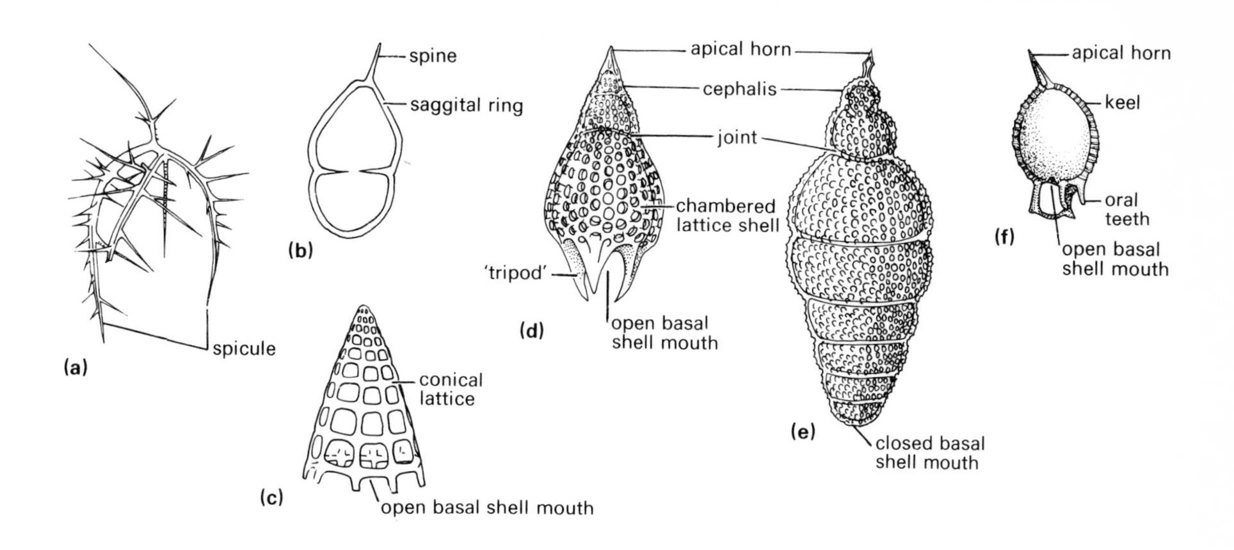

Figure 12.7 Nasellarian and phaeodarian Radiolaria. (a) *Campylacantha* ×200; (b) *Acanthocircus* ×40; (c) *Bathropyramis* ×133; (d) *Podocyrtis* ×100; (e) *Cyrtocapsa* ×200; (f) *Challengerianum* ×187. ((a) after Campbell 1954a from Jorgensen; (b) after Campbell 1954a from Squinabol; (c), (d) & (e) after Campbell 1954a from Haeckel; (f) after Reshetnjak Fig. 24.19b *in* Funnell & Riedel 1971)

has a double wall rather than the single wall found in the former groups, and a basal shell mouth as in the Nassellaria. Only the more robust shells are known as fossils, such as *Challengerianum* (Mioc.–Rec., Fig. 12.7f). This has an ovate shell with an apical horn, a marginal keel, an open basal shell mouth surrounded by oral teeth and a skeleton wall with a fine hexagonal, diatom-like mesh.

General history of radiolarians

Many of the Precambrian and Palaeozoic records of radiolarians appear in fact to have been of acritarchs. The earliest acceptable examples of spherical Spumellaria are of middle Cambrian age (Nazarov 1975). A variety of distinct Spumellaria flourished in the Palaeozoic, joined by the Albaillellaria in the late Devonian to early Permian (Holdsworth, pp. 167–84 *in* Swain 1977). The dramatic reduction of species during the Permian and Triassic periods (Tappan & Loeblich 1973) may relate to the tectonic closure of some late Palaeozoic ocean basins at this time, but it could also reflect the lack of published information. The opening of new ocean basins in the Jurassic and Cretaceous may also have been responsible for radiations seen in the group, with Nessellaria appearing for certain by late Triassic times.

The earliest Phaeodaria are of Miocene age, whilst fossil Acantharia have been reported from Eocene and younger strata (Campbell 1954a), Riedel (pp. 291–8) *in* Harland *et al.* 1967), tacitly, considered these records as dubious.

The fossil record suggests that, unlike diatoms and silicoflagellates, the Radiolaria did not flourish in the cooler Cainozoic Era (Fig. 12.8), the equatorial belt which they prefer having contracted steadily during this time. The competitive pressure for silica brought about by the diatom radiation may also have encouraged a more economic use of silica in the skeleton. Hence many Recent Nassellaria and Phaeodaria are delicately constructed and do not occur as fossils. If this is so then the apparent drop and rise in diversity may be misleading, with the data merely recording a decrease in the preservation potential of radiolarians, of oceanic environments, or both.

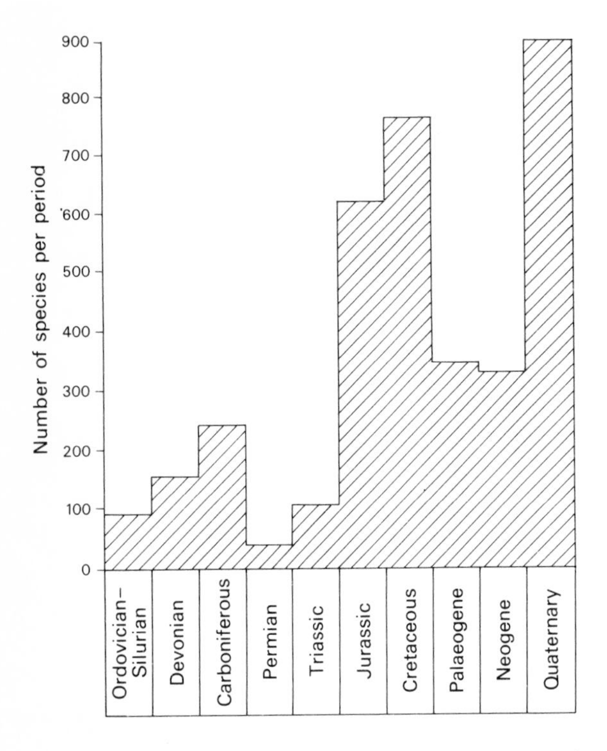

Figure 12.8 Apparent changes in the species diversity of polycystine Radiolaria through time. (Based on Tappan & Loeblich 1973)

Applications of radiolarians

Most of the studies of fossil Radiolaria emphasise their value to biostratigraphic correlation of oceanic sediments, as almost any *Initial report of the deep-sea drilling project* will demonstrate (see Riedel & Sanfilippo 1971, Goll 1972, Petrushevskaya 1975, Foreman 1975). Zonal schemes are reviewed for the Mesozoic by Foreman (pp. 305–20 *in* Swain 1977) and by Pessagno (pp. 913–50 *in* Ramsay 1977) and for the Cainozoic by Casey and McMillen (pp. 521–44 *in* Swain 1977) and Riedel and Sanfilippo (pp. 847–912 *in* Ramsay 1977). Radiolarians are particularly useful where the calcareous microfossils have suffered dissolution. Their often complex structure allows the charting of minute evolutionary changes (see Moore 1972, Knoll & Johnson 1975, Foreman 1975). Comparison of their evolutionary history with changes in the Earth's magnetic field over the past three million years has also revealed a curious correlation between magnetic reversal events and the rates of evolution in certain lineages (Hays 1970).

Limited knowledge concerning the ecology of Recent depth assemblages has so far curtailed the use of radiolarians as markers of ancient climates and water masses but attempts to do this have been made by Nigrini (1970) and Riedel (pp. 567–94 *in* Funnell & Riedel 1971). Their value as depth indicators in marine sediments is examined by Funnell (1967), and Ramsay (1973) outlines their value as indicators of changing calcium carbonate compensation depths with time. They have also been used to indicate palaeogeographic and tectonic changes in ocean basins. For example, radiolarian stratigraphy gave early support to the hypothesis of sea-floor spreading (Riedel 1967), and the closure of the Panama isthmus about 3.5 Ma ago is reflected in changing radiolarian assemblages in the Atlantic (Coney & McMillen, pp. 521–4 *in* Swain 1977).

SUBCLASS HELIOZOA

The Heliozoa closely resemble the Radiolaria but they occupy a freshwater planktonic niche and lack the distinctive central capsule membrane between ectoplasm and endoplasm. Their skeletons may comprise a spherical lattice of chitinoid matter weakly impregnated with silica, or isolated siliceous spicules and plates embedded in the mucilage near the outer ectoplasm. A few can agglutinate a skeleton of sand grains or diatom frustules or even survive without a skeleton at all. Unfortunately, these delicate structures tend to fall apart after death, thereby obscuring their heliozoan origin. Heliozoans are, none the less, known as fossils from a few Pleistocene lake sediments (Moore 1954).

Further reading

An introduction to the modern classification with helpful sections on morphology and distribution, is the paper by Petrushevskaya (1971). Further information on the ecology of both living and fossil radiolarians can be found in reviews within Funnell and Riedel (1971), Swain (1977) and Ramsay (1977). Identification of specimens should be assisted by reference to Campbell (1954a) and Foreman and Riedel (1972 to date).

Hints for collection and study

Fossil Radiolaria can be extracted from mudstones, shales and marls using methods A to E, or from limestones using method F (see Appendix). The residues should then be washed over a 200 mesh sieve, dried and concentrated with CCl_4 (methods I and J) and viewed with reflected light (method O) or with well condensed transmitted light, as with diatoms (*q.v.*). Radiolarian cherts are usually studied in relatively thick petrographic thin sections, viewed with transmitted light at about 400 × magnification or higher. Further information on preparatory techniques is given by Burma (pp. 7–14 *in* Kummel & Raup 1965) and by Riedel and Sanfilippo (pp. 852–8) and Pessagno (pp. 918–20) *in* Ramsay (1977).

13 Phylum Sarcodina – Foraminifera

The Foraminiferida are an order of single-celled protists that live either on the sea floor or amongst the marine plankton. The soft tissue (protoplasm) of the foraminiferid cell is largely enclosed within a shell (**test**) variously composed of secreted organic matter (**tectin**), secreted minerals (calcite, aragonite or silica) or of agglutinated particles. This test consists of a single **chamber** or several chambers mostly less than 1 mm across and each interconnected by an opening (**foramen**) or several openings (foramina). The group, which takes its name from these foramina, is known from early Cambrian times through to Recent times.

Foraminiferid tests can be very abundant in marine sediments, sometimes making up the bulk of the rocks, as in *Globigerina* marls and oozes or in Nummulitic limestones. 'Foraminifera' (as they are informally referred to) are important as biostratigraphic indicators in marine rocks of late Palaeozoic, Mesozoic and Cainozoic age, largely because they are abundant, diverse and easy to study. Planktonic foraminifera are widespread and have had rapidly evolving lineages, factors which greatly aid inter-regional correlation of strata. Because the developmental stages of foraminiferid life history are encapsulated in the test, they also lend themselves well to evolutionary studies. Ecological sensitivity renders the group particularly useful in studies of recent and ancient environmental conditions.

The living foraminiferid

The protoplasm of a foraminiferid comprises a single cell differentiated into an outer layer of clear **ectoplasm** and an inner layer of darker, coloured **endoplasm**.

The ectoplasm surrounds the test and gives rise to numerous thread-like (**filose**) or branching (**reticulose**) **pseudopodia** along which the food material is drawn and the debris expelled (Fig. 13.1a). These pseudopodia are also employed as a means of pulling the test along and for anchorage (Fig. 13.1b).

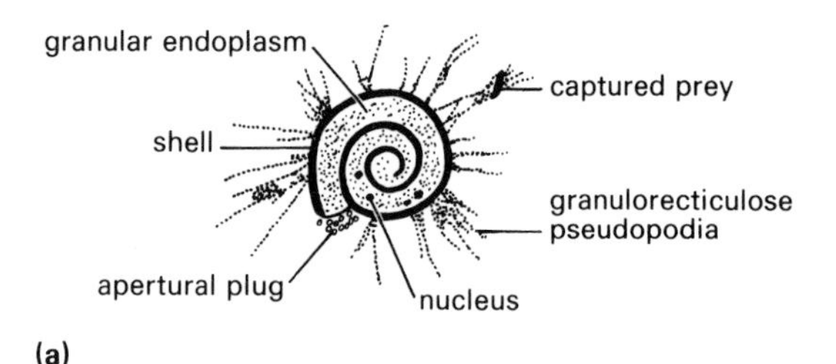

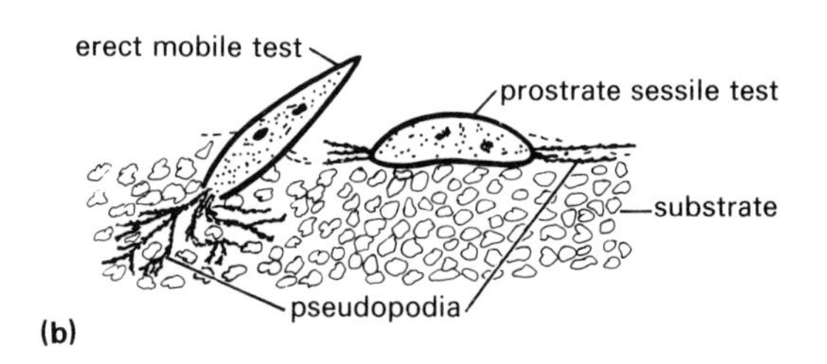

Figure 13.1 (a) Living foraminiferid (*Spirillina vivipara*) as seen in transmitted light. (b) Attitude of benthic foraminiferid tests in life (diagrammatic)

The endoplasm is found within the test and contains food vacuoles and the nucleus. Although a **uninucleate** condition may be found in the simplest forms, many nuclei occur together within the single cells of many-chambered foraminifera (i.e. a **multinucleate** condition). Cells of algal symbionts are also found in the endoplasm of some shallow water species.

Foraminifera feed by engulfing or trapping small organisms and particles with their sticky pseudopodia. Food requirements vary between species but

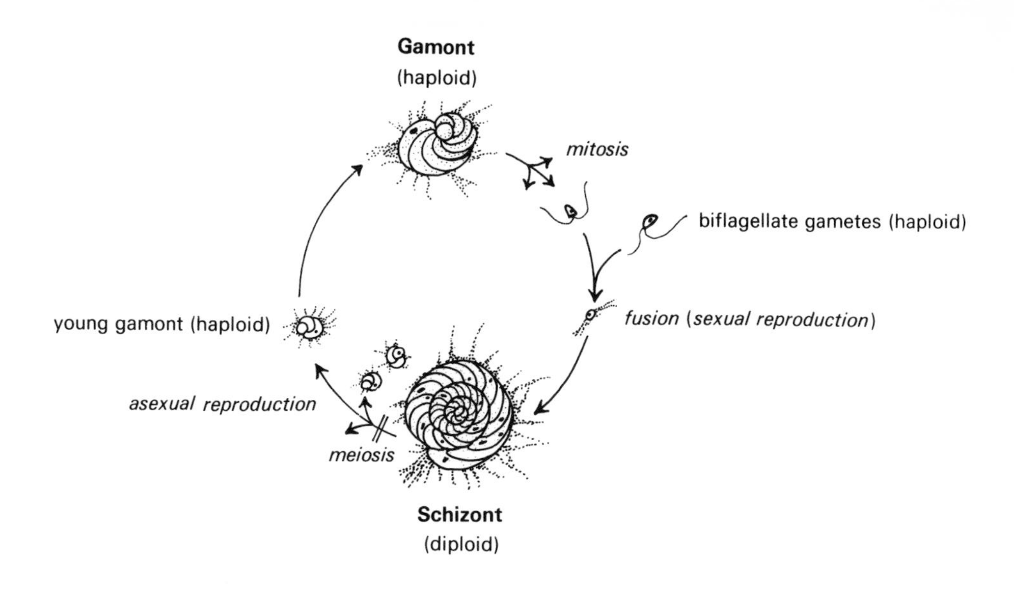

Figure 13.2 Foraminiferid life history (diagrammatic)

include bacteria, diatoms and other protists, small crustaceans, molluscs, nematodes and invertebrate larvae. A few are thought to be parasitic. This engulfed food is digested outside the test in a **food vacuole**, the digestion products and food vacuoles then passing to the endoplasm through an aperture in the test.

Foraminiferid life history

The life history of a foraminiferid is characterised by an alternation between two generations: a **gamont** generation which reproduces sexually, and a **schizont** generation which reproduces asexually (Fig. 13.2). Although this cycle may be completed within a year in tropical latitudes, it can take two or more years at higher latitudes. This alternation, however, is not always strictly followed.

The schizont usually undergoes asexual reproduction during the winter months, the process beginning with withdrawal of the protoplasm into the test. This protoplasm then splits into numerous tiny daughter cells, each with a nucleus or several nuclei containing only half the number of chromosomes originally found in the parent nucleus: a reduction division (meiosis) from a diploid to a haploid chromosome number has taken place. Chamber formation then begins and this new gamont generation is released into the water to disperse. When mature, generally in the summer months, the protoplasm is again withdrawn and divides into daughter cells containing the same, haploid chromosome number of the parent. In most cases these sexual cells (**gametes**) bear two whip-like **flagella**. When released from the parent test, two gametes may fuse (sexual reproduction) to form the next, schizont generation with a diploid chromosome number. The parent test is often left empty after the dispersal of the juveniles.

Tests of these two generations are slightly different in appearance (**dimorphic**). Those of the gamont are more common, are generally smaller and have a relatively large **megalospheric** initial chamber, called a **proloculus**. Those of the schizont are larger, reveal more developmental stages and have a relatively small, **microspheric** proloculus (e.g. Figs 13.2, 13.2d). This phenomenon has led unwittingly to the description of many dimorphs as distinct species or even as distinct genera.

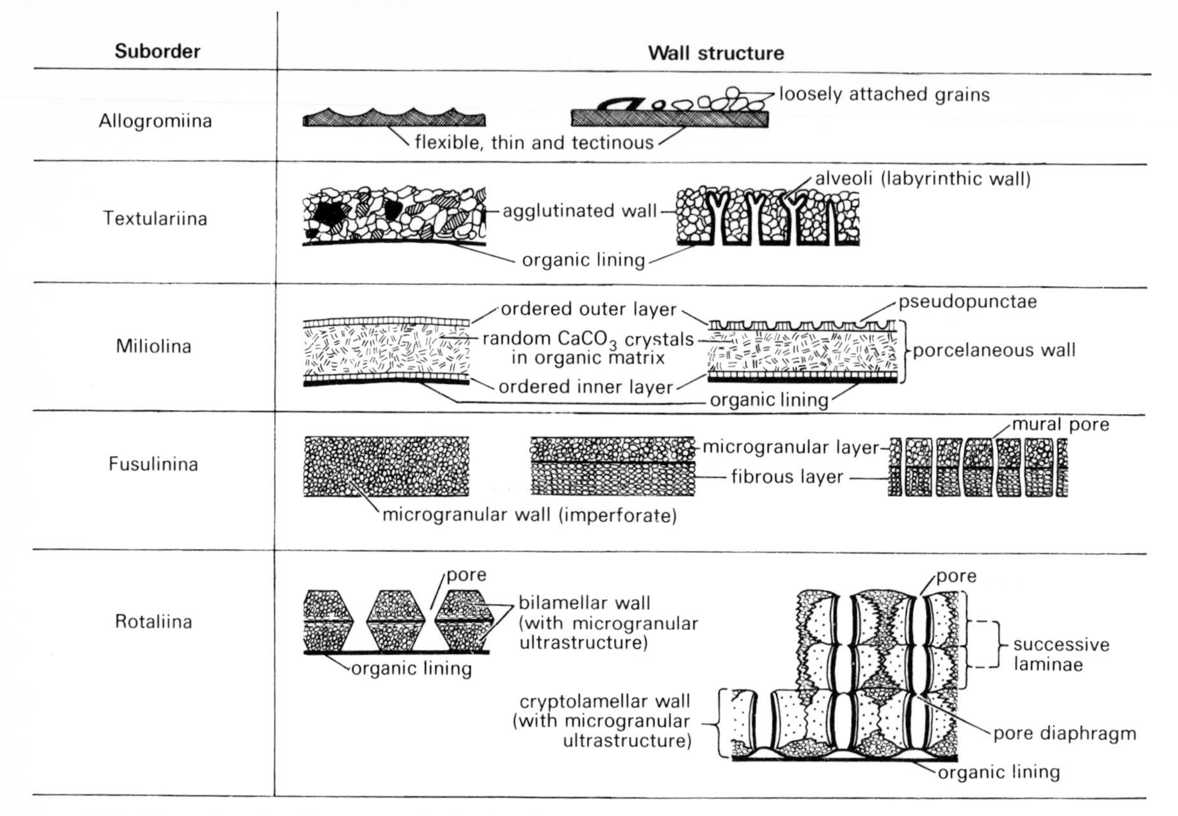

Figure 13.3 Examples of wall structures in the five foraminiferid suborders (diagrammatic, mainly based on studies using electron microscopy)

The test

Wall structure and composition. The structure and composition of the test wall is important to the classification of the group.

Organic-walled forms belong to the Suborder Allogromiina. These have a thin, non-rigid test of proteinaceous or pseudochitinous matter generally termed tectin (Fig. 13.3). Similar material is also present as a thin lining to the chambers of most hard-tested foraminifera, where it may act as a template for mineralisation.

The Suborder Textulariina encompasses forms with **agglutinated** tests. In these, organic and mineral matter from the sea floor is bound together by an organic, calcareous or ferric oxide cement (Fig. 13.3). The grains are commonly selected for size, texture or composition (e.g. coccoliths, sponge spicules and heavy minerals).

Calcareous tests are by far the most abundant and fall into three suborders, each with a different wall structure: the Miliolina, Fusulinina and Rotaliina. **Porcelaneous** tests of the Miliolina appear a distinctive milky white in reflected light and an amber colour in transmitted light. They are constructed of tiny needles of high magnesian calcite randomly arranged for the most part, but the outer surfaces are built with horizontally or vertically arranged needles (Fig. 13.3).

Tests of the Fusulinina appear dark in thin sections viewed with transmitted light and opaque (usually brown or grey) in reflected light. They are built of minute granules of calcite which may be arranged randomly or aligned normal to the surface of the test, giving the wall a fibrous appearance (Fig. 13.3). These granular and pseudo-fibrous layers of **microgranular** calcite are often combined in the structure of a single, multi-layered wall.

The Rotaliina have tests which are generally glassy (**hyaline**) when viewed with reflected light and grey to clear in transmitted light. However, thick walls, fine dense perforations, granules, spines and pigments may obscure this clarity. Rotaliine walls are further distinguished as being radial hyaline, granular hyaline, monocrystalline or spicular when seen in thin section. **Radial** walls are constructed of

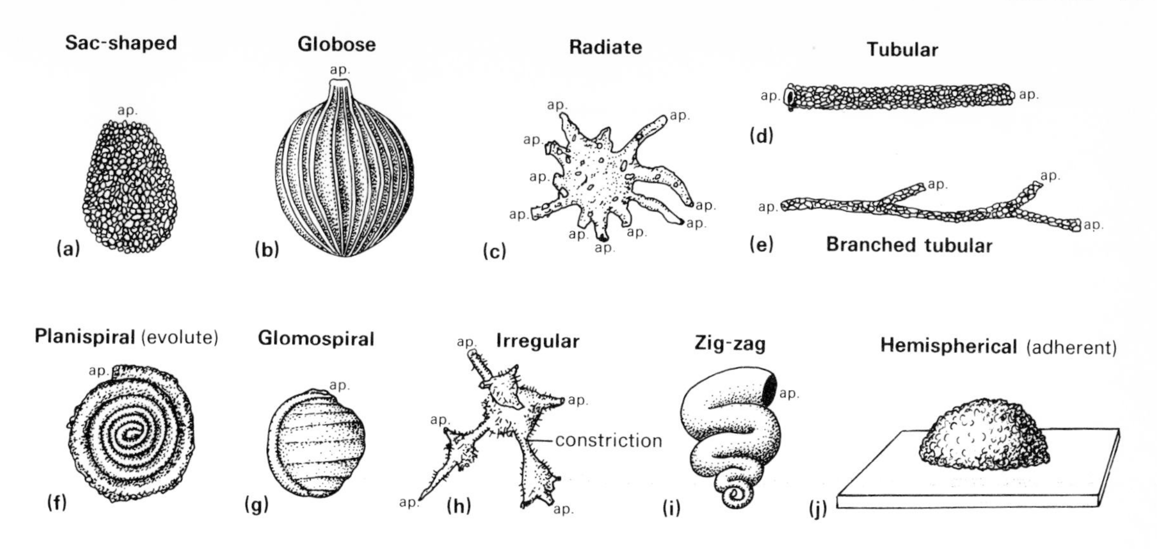

Figure 13.4 Examples of unilocular tests. (a) *Pleurophrys* ×200; (b) *Lagena* ×53; (c) *Astrorhiza* ×49; (d) *Bathysiphon* ×7; (e) *Rhizammina* ×12; (f) *Ammodiscus* ×17; (g) *Usbekistania* ×66; (h) *Aschemonella* ×3; (i) *Ammovertella* about ×20; (j) *Hemisphaerammina* about ×16; ap. = aperture. ((c), (h), (i) & (j) after Loeblich & Tappan 1964; (a) modified from Loeblich & Tappan 1964 after Saedeleer; (b) after Loeblich & Tappan 1964 from H. B. Brady; (g) after Loeblich & Tappan 1964 from Suleymanov)

calcite crystals whose *c*-axes are perpendicular to the test surface giving a black-cross polarisation figure with coloured rings when viewed with crossed nicols. Optically '**granular**' walls are constructed of crystals with *c*-axes oriented obliquely or randomly to the test surface. These give no polarisation figure except for minute flecks of colour when viewed under crossed nicols.

The walls of many foraminifera are traversed by small straight pores or branched alveoli through which the pseudopodia pass, linking ectoplasm and endoplasm (Fig. 13.3). Perforate organic diaphragms across these pores may act as semi-permeable membranes. Such radial pores are typical of the Rotaliina but are also found in certain of the more complex Fusulinina and Textulariina. They give to the wall a pseudo-radial or pseudo-fibrous appearance in thin section.

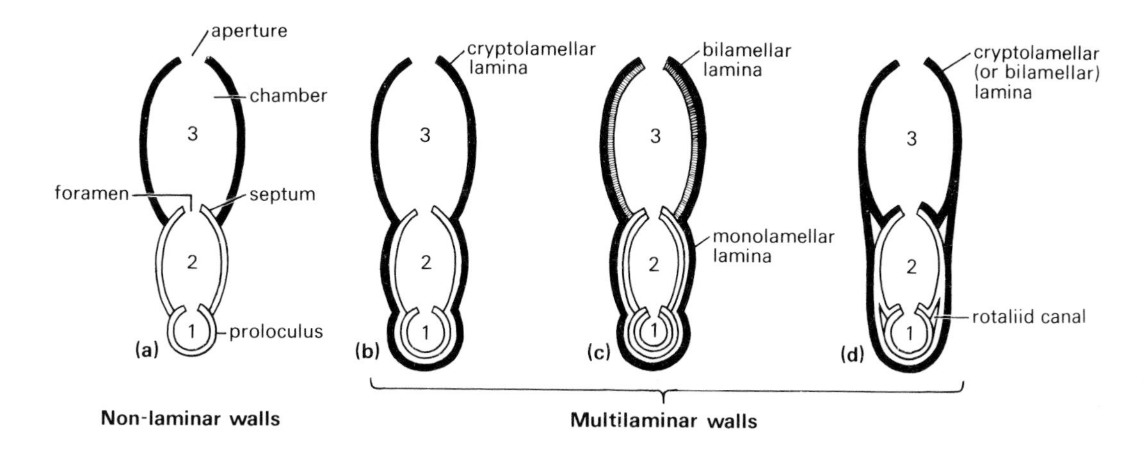

Figure 13.5 Diagrammatic axial sections illustrating different modes of chamber addition: (a) non-laminar; (b) multilaminar, cryptolamellar; (c) multilaminar, monolamellar–bilamellar; (d) multilaminar, cryptolamellar, with septal flaps and canals. The uniserial growth is shown here for simplicity.

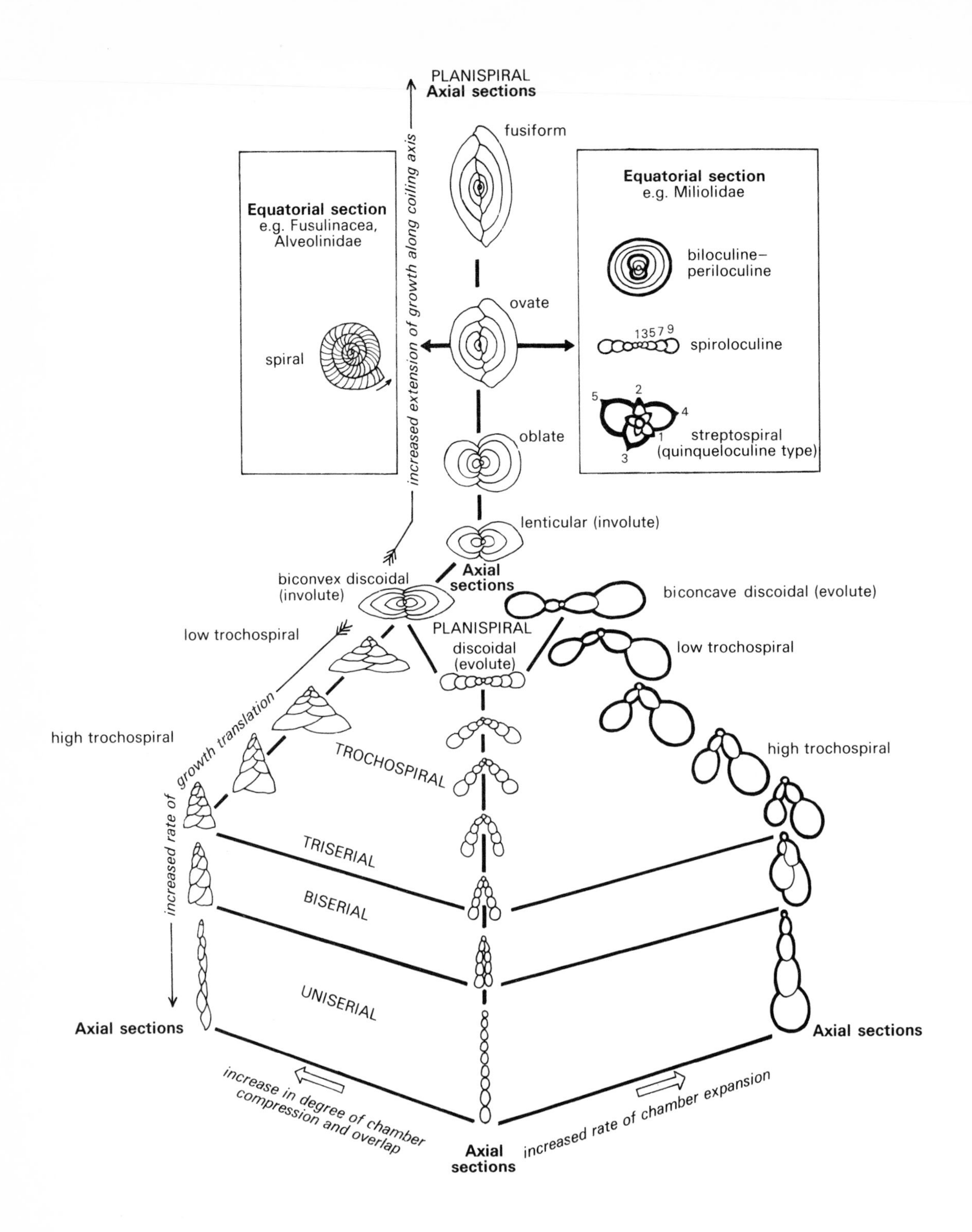

Figure 13.6 Diagram to illustrate the main growth forms in multilocular tests of foraminifera. Axial sections are those cut parallel to and including the main axis of symmetry and growth. Equatorial sections (*sensu lato*) are cut at right angles to this axis, at the widest point on the test.

Chamber development. Foraminiferid tests may consist of a single chamber (**unilocular**) or of two or more chambers (**multilocular**). In unilocular forms (Fig. 13.4), chamber growth proceeds gradually along with protoplasmic growth. In multilocular forms (Fig. 13.5) protoplasmic growth is gradual but test growth is periodic, a new and larger chamber being added at regular intervals. Chamber addition begins with the construction of a loosely bound growth cyst, composed largely of food debris. The pseudopodia are then withdrawn to occupy the space of the new chamber, building first a thin organic wall and then an agglutinated or calcareous one on the outer side, or on both sides.

Where there is no overlap of previous chamber walls by the new wall, the arrangement is termed **non-laminar** (or, incorrectly, non-lamellar; see Fig. 13.5a). If overlap occurs, thin sections will reveal the layers of successive walls (i.e. **multilaminar**, Fig. 13.5b–d). Each lamina may also be composed of two distinct lamellae (i.e. **bilamellar**, Fig. 13.5c). Different types of lamellar walls have been used in classification of the non-laminar Fusulinina and multilaminar Rotaliina (see Table 13.1). Where the lamellar structure is only visible with electron microscopy, the walls may be called **cryptolamellar**.

Chamber architecture and shape. Foraminiferid tests may appear to represent a bewildering array of modes of growth. Although the variation is remarkable it is possible to impose a degree of order by recognising that most multilocular test types arise as the result of interaction between three variables during growth: the rate of translation (i.e. the net rate of movement along the growth axis to the net movement away from the growth axis), the rate of chamber expansion, and the chamber shape (Fig. 13.6).

Different rates of translation produce the four common growth plans of the foraminiferid test: planispiral, trochospiral, biserial and uniserial. In **Planispiral** tests the rate of translation is zero, the chamber or chambers being arranged more or less symmetrically in a plane coil about the growth axis. This growth plan may be further modified by different rates of chamber expansion (e.g. discoidal to biconcave form) and by an extension of growth along the coiling axis (e.g. biconvex to fusiform).

Where material is added in a helical coil the test is called **trochospiral**. Such tests have a **spiral** side and an **umbilical** side, which are more evolute and more involute respectively (Fig. 13.8a). In multilocular tests, a successive decrease in the spiral angle may ultimately bring about a reduction in the number of chambers per whorl to three (**triserial**) although triserial and biserial forms with wide spiral angles are known. Further reduction may obscure or eliminate the spiral component, resulting generally in **biserial** and **uniserial** growth plans (i.e. two and one chamber per whorl respectively, Fig. 13.6). Not infrequently, some of these arrangements may be found together in one test, with developmental changes from planispiral or trochospiral to uniserial.

The rate of chamber expansion may be defined as the rate of increase in volume (or of width, length or depth) from one chamber to the next. In most foraminifera this remains a fairly constant logarithmic trend, at least through early ontogeny. However, the number of chambers per whorl in a species can change through life or between localities and is therefore an unreliable taxonomic character.

Chamber shape varies widely. Unilocular tests may be flask-shaped, globose, tubular, branched, radiate or irregular (Fig. 13.4). Although the chambers of multilocular forms generally remain of constant shape through ontogeny, their arrangement and ornament can vary. Common shapes include globular, tubular, compressed, lunate and wedge-shaped.

Expansion rate and chamber shape are closely linked. Figure 13.7 demonstrates tests with identical rates of volumetric expansion but differing chamber proportions and growth plans. In forms with globular chambers, the length (l), and width (w_1 and w_2) increase equally in logarithmic proportion. Those with compressed chambers often have one virtually constant parameter, while those with tubular chambers may have both the width components fixed. The latter is also the case with **cyclical** chambers in which ring-shaped chambers are added to the periphery of the test.

Apertures and foramina. The **aperture** is found in the wall of the final chamber and serves to connect the external pseudopodia with the internal endoplasm, allowing passage of food and contractile vacuoles, nuclei and release of the daughter cells. Its position remains more or less constant through ontogeny so that each chamber is linked to the next by a foramen or several foramina (see Fig. 13.5). In forms that

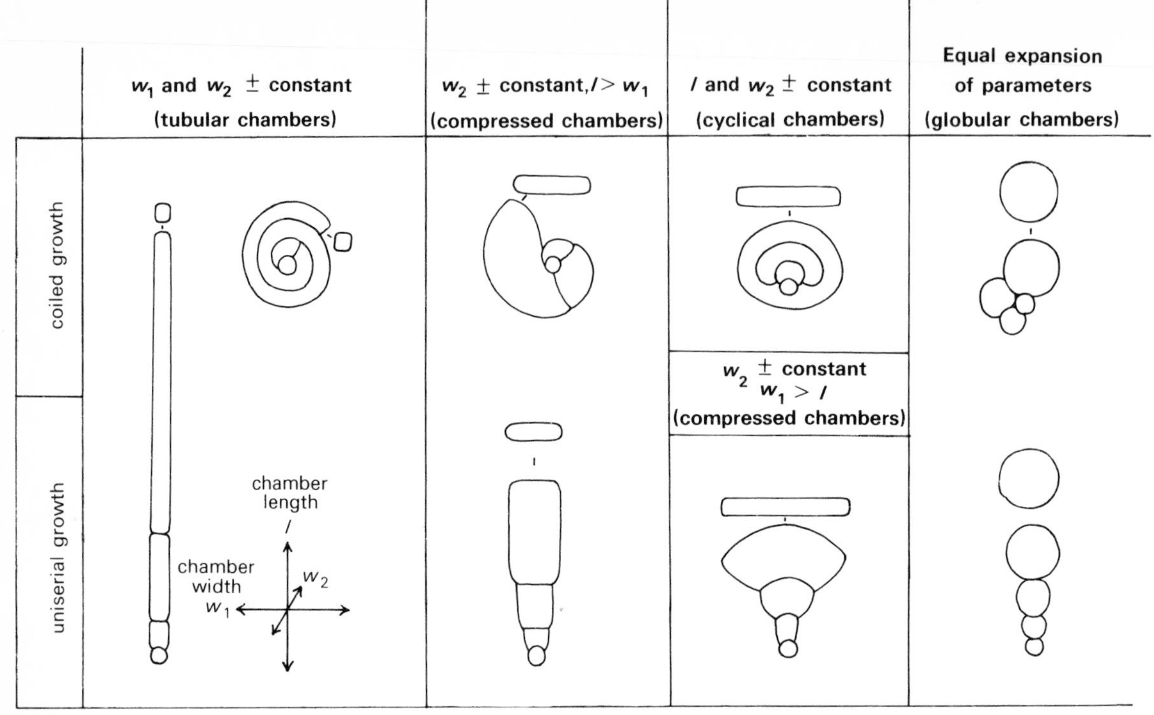

Figure 13.7 Diagrammatic illustration of tests bearing chambers of similar volume but with different growth coordinates (see text)

lack apertures, foramina may be secondarily developed by resorption of the chamber wall.

Apertures may be single or multiple in number and terminal, areal, basal, extraumbilical, umbilical or sutural in position (Figs 13.8, 13.12). Their shape varies widely, e.g. rounded, bottle-necked (phialine), radiate, dendritic, sieve-like (cribrate), cruciform, slit- or loop-shaped. Apertures can be further modified by the presence of an apertural lip or flap (termed a labiate aperture, Fig. 13.8c), teeth (dentate aperture, Fig. 13.8e), a cover plate (bullate aperture, Fig. 13.8f) or an umbilical boss (Fig. 13.8g). The foramina of the rotaliine Rotaliacea and Robertinacea have secondarily formed **septal flaps** (Fig. 13.5d). Such apertural and foraminal structures are used for classification below the subordinal level.

Sculpture. The external surface of the test may bear spines (termed spinose), keels (carinate), rugae (rugose), fine striae (striate), coarser costae (costate), granules (granulate) or a reticulate sculpture. These features should be used with caution in distinguishing certain genera and species for they vary through ontogeny and with environment.

Test function. Much work remains to be done on the function of the test but, principally, it appears to reduce biological, physical and chemical stresses. Biological pressures include, for example, the risk of ingestion by deposit feeders or infestation by parasitic nematode worms. Physical stresses include harmful radiation (including ultra-violet light) from the Sun, water turbulence and abrasion. Chemical stresses encompass fluxes in salinity, pH, CO_2, O_2 and inorganic nutrient levels in the water. In these cases the protoplasm can withdraw into the inner chambers, leaving the outer ones as protective 'lobbies', or a detrital plug may close the aperture (see Marszalek *et al.* 1969). $CaCO_3$ shells may also help to buffer the acidity of nutrient-rich, oxygen-deficient environments.

Additional advantages of the test include the stability it gives to the organism (especially if spinose or thick-walled and calcareous or agglutinated). Without a shell, the steady build up of biomass and regular control of cell division found in the group would also prove difficult. Surface sculpture may variously assist buoyancy (e.g. spines and keels of planktonic forms), improve adherence, strengthen the test against crushing and help to

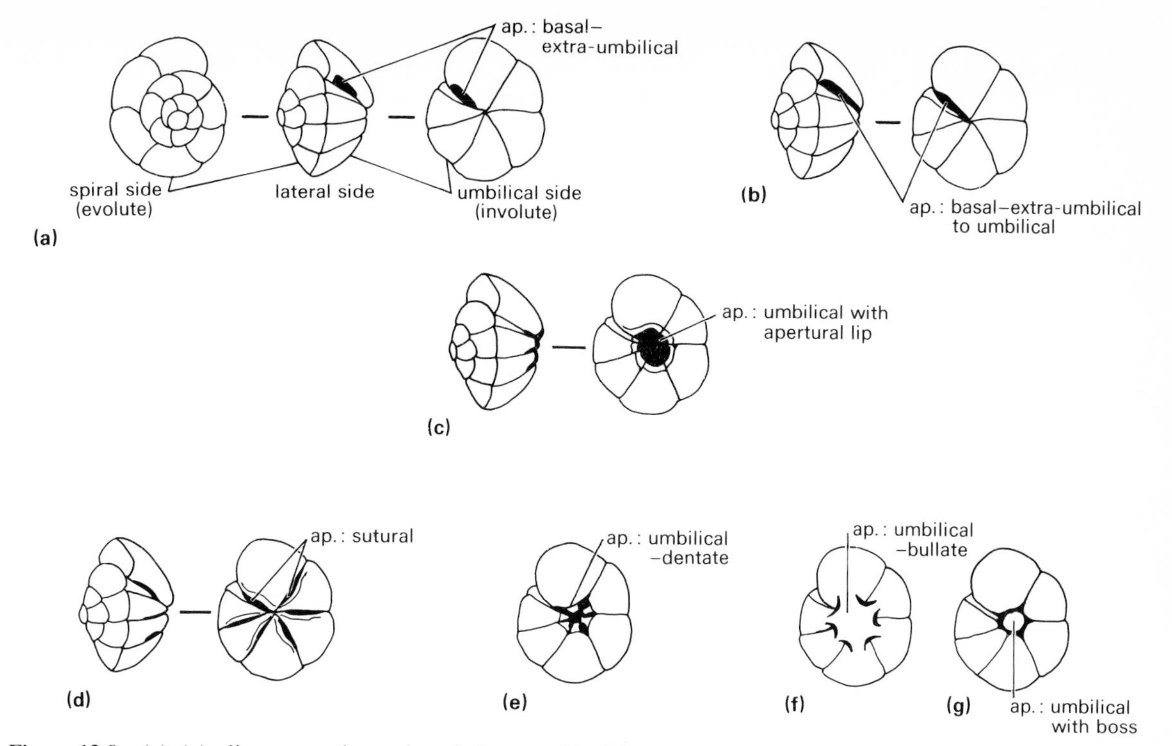

Figure 13.8 (a)–(g), diagrammatic trochospiral tests with different kinds of aperture; ap. = aperture.

channel ectoplasmic flow to and from the apertures, pores and umbilicus.

Foraminiferid ecology

Studies of the ecology of Recent foraminifera have greatly enhanced their value as palaeoenvironmental indicators. Some aspects of these ecological studies are reviewed by Phleger (1960), Murray (1973), Boltovskoy and Wright (1976) and many authors in Schafer & Pelletier (1976). The latter volume also brings together papers on foraminiferid palaeoecology.

Food. Foraminifera play a prominent role in marine ecosystems as micro-omnivores, i.e. they feed on small bacteria, algae, protists and invertebrates (Lipps & Valentine 1970). Some are scavengers, feeding on dead organic particles. Certain foraminifera from reef and carbonate shoal environments appear to benefit from endosymbiotic algae in much the same way as do the hermatypic corals, e.g. *Archaias* (Fig. 13.20b), *Elphidium* (Fig. 13.28b). It is possible that the fossil 'larger foraminifera' achieved their great size in this way, the algae providing nutrients from photosynthesis and favouring maximum $CaCO_3$ precipitation by the uptake of CO_2.

High-diversity foraminiferid assemblages strongly suggest a wide range of available food resources. Such partitioning of resources among species is typical of stable habitats, even where food is scarce. Conversely, seasonal fluctuations in food supply, as in boreal waters, may result in foraminiferid blooms of great abundance but low diversity. These 'opportunistic' species must reach maturity quickly, hence they are relatively small in size. Planktonic foraminifera tend to thrive in regions of oceanic upwelling because their primary food supply, the phytoplankton, are able to prosper in such nutrient-enriched waters.

Predation. Benthic foraminifera stand a very high chance of being ingested by creatures such as worms, crustaceans, gastropods, echinoderms and fish that browse on the sediments and organisms upon the sea floor. As yet, the effects of such predation on living foraminiferid populations is little known and may serve either to raise or lower diversity. It should be noted, however, that the fossil record may be distorted by the selective destruction of tests in the gut of many deposit feeders (Mageau & Walker, pp. 89–105 *in* Schafer & Pelletier 1976).

Substrate. Silty and muddy substrates are often rich in organic debris and the small pore spaces contain bacterial blooms. Such substrates are therefore attractive to foraminifera and support large populations. Many of these species are thin-shelled, delicate and elongate forms (see Figs 13.21a, d; 13.30a, b).

The larger pore spaces of sands and gravel contain fewer nutrients and therefore support sparser populations. Foraminifera from these coarser substrates may be thicker-shelled, heavily ornamented and of biconvex or fusiform shape (see Figs 13.18a, 13.19c, 13.23g, 13.29). Although foraminifera have been found living up to 200 mm below the sediment surface, the majority are feeding within the top 10 mm or so, the depth of burial varying between species.

Those foraminifera which prefer hard substrates (i.e. rock, shell, sea grasses and algae) are normally attached, either temporarily or permanently by a flat or concave lower surface. Typical growth forms include discoidal, annular, flabelliform, concavo-convex, plano-convex, dendritic and irregular (see Figs 13.4j, 13.12g, 13.19a, 13.20c, 13.23a, e, f, 13.25a, b). They often develop a relatively thin test and exhibit greater morphological variability than seen in sediment-dwelling and planktonic forms (see Brasier 1975).

Light. The zone of light penetration in the oceans (the **photic zone**) is affected by water clarity and the incident angle of the Sun's rays. Hence the photic zone is deeper in tropical waters (<200 m) and decreases in depth towards the poles where it also varies with marked seasonality. Primary production of nutrients by planktonic and benthic algae render this zone attractive to foraminifera, especially the porcelaneous Miliolina and the larger forms (see Haynes 1965). This may be due not only to their preference for algal food sources or even to their association with endosymbiotic algae. Many foraminifera also thrive in the protected and food-rich niches around and on algal fronds, roots and sea-grass meadows.

Temperature. Each species is adapted to a certain range of temperature conditions, the most critical being that range over which successful reproduction can take place. Generally, this range is narrowest for low-latitude faunas adapted to stable, tropical climates. However, stratification of the oceans results in the lower layers of water being progressively cooler, as for example in tropical waters where the surface may average 28°C but the bottom waters of the abyssal plains may average less than 4°C. These cooler, deeper waters may be characterised by cool-water benthic assemblages that otherwise are found at shallower depths nearer the Poles (Fig. 13.9).

Planktonic foraminifera are also adapted to different oceanic layers of particular temperatures and densities. Forms adapted to the cooler, denser water bodies have fewer buoyancy problems and consequently a lower test-porosity than those from warmer or shallower waters (see Frerichs *et al.* 1972). The latter improve their buoyancy with a high test-porosity and prominent spines (Fig. 13.10). None the less, the deep-water planktonic forms have to cope with the effects of $CaCO_3$ solution (due to higher pressure, lower pH and other factors) which may account for their extra 'crust' of radial, hyaline calcite (see Fig. 13.27h). In several planktonic species (e.g. *Globigerina pachyderma*) warm and cool populations can be distinguished by a predominance of right-hand (dextral) or left-handed (sinistral) coiling (Fig. 13.9). The sequence of Pleistocene temperature fluctuations has been determined from studies of these and similar foraminifera obtained in deep-sea cores. The distribution and ecology of Recent planktonic forms are discussed in more detail by various authors in Funnell and Riedel (1971) and by Bé *in* Ramsay (1977, pp. 1–100).

Oxygen. Oxygen concentrations do not vary greatly in present seas and oceans, with a few exceptions such as the Black Sea. Oxygen deficiency is usually caused by high organic productivity at the sea's surface leading to anaerobic bacterial blooms on the sea floor and the production of H_2S. However, it appears that O_2 deficiency does not greatly affect small organisms such as foraminifera, for their O_2 requirements are very modest and high standing crops are known from these conditions. Anaerobic assemblages are typified by small, thin-shelled,

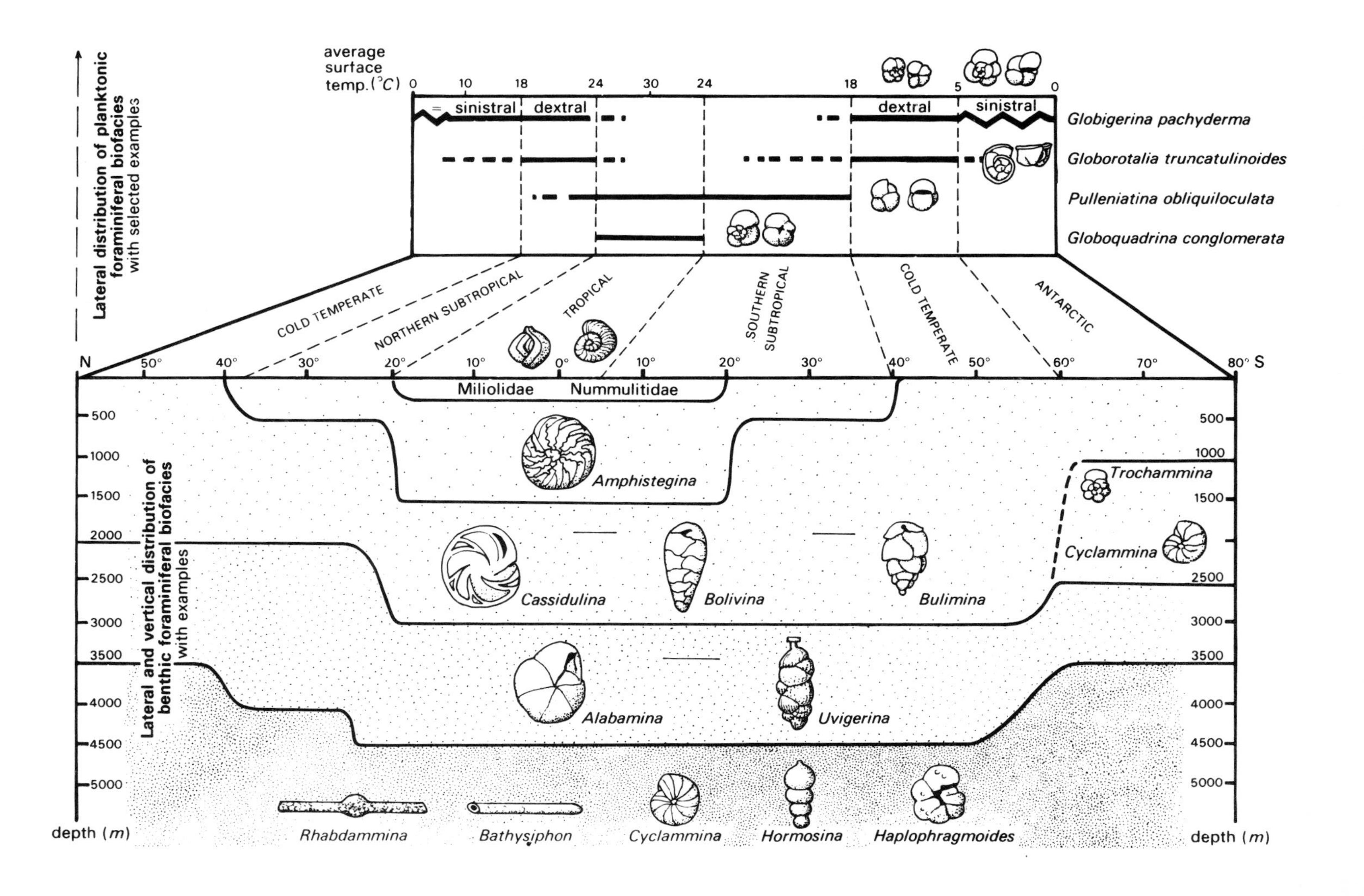

Figure 13.9 Diagram illustrating how benthic and planktonic foraminiferid assemblages (and some of the typical taxa) change with depth and latitude in the Pacific Ocean, especially in relation to temperature. (Based partly on Saidova 1967)

unornamented calcareous or agglutinated assemblages. Although low O_2 decreases the ability to secrete $CaCO_3$ it can increase its subsequent chances of preservation, unless conditions are also acidic (see Phleger & Soutar 1973).

Salinity. The majority of foraminifera are adapted to normal marine salinities (about $35^0/_{00}$) and it is in such conditions that the highest diversity assemblages are found. The low salinity of brackish lagoons and marshes favours low-diversity assemblages of agglutinated foraminifera (mostly with non-labyrinthic walls and organic, siliceous or ferruginous cements, e.g. *Reophax*, Fig. 13.10) and certain Rotaliacea (e.g. *Ammonia*, Fig. 13.10). The soft, tectinous Allogromiina are found in fresh and brackish waters, but they are rarely encountered as fossils. The high $CaCO_3$ concentrations of hypersaline waters favour the porcelaneous Miliolina (especially the Nubecularidae and Miliolidae, e.g. *Triloculina*, Fig. 13.10) but deter most other groups.

Triangular plots of the relative proportions of Textulariina, Miliolina and Rotaliina have proved useful as indices for palaeosalinity. Samples from certain habitats usually fall within the proscribed fields (Fig. 13.10; see Murray 1973). However, the method can give misleading results where there has been selective post-mortem reworking, solution or disaggregation of tests.

$CaCO_3$. The solubility of $CaCO_3$ is less in warm than in cool waters. This in part accounts for the thicker tests and the occurrence of foraminiferid limestones and oozes at low latitudes. $CaCO_3$ solubility also increases with pressure (i.e. depth). Furthermore, the ratio of CO_2 to O_2 increases with depth because algae cannot photosynthesise below the photic zone, although animals continue to respire. This leads to a decrease in pH with depth, from about 8·2 to as low as 7·0. The level at which $CaCO_3$ solution equals $CaCO_3$ supply is called the calcium carbonate compensation depth (or CCCD). As this can be difficult to locate, the concept of the **lysocline** (i.e. the level of maximum change in the rate of solution) is often preferred. Either way, the net result is a drop in the number of calcareous organisms with depth, there being few below 3000 m (Figs 13.9 and 13.10). For this reason, the agglutinated foraminifera dominate populations from abyssal depths.

Foraminifera and sedimentology

Planktonic foraminifera are important contributors to deep-sea sedimentation and, with coccoliths, account for more than 80% of modern carbonate deposition in seas and oceans. At present the foraminifera contribute more than the coccolithophores, although this was not the case with earlier chalks and oozes (Bramlette 1958). Three factors are important in controlling the deposition of *Globigerina* ooze (i.e. ooze in which over 30% of sediment is globigerinacean): climate, depth of the lysocline and terrigenous sediment supply. The position and strength of currents, especially diverging and upwelling currents, are greatly affected by climate and themselves affect the plankton productivity. Berger (1971) estimates than from 6 to 10% of the living population of planktonic foraminifera leave empty tests every day, mostly as a result of reproduction. These tests settle quite rapidly and are less susceptible to dissolution than coccoliths (which lack organic outer layers), except when they approach the lysocline which usually lies between 3000 and 5000 m depth. Fluctuations in the depth of the lysocline during Mesozoic and Cainozoic times are now known to have caused cycles of deposition and dissolution (see Berger 1973), selectively removing some of the smaller or more delicate forms and rendering the fossil record of the deep sea much less complete than might be wished. Even where the conditions are otherwise favourable, *Globigerina* oozes cannot accumulate where there is an influx of terrigenous clastics, hence they are rarely found on continental shelves. At present such oozes are accumulating mainly between 50° N and 50° S at depths between about 200 and 5000 m, especially along the mid-oceanic ridges (Fig. 13.11). In many cases, though, they are diluted with the siliceous remains of diatoms and radiolarians.

Classification

Kingdom PROTISTA
Phylum SARCODINA
Class RHIZOPODA
Order FORAMINIFERIDA

The basis for subdivision of the Order Foraminiferida is open to discussion, but the generally

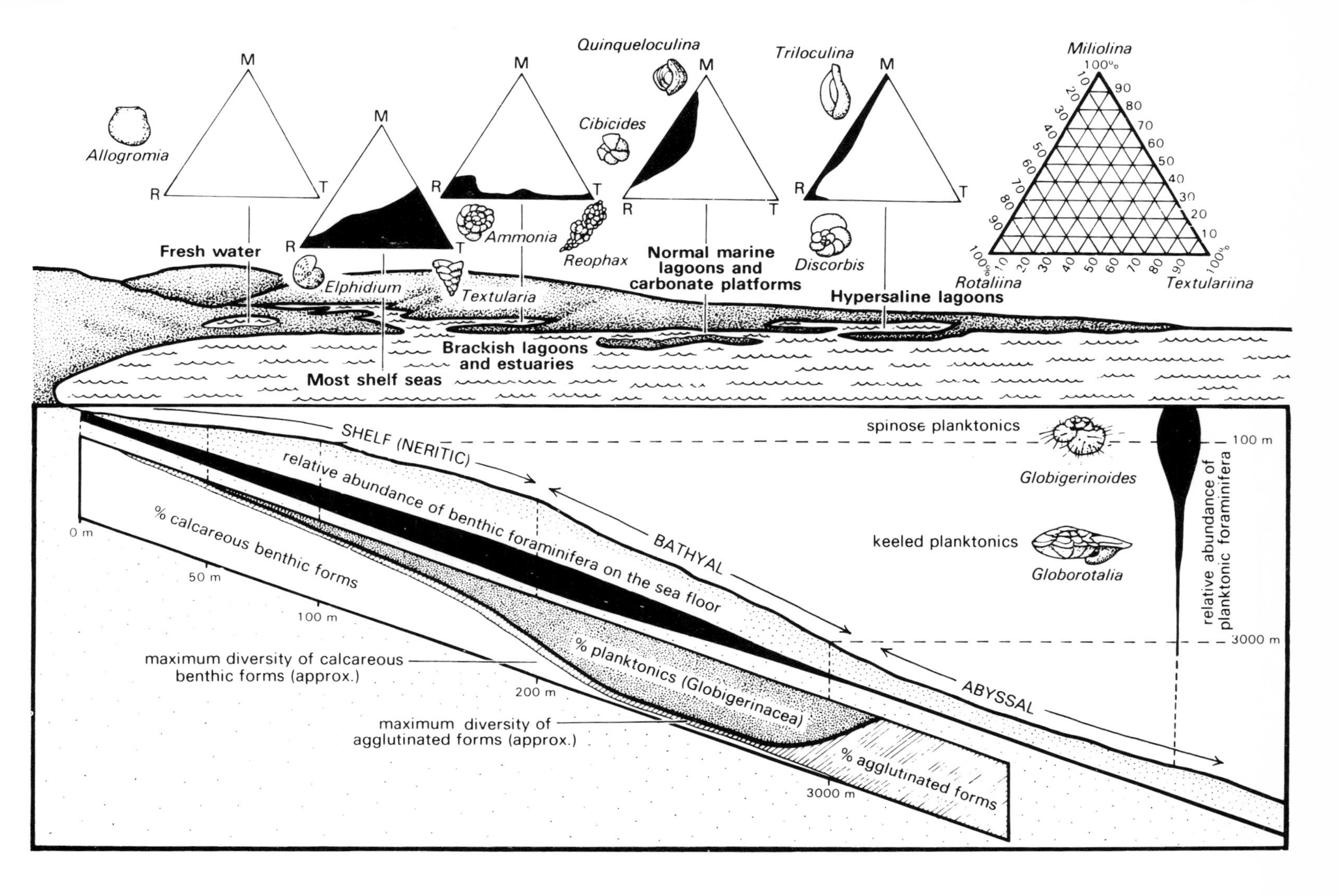

Figure 13.10 Diagram illustrating how benthic and planktonic foraminiferid abundance and general composition change with depth and salinity. Some typical genera are shown.

Table 13.1 A guide to the morphological character of foraminiferid suborders and superfamilies.

Suborder	*Superfamily*	*Character of wall*		*Septation*	*Chamber architecture*	*Aperture*
ALLOGROMIINA	Lagynacea	Non-laminar, thin tectinous; mostly imperforate		Unilocular	Irregular, sac, flask or tube-shaped	Single or multiple, simple terminal or absent
TEXTULARIINA	Ammodiscacea	Non-laminar, agglutinated	Agglutinated outer layer with tectinous inner layer; imperforate and perforate	Unilocular, simple or labyrinthic	Tubular, branching, radiate, globular or enrolled	Single or multiple, simple terminal or absent
	Lituolacea		Siliceous or agglutinated with $CaCO_3$, SiO_2 or ferruginous cement; mostly imperforate but may have labyrinthic walls	Multilocular, simple, labyrinthic or with chamberlets	Planispiral, trochospiral, triserial or uncoiled straight biserial and uniserial	Single or multiple, terminal or basal
FUSULININA	Parathuram-minacea	Non-laminar, calcareous, microgranular	$CaCO_3$ granules in $CaCO_3$ cement; imperforate	Unilocular or multilocular, simple	Single globular, irregular or multiple uniserial (straight or branched)	Single or multiple, simple terminal or absent
	Endothyracea		Fibrous or granular often in 2 layers; may be some agglutinated matter; finely perforate	Multilocular, simple or with chamberlets	Planispiral, trochospiral, uniserial or biserial	Single or multiple, basal, terminal or absent
	Fusulinacea		Up to 4 layers, including dark outer tectum; perforate	Multilocular, mostly with chamberlets, may be simple	Planispiral, mostly fusiform, some ovate or discoidal	Absent; septal and mural pores only
MILIOLINA	Miliolacea	Non-laminar, calcareous, porcelaneous (high Mg calcite) with tectinous inner lining; imperforate adult stage; may be agglutinated outer layer or pseudopunctae		Unilocular or multilocular, simple or with chamberlets	Spiral tube of early stages followed by planispiral and streptospiral growth, by uncoiled growth, irregular branched growth or uniserial growth	Single or multiple, basal, areal or terminal; often with tooth in streptospiral forms
ROTALIINA (part)	Spirillinacea	Multilaminar, calcareous, hyaline, perforate	Single crystal of high Mg calcite	Unilocular or multilocular, simple	Planispiral, trocho-spiral, biserial or cyclical	Simple terminal or umbilical
	Nodosariacea		Finely perforate, optically radial, cryptolamellar calcite	Unilocular or multilocular, simple	Planispiral, uncoiled straight biserial uniserial, irregular or streptospiral	Rarely simple, usually radiate, peripheral or terminal
	Buliminacea		Fine to coarsely perforate, optically radial cryptolamellar calcite	Multilocular, simple	Trochospiral, straight biserial or uniserial	Basal slit or terminal; may have tooth plate
	Duostominacea		Optically radial and granular crypto-lamellar calcite or aragonite	Multilocular, simple	Planispiral or trochospiral	Single, or double, basal

Suborder	*Superfamily*	*Character of wall*		*Septation*	*Chamber architecture*	*Aperture*
ROTALIINA (part)	Robertinacea	Multilaminar, calcareous, hyaline, perforate	Optically radial bilamellar aragonite	Multilocular, simple or with chamberlets; septal flaps present	Trochospiral	Basal slit and supplementary sutural apertures
	Discorbacea		Optically radial crypto- and bilamellar low Mg calcite	Unilocular or multilocular, simple	Planispiral, trochospiral, uncoiled biserial or irregular	Basal, areal or peripheral
	Globigerinacea		Optically radial bilamellar low Mg calcite	Multilocular, simple	Planispiral, trochospiral, uncoiled biserial or uniserial	Basal; may be supplementary apertures
	Rotaliacea		Optically radial bilamellar calcite (mostly low Mg) with canals ('canaliculate')	Multilocular, simple or divided into median chambers with chamberlets and lateral chambers; septal flaps	Planispiral to trochospiral; test often biconvex or conical in profile	Primary apertures absent; may be secondary basal foramina
	Orbitoidacea		Optically radial bilamellar calcite	Multilocular, simple or divided into median and lateral chambers	Planispiral; embryonic chambers thick and bilocular; test often discoidal or biconvex	Usually absent. Median chambers with connecting canals (stolons)
	Cassidulinacea		Optically granular cryptolamellar low Mg calcite	Multilocular, simple	Planispiral, trochospiral, or straight biserial, uniserial or cyclical	Slit- or loop-shaped or multiple
	Nonionacea		Optically granular crypto- and bilamellar low Mg calcite	Multilocular, simple	Planispiral or trochospiral; test often biconvex	Single, basal or areal
	Carterinacea		Parallel orientated $CaCO_3$ spicules in $CaCO_3$ matrix	Multilocular, simple and with chamberlets	Trochospiral	Umbilical

accepted taxonomy of Loeblich and Tappan (1964) takes the following features into account, in order of importance: wall structure and composition, chamber shape and arrangement, aperture, and ornament. The salient features of the currently recognised suborders and superfamilies are noted in Table 13.1. The extent to which wall structure indicates evolutionary relationships should be called into question, however, for there is the likelihood that different kinds of hyaline wall structure may represent grades of development through which several distinct lineages have passed (see Loeblich & Tappan, pp. 1–54 *in* Hedley & Adams 1974). Evidence from scanning electron microscopy led Loeblich and Tappan to modify their original classification by extending the scope of the Discorbacea, restricting the scope of the Orbitoidacea and erecting the Duostominacea and Nonionacea (see also Tappan, pp. 301–33 *in* Schafer & Pelletier 1976). This classification is followed here but the emphasis is placed on features visible with an optical microscope.

Three artificial groups of foraminifera, all widely used for stratigraphic correlation, deserve a prior mention: the 'larger' benthics, the 'smaller' benthics and the 'planktonics'. The term 'larger forami-

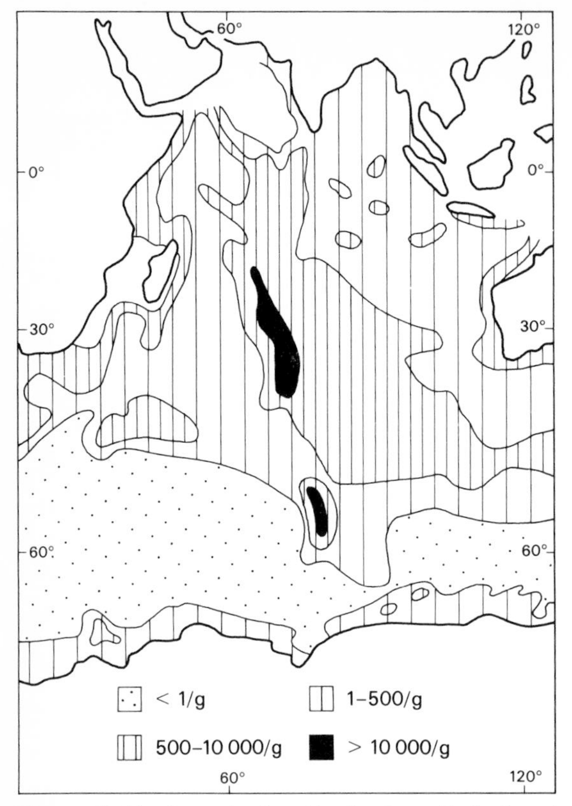

Figure 13.11 Quantitative distribution of planktonic foraminifera in the surface layer of bottom sediments from the Indian Ocean. (Based on data in Belyaeva 1963)

nifera' is loosely applied to groups characterised by large individuals (> 1 mm diameter) with complex endoskeletal structures. Both the living and extinct examples have usually evolved in tropical or subtropical habitats. Three basic growth plans are characteristic of larger foraminifera: fusiform planispiral, discoidal planispiral (including cyclical) and low trochospiral–uniserial. Chambers are commonly further subdivided into smaller units called **chamberlets** by secondary septa (**septulae**) or by **pillars**. The walls may be filled with small branched caverns (**labyrinthic walls**). Each of these basic growth plans has been occupied at some time and place by the four suborders of hard-tested foraminifera.

'Smaller' benthic foraminifera comprise the majority of individuals and species. It is from certain of these that the 'larger' foraminifera have evolved, hence they may share some characteristics.

The majority of benthic foraminifera have a short planktonic juvenile stage in their life cycle, to aid outbreeding and dispersal. Entirely planktonic forms, however, belong mostly to the Superfamily Globigerinacea (M. Jur.–Rec.). These are notable amongst the smaller foraminifera for their value in inter-regional biostratigraphic correlation and their significance as rock builders: for example, of chalk, *Globigerina* marls and *Globigerina* oozes. Planktonic adult stages are known in several other rotaliine groups, as in the discorbacean *Tretomphalus* which leads a normal benthic life until it reaches maturity, at which point a final, spherical float chamber is formed and the individual drifts up to the water's surface (Fig. 13.23b). Planktonic agglutinated forms have also been reported.

Suborder Allogromiina. SUPERFAMILY LAGYNACEA. These foraminifera have an entirely organic test with only one chamber. They are rarely encountered as fossils, being found mostly in Recent fresh or brackish water sediments. None the less, they are known in marine sediments since late Cambrian times. *Allogromia* (Rec., Fig. 13.12a) has an ovate test with a rounded terminal aperture. *Pleurophrys* (Rec., Fig. 13.4a) is similar but smaller. *Shepheardella* (Rec., Fig. 13.12b) has a long tubular test with an aperture at each end. Both larger and planktonic types are unknown in this suborder.

Suborder Textulariina. The Textulariina, characterised by non-laminar, agglutinated tests, are divisible into two superfamilies: the Ammodiscacea, which are mostly unilocular and the Lituolacea, which are multilocular.

SUPERFAMILY AMMODISCACEA. The Ammodiscacea range from early Cambrian to Recent times. All would be considered 'smaller' benthic foraminifera although *Astrorhiza* can be up to ten millimetres in diameter. *Saccammina* (Sil.–Rec., Fig. 13.12f) is a simple globular form with a terminal aperture. Irregularly arranged chambers of similar type are found in the multilocular *Sorosphaera* (Sil.–Rec., Fig. 13.12e). In *Technitella* (Olig.–Rec., Fig. 13.12d) the test is fusiform and built of carefully selected sponge spicules. Tubular tests generally have several apertures and may be simple and unbranched as in *Bathysiphon* (?Camb., Ord.–Rec., Fig. 13.4d), branched as in *Rhizammina* (Rec., Fig. 13.4e) or radiating from a central point as in *Astrorhiza* (?M. Ord.–Rec., Fig. 13.4c), *Aschemonella* (U. Dev.–

Rec., Fig. 13.4h) and *Rhabdammina* (Ord.–Rec., Fig. 13.12c).

Planispiral coiling is seen in *Ammodiscus* (Sil.–Rec., Fig. 13.4f) and glomospiral coiling (like a skein of wool) in *Usbekistania* (Jur.–Rec., Fig. 13.4g). Adherent forms are irregularly branched or may meander and zig-zag across the substrate (e.g. *Ammovertella*, L. Carb.–Rec., Fig. 13.4i; *Tolypammina*, U. Ord.–Rec., Fig. 13.12g).

SUPERFAMILY LITUOLACEA. Tests of the Lituolacea are more complex than those of the Ammodiscacea. The simplest of the 'smaller' benthic forms are commonly straight uniserial (e.g. *Reophax*, U. Dev.–Rec., Fig. 13.13a; *Hormosina*, Jur.–Rec., Fig. 13.13b) or biserial (*Textularia*, U. Carb.–Rec., Fig. 13.14b). Both kinds of growth are combined in different stages of *Bigenerina* (U. Carb.–Rec., Fig. 13.14c). Triserial tests are also common in the group (e.g. *Verneuilina*, Jur.–Rec., Fig. 13.14d), and *Miliammina* (L. Cret.–Rec., Fig. 13.13c) is coiled like a miliolid (see below, and fig. 13.19e).

Coiled growth plans are also common, as in the planispiral *Cyclammina* (Cret.–Rec., Fig. 13.13d) and the trochospiral *Trochammina* (L. Carb.–Rec., Fig. 13.14e). A combination of planispiral and uniserial growth is seen in the uncoiled test of *Ammobaculites* (L. Carb.–Rec., Fig. 14.13a).

The 'larger' agglutinated foraminifera have tests mostly constructed of calcareous particles with a mineral cement. Examples in the Lituolidae are found in rocks formed in warm shallow facies of Jurassic and Cretaceous age. *Spirocyclina* (U. Cret.) and its relatives had almost planispiral, compressed tests and labyrinthic walls (Fig. 13.13f). *Loftusia* (U. Cret) resembles the more ancient fusulines in having a planispiral fusiform test with a labyrinthic wall, irregular septa and chamberlets (Fig. 13.13e). The Dicyclinidae were long ranged (U. Trias.–M. Eoc.) and comprise discoidal or low conical forms with cyclical chambers that may be further subdivided into chamberlets (e.g. *Cyclolina*, U. Cret., Fig. 13.15b; *Cyclopsinella*, U. Cret., Fig. 13.15c; *Dicyclina*, U. Cret., Fig. 13.15d). Conical forms belonging to the Orbitolinidae (L. Cret.–U. Eoc.) have uniserial stacks of saucer-shaped chambers following an early trochospiral stage (e.g. *Coskinolina*, L. Cret.–U. Eoc., Fig. 13.15a; *Orbitolina*, L.–U. Cret., Fig. 13.15e). Radial septulae subdivide these chambers into an outer radial zone of tubular chamberlets. Smaller horizontal and vertical plates may form, within these chamberlets, a marginal zone of minute **cellules**. In the centre of the chambers is a reticulate zone in which the radial chamberlets are further subdivided by vertical pillars.

Suborder Fusulinina. The Fusulinina contains those

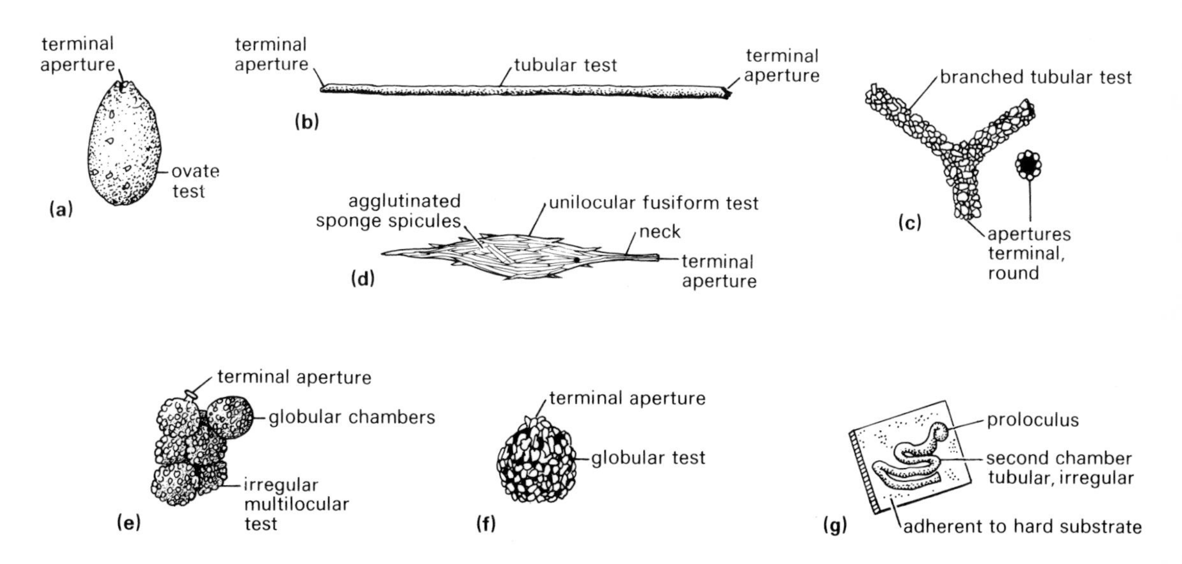

Figure 13.12 Suborder Allogromiina. (a) *Allogromia* ×23; (b) *Shepheardella* ×8. Suborder Textulariina, Superfamily Ammodiscacea. (c) *Rhabdammina* ×10; (d) *Technitella* ×17; (e) *Sorosphaera* ×7·5; (f) *Saccammina* ×10·5; (g) *Tolypammina* ×12·5. ((e) after Loeblich & Tappan 1964)

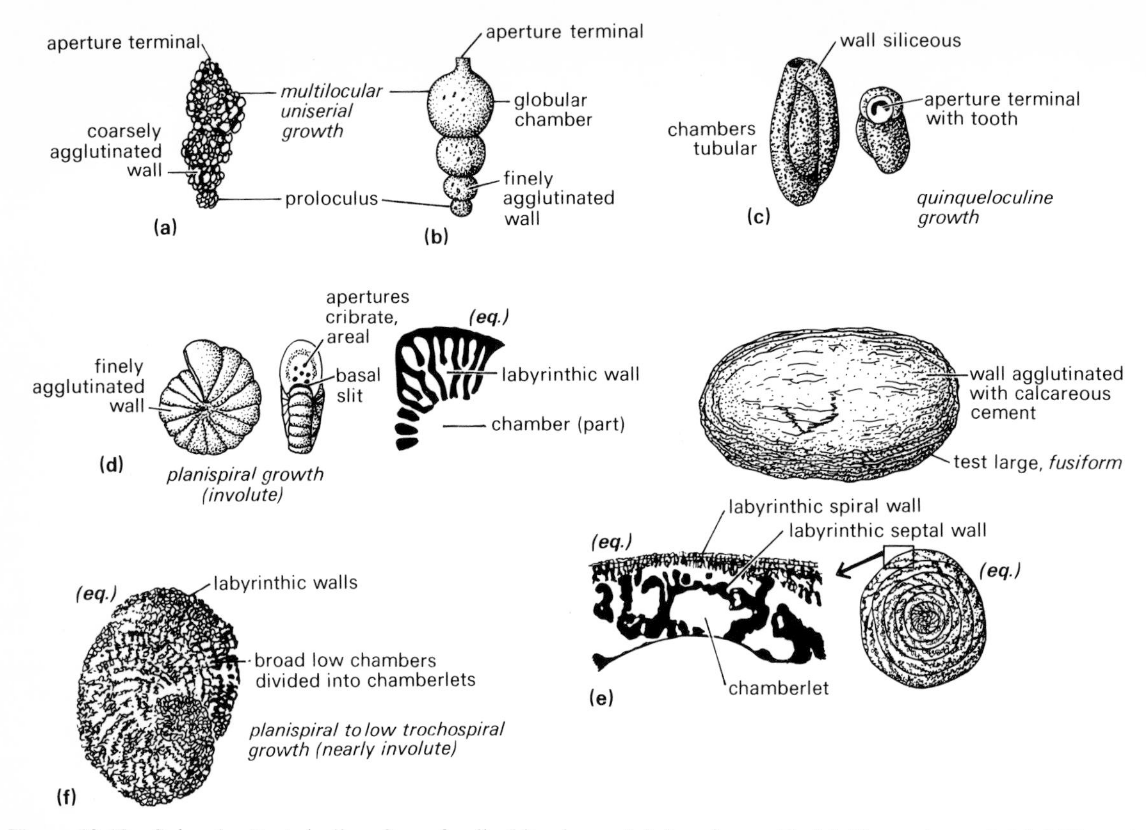

Figure 13.13 Suborder Textulariina, Superfamily Lituolacea. (a) *Reophax* ×18; (b) *Hormosina* ×6; (c) *Miliammina* ×33; (d) *Cyclammina* ×4; (e) *Loftusia*, above ×0·7, lower left ×22, lower right ×3·5; (f) *Spirocyclina* ×9·5. ((a) & (b) modified from Loeblich & Tappan 1964; (e) adapted from Loeblich & Tappan 1964 after Carpenter & Brady; (f) adapted from Loeblich & Tappan 1964 after Maync)

The following abbreviations are used on this and the following figures in this chapter: (*eq*) = equatorial section; (*ax*) = axial section.

foraminifera with non-laminar, calcareous, microgranular walls. The group was largely Palaeozoic in age, becoming extinct in the Triassic, although its descendants are extant in the hyaline Suborder Rotaliina.

SUPERFAMILY PARATHURAMMINACEA. The Parathuramminacea were small benthic forms with simple microgranular walls. The architecture was also simple, ranging from unilocular to straight uniserial (e.g. *Saccaminopsis*, Ord.–Carb., Fig. 13.16a; *Earlandinita*, L.–U. Carb., Fig. 13.16b). This group is known with certainty from Ordovician through to Carboniferous times.

SUPERFAMILY ENDOTHYRACEA. The Endothyracea were small, multilocular foraminifera with walls generally differentiated into an outer granular layer and an inner fibrous layer, also microgranular but of fibrous appearance owing to the perforations. The architecture was variable and included uniserial forms (e.g. *Nodosinella*, U. Carb.–Perm., Fig. 13.16c, a possible ancestor of certain Rotaliina), biserial (e.g. *Palaeotextularia*, L. Carb.–Perm., Fig 13.16f), high trochospiral (e.g. *Tetrataxis*, L. Carb.–Trias., Fig. 13.16e) and planispiral forms (e.g. *Endothyra*, L. Carb.–Perm., Fig. 13.17a). This superfamily lived from late Silurian to Triassic times.

SUPERFAMILY FUSULINACEA. The extinct fusulines were 'larger' forms which also had microgranular perforate tests, but with chambers arranged planispirally in a discoidal to fusiform plan. Two kinds of wall structure are found. The ancestral, **fusulinid** wall is primarily two-layered with a dark, partly organic outer **tectum** and an inner, clear **diaphano-**

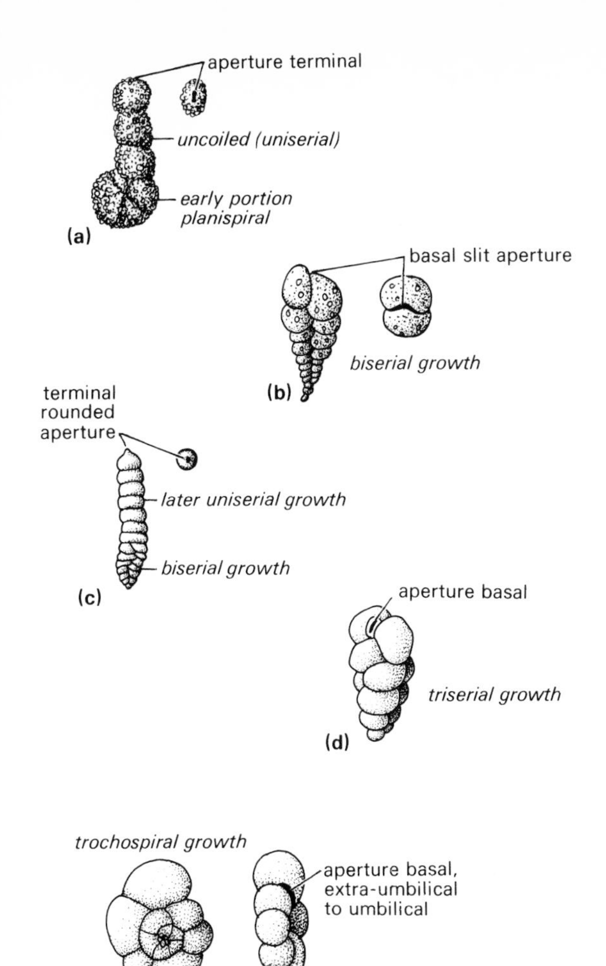

Figure 13.14 Suborder Textulariina, Superfamily Lituolacea. (a) *Ammobaculites* ×20; (b) *Textularia* ×12·5; (c) *Bigenerina* ×11·5; (d) *Verneuilina* ×13·5; (e) *Trochammina* ×29. ((a) after Pokorný 1963 from d'Orbigny; (b) & (d) after Morley Davies 1971 from H.B. Brady; (c) & (e) after Loeblich & Tappan 1964)

theca (Fig. 13.18b). Secondary deposition of a dark **epitheca** within the chamber may give the inner walls a four-layered appearance. The **schwagerinid** wall lacks this secondary thickening and the mural pores are much enlarged to form **alveoli** (Fig. 13.18c). This gives the clearer inner layer a fibrous appearance termed the **keriotheca**. The schwagerinid wall is typical of the larger fusulines of the later Pennsylvanian (U. Carb.) and Permian periods.

The early chambers of microspheric fusulines indicate they had an ancestor like the small planispiral *Endothyra* (Fig. 13.17a). Evolutionary trends included changes in shape, size and wall structure. For example, there was a progressive folding of the septa in some lineages, the forward folds of one septum generally meeting the backward folds of the next (e.g. *Profusulinella*, U. Carb., Fig. 13.17b; *Fusulina*, U. Carb., Fig. 13.17c). A small passage (**cuniculus**) connected adjacent chamberlets (Fig. 13.18a). In some forms a **tunnel** was formed by selective resorption of the septa and secretion of two bordering ridges called **chomata**, (see Figs 13.17b, 13.18a), thereby connecting the mid-floor of each chamber. In the Permian schwagerinids there was a tendency to fill the central, axial chambers with secondary calcite (e.g. *Schwagerina*, Perm., Fig. 13.17d). The late Permian verbeekinids had flat septa with foramina and spiral walls bearing axial and transverse projections (septulae) into the chambers (e.g. *Neoschwagerina*, U. Perm., Fig. 13.17e).

These highly specialised foraminifera were adapted to carbonate and reefal facies in the late Carboniferous and Permian but became extinct at the end of that period. Their evolutionary trends are reviewed in more depth by Dunbar (pp. 25–44 *in* von Koenigswald *et al.* 1963).

Suborder Miliolina. The Miliolina have imperforate calcareous tests of porcelaneous appearance with a planispirally coiled proloculus. Subsequent growth may continue planispirally (e.g. *Cyclogyra*, Carb.–Rec., Fig. 13.19a), uncoil and develop uniserially (e.g. *Nubeculinella*, U. Jur., Fig. 13.19b) or coil streptospirally. **Streptospiral** coiling here involves the addition of tubular chambers (generally half a whorl in length) arranged lengthwise about a growth axis. When added in the same plane (i.e. at 180° to one another) the arrangement is called **spiroloculine** if the chambers are evolute and **biloculine** if they are involute (see Fig. 13.6). More usually, however, chambers are added at angles of 144° leaving five chambers visible from the outside (**quinqueloculine**, e.g. *Quinqueloculina*, Jur.–Rec., Fig. 13.6 and Fig. 13.19e). In *Triloculina*, the chambers are added at angles of 120° and only three chambers are visible from outside the test (**triloculine**). Such streptospiral growth forms may later 'unroll' to uniserial, as in *Articulina* (M. Eoc.–Rec., Fig. 13.19d).

'Larger' porcelaneous foraminifera fall mainly into two families: the Soritidae and the Alveolinidae. The Soritidae have thrived in reefal and

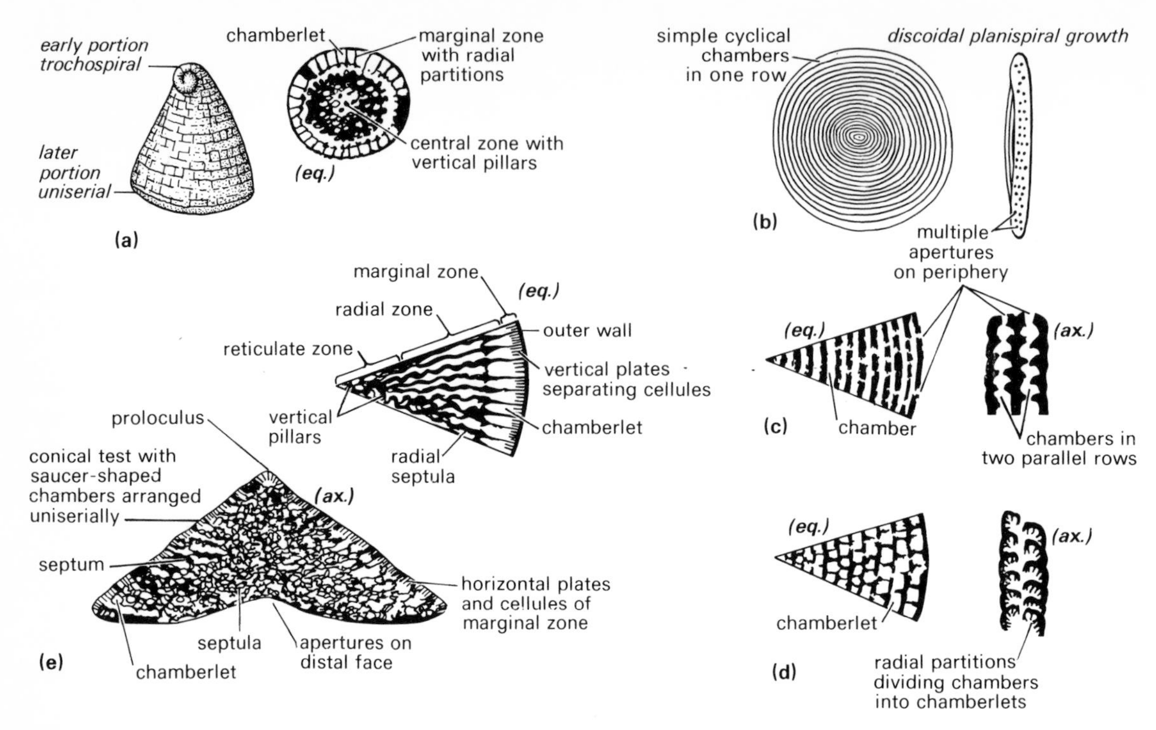

Figure 13.15 Suborder Textulariina, Superfamily Lituolacea. (a) *Coskinolina* ×9·5; (b) *Cyclolina* ×11·5; (c) *Cyclopsinella* ×16; (d) *Dicyclina* ×16; (e) *Orbitolina* left ×13, above right ×21. ((a) after Morley Davies 1971; (e) after Loeblich & Tappan 1964 from Egger)

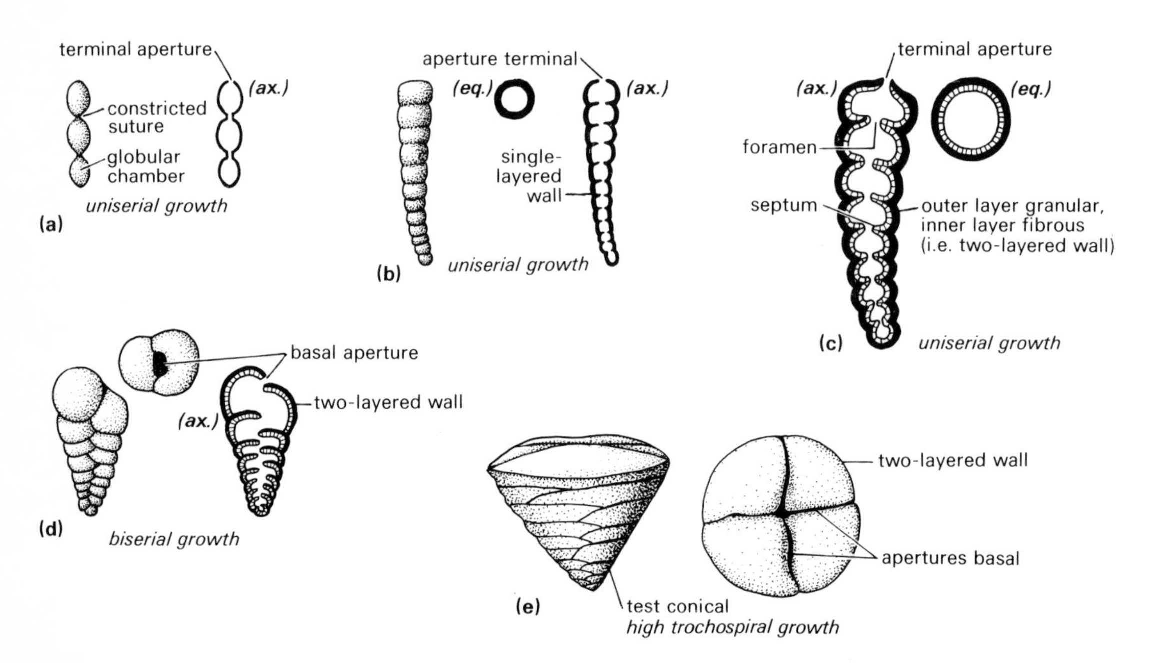

Figure 13.16 Suborder Fusulinina, Superfamily Parathuramminacea. (a) *Saccaminopsis* ×1·5; (b) *Earlandinita* ×40. Superfamily Endothyracea. (c) *Nodosinella* ×16·5; (d) *Palaeotextularia* ×23; (e) *Tetrataxis* ×34. ((a) after Loeblich & Tappan 1964 from H. B. Brady; (b) & (c) after Cummings 1955; (d) after Loeblich & Tappan 1964 from Galloway & Ryniker; (e) after Loeblich & Tappan 1964)

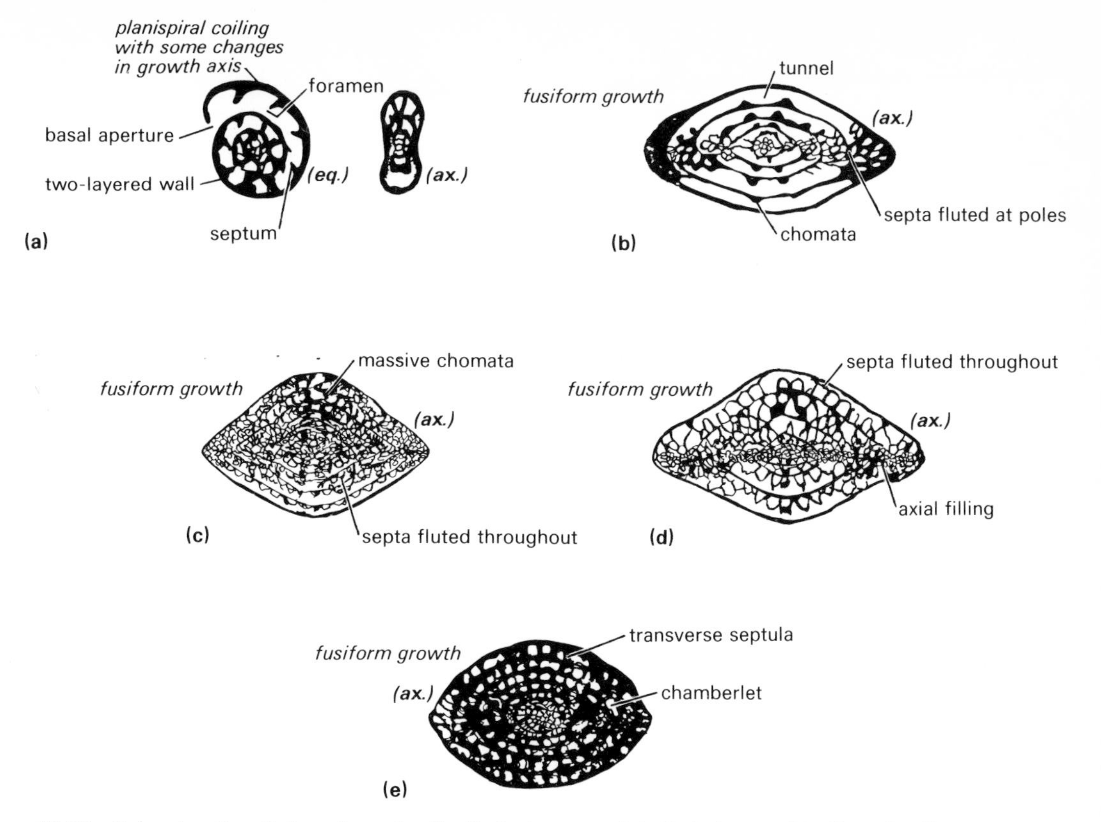

Figure 13.17 Suborder Fusulinina, Superfamily Endothyracea. (a) *Endothyra* ×22. Superfamily Fusulinacea: (b) *Profusulinella* ×20; (c) *Fusulina* ×7; (d) *Schwagerina* ×7; (e) *Neoschwagerina* ×13. (All after Loeblich & Tappan 1964; (a) from Zeller, (b) from Rauzer-Chernousova, (c), (d) & (e) after Thompson)

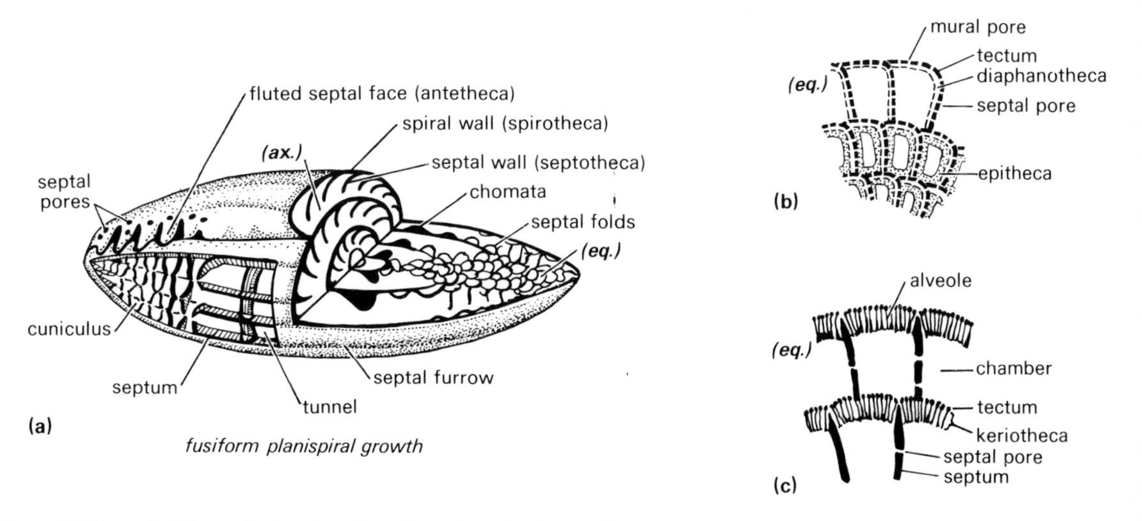

Figure 13.18 (a) Schematic fusuline, based on *Parafusulina* and *Fusulinella*; (b) 'fusulinid' wall; (c) 'schwagerinid' wall.

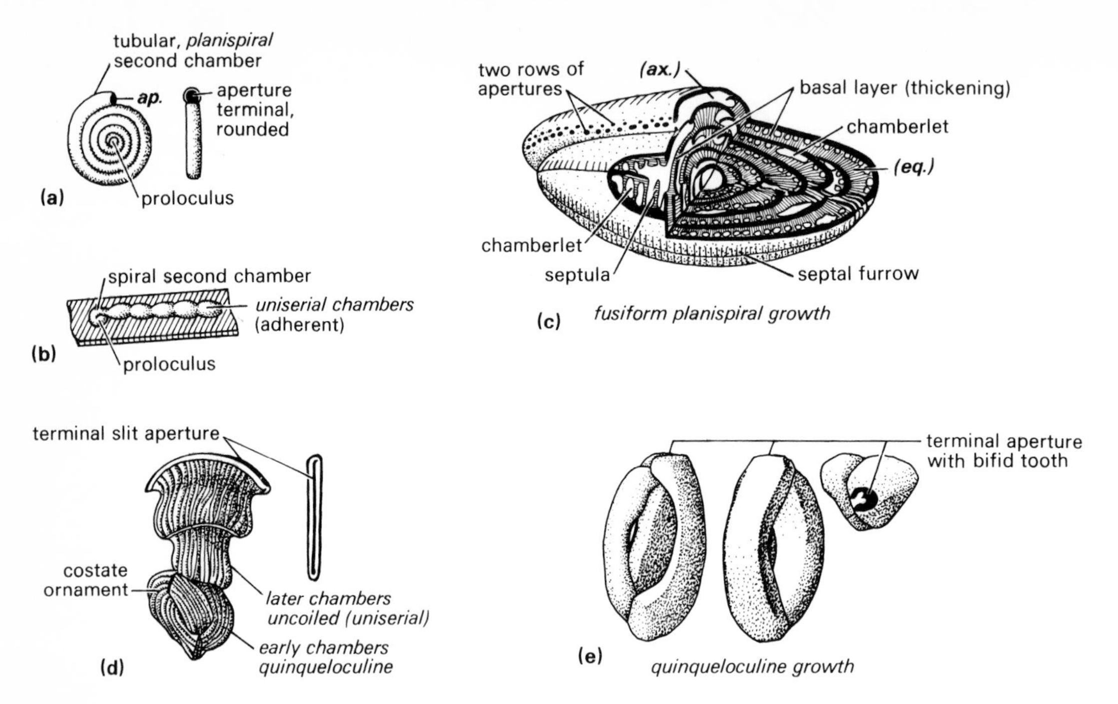

Figure 13.19 Suborder Miliolina. (a) *Cyclogyra* ×40; (b) *Nubeculinella* ×37; (c) schematic diagram of the alveolinid *Fasciolites* ×21·5; (d) *Articulina* ×33; (e) *Quinqueloculina* ×23. ((c) modified from Loeblich & Tappan 1964 after Neumann; (e) Loeblich & Tappan 1964; (d) after Pokorný 1963 from H. B. Brady)

carbonate habitats since the late Triassic period. These have a test which is perforate in the earliest stages and may be pseudopunctate throughout (see Fig. 13.3), but like all other milioline tests they are properly regarded as imperforate. Coiling is basically discoidal planispiral, further modified to cyclical, fan-shaped (flabelliform) or straight uniserial in the later stages of growth (e.g. *Peneroplis*, Eoc.–Rec., Fig. 13.20a). Interseptal buttresses or septulae subdivide the chambers into chamberlets in genera such as *Archaias* (M. Eoc.–Rec., Fig. 13.20b). The all-embracing, annular chamber addition in forms like *Orbitolites* (U. Palaeoc.–Eoc., Fig. 13.20c) is called cyclical.

The Alveolinidae also have imperforate tests with a perforate proloculus (e.g. *Fasciolites*, L. Eoc., Fig. 13.19c). Coiling is fusiform to ovate planispiral. The chambers are divided by septulae into numerous tubular chamberlets arranged in one or more rows. This group exhibits remarkable convergence with the Palaeozoic fusulines but is much younger, ranging from early Cretaceous to Recent times.

Suborder Rotaliina. Rotaliine foraminifera have a calcareous hyaline test which is both multilaminar and perforate. Subdivision into superfamilies is based largely on knowledge of wall structure. 'Larger' forms are found mainly in the Rotaliacea and Orbitoidacea and planktonics in the Globigerinacea.

SUPERFAMILY NODOSARIACEA. The Nodosariacea have walls of optically radial calcite known to be of bilamellar ultrastructure under the electron microscope but monolamellar when viewed optically. Such a hidden ultrastructure should be called cryptolamellar. An aperture of radially arranged slits is typical, except in the unilocular genus *Lagena* (Jur.–Rec., Fig. 13.21d). *Nodosaria* has a simple uniserial test (Perm.–Rec., Fig. 13.21a). In *Frondicularia* (Perm.–Rec., Fig. 13.21b) the test is also uniserial but the chambers are compressed and V-

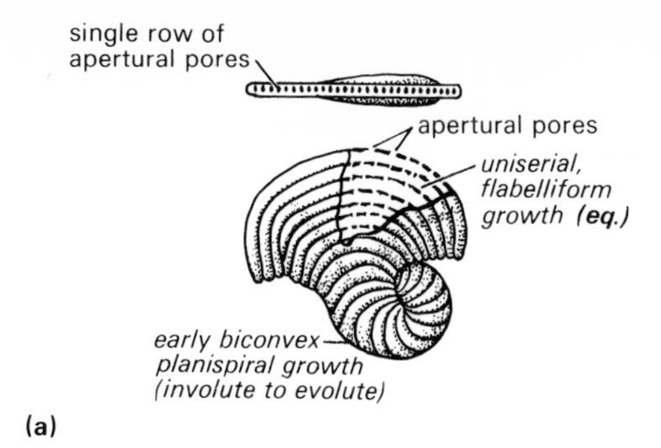

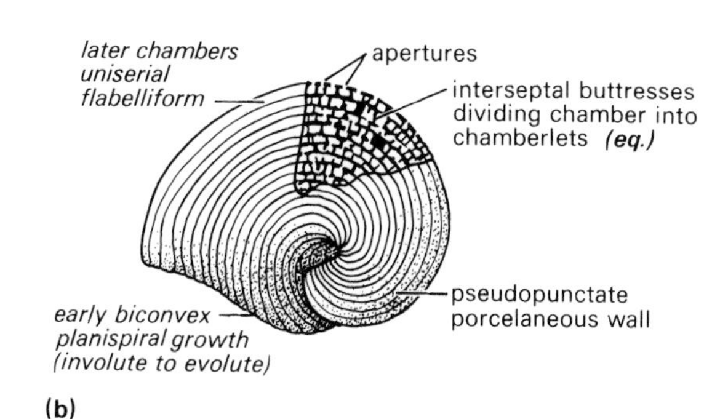

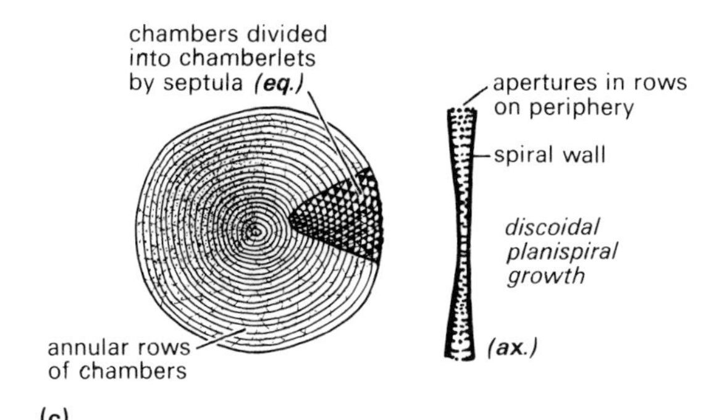

Figure 13.20 Suborder Miliolina. (a) *Peneroplis* ×20; (b) *Archaias* ×12·5; (c) *Orbitolites* ×7.

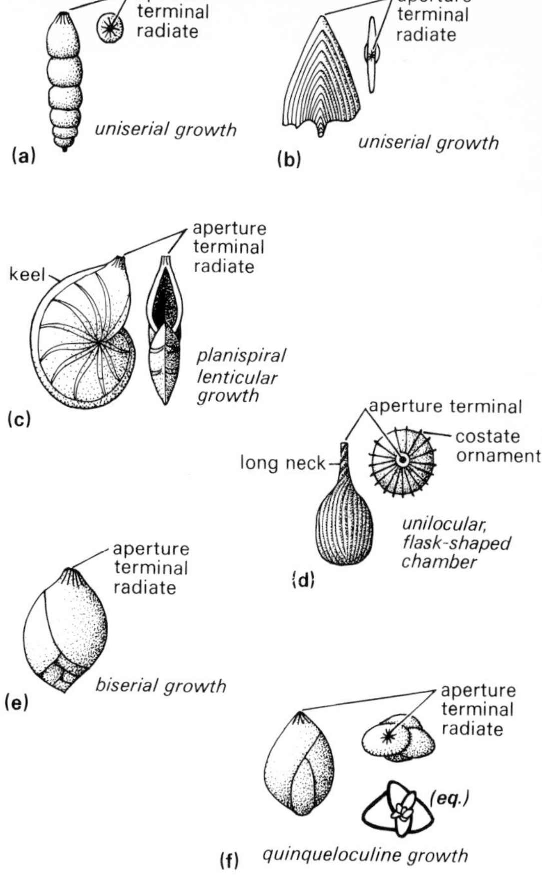

Figure 13.21 Suborder Rotaliina, Superfamily Nodosariacea. (a) *Nodosaria* ×10; (b) *Frondicularia* ×5; (c) *Lenticulina* ×8; (d) *Lagena* ×30; (e) *Polymorphina* ×12·5; (f) *Guttulina* ×21·5. ((a) after Morley Davies 1971; (b), (d) & (e) after Morley Davies 1971 from H. B. Brady; (c) after Morley Davies 1971 from von Hantken; (f) after Loeblich & Tappan 1964 from d'Orbigny)

shaped. *Lenticulina* (Trias.–Rec., Fig. 13.21c) is a common involute planispiral form. Biserial growth is seen in *Polymorphina* (Palaeoc.–Rec., Fig. 13.21e) and streptospiral (quinqueloculine) growth in *Guttulina* (Cret.–Rec., Fig. 13.21f).

SUPERFAMILY BULIMINACEA. The Buliminacea also have optically radial, cryptolamellar calcite walls, but the aperture is generally a basal, comma-shaped slit. Biserial growth is very common, as in *Bolivina* (U. Cret.–Rec., Fig. 13.22c) or triserial, as in *Bulimina* (Palaeoc.–Rec., Fig. 13.22b). In *Rectobolivina* (M. Eoc.–Rec., Fig. 13.22d) and *Pavonina* (Mioc.–Rec., Fig. 13.22e) these plans are modified in later stages to uniserial growth, the former with globular and the latter with C-shaped, flaring flabelliform) chambers. In *Islandiella* (Palaeoc.–Rec., Fig. 13.22f) the biserial arrangement is even planispirally enrolled. The long chambers of *Buliminella* (U. Cret.–Rec., Fig. 13.22a) are arranged in a high trochospiral coil.

SUPERFAMILY DISCORBACEA. The Discorbacea now contains genera known to have walls of either cryptolamellar or bilamellar, optically radial calcite. The tests of discorbaceans are often trocho-

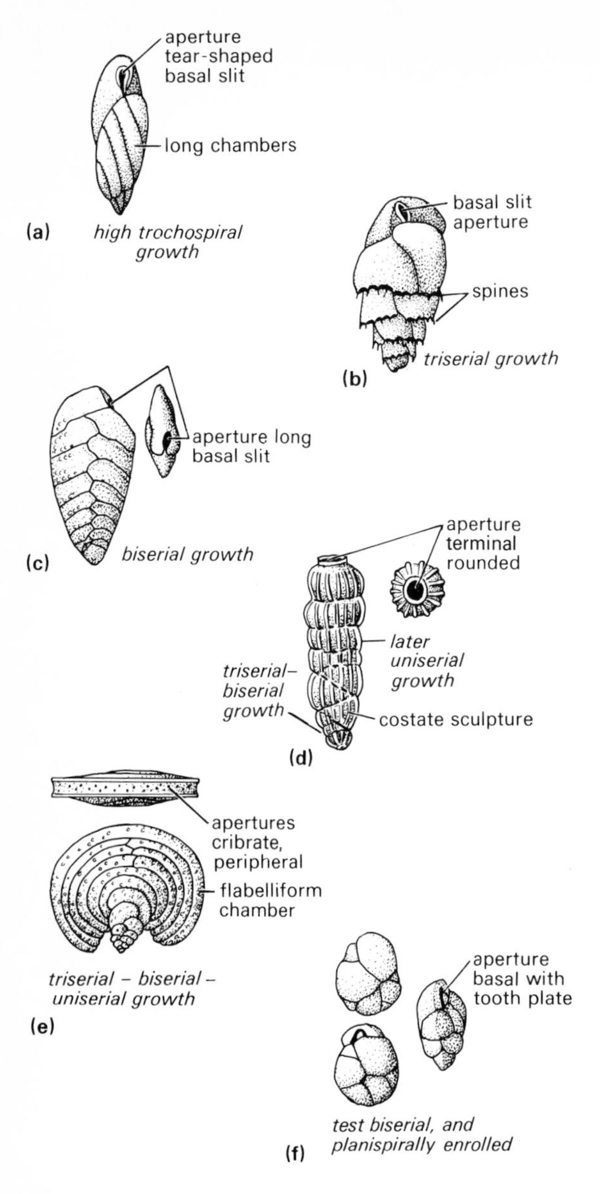

Figure 13.22 Suborder Rotalina, Superfamily Buliminacea. (a) *Buliminella* ×90; (b) *Bulimina* ×33; (c) *Bolivina* ×20; (d) *Rectobolivina* ×38; (e) *Pavonina* ×55; (f) *Islandiella* ×11. ((a) after Pokorný 1963 from d'Orbigny; (b) after Loeblich & Tappan 1964 from Cushman & Parker; (c) after Pokorný 1963 from Cushman; (d) after Loeblich & Tappan 1964; (f) after Loeblich & Tappan 1964 from Nørvang)

spiral and fresh specimens may be coloured brown. *Discorbis* (Eoc.–Rec., Fig. 13.23a) has a plano-convex profile, as does the juvenile stage of *Tretomphalus* (Rec., Fig. 13.23b), but the final chamber is a globular **float chamber** for planktonic dispersal in the latter genus. *Siphonina* (Eoc.–Rec., Fig. 13.23c) has a biconvex profile and an areal aperture borne on a short neck. The umbilical boss seen in many discorbaceans is covered by a rosette of **secondary** chambers in *Asterigerina* (Cret.–Rec., Fig. 13.23d). A similar development occurs in *Amphistegina* (Eoc.–Rec., Fig. 13.23g) but the sutures are more angular and the trochospiral growth is hidden by overlap of the chambers. *Cibicides* (Cret.–Rec., Fig. 13.23e) is a common genus that deviates from the normal in having a basal aperture that extends from the umbilical side to the spiral side. This spiral side is flat or concave whilst the umbilical side is convex. *Planorbulina* (Eoc.–Rec., Fig. 13.23f) has a *Cibicides*-like early stage followed by more irregular addition of chambers in a planispiral manner. The essentially discoidal, planispiral growth of chambers in *Linderina* (Eoc.–Mioc., Fig. 13.23h) is rendered into a stronger, lenticular test by the lateral secretion of layers of calcite.

SUPERFAMILY ORBITOIDACEA. The orbitoids are a late Cretaceous to Miocene group of 'larger' foraminifera which originated in the tropical Americas. Their tests are radial hyaline and perforate, with a discoidal mode of growth, the chambers being arranged in annular cycles rather than plane spirals (Fig. 13.24). A median (**equatorial**) layer of chambers is differentiated from the **lateral** chambers, seen most clearly in axial thin sections (e.g. *Discocyclina*, Eoc., Fig. 13.24a). Radiating calcite pillars give rise to granules on the outer surface. Equatorial sections are important both for taxonomy and biostratigraphic zoning, note being made of the form of the embryonic chambers and the shape of the median chambers (e.g. *Lepidocyclina*, Eoc.–M. Mioc., Figs 13.24b, c).

SUPERFAMILY SPIRILLINACEA. In the Spirillinacea the wall consists of a single crystal of calcite. They are small benthic forms often found adhering to algae and other hard substrates. It is possible that they developed independently from the other hyaline superfamilies (Fig. 13.32). *Spirillina* (Jur.–Rec., Fig. 13.25a) has a long, planispiral second chamber and terminal aperture. *Patellina* (L. Cret.–Rec., Fig. 13.25b) has a trochospiral to biserial test in which the chambers are subdivided by a scroll-like median

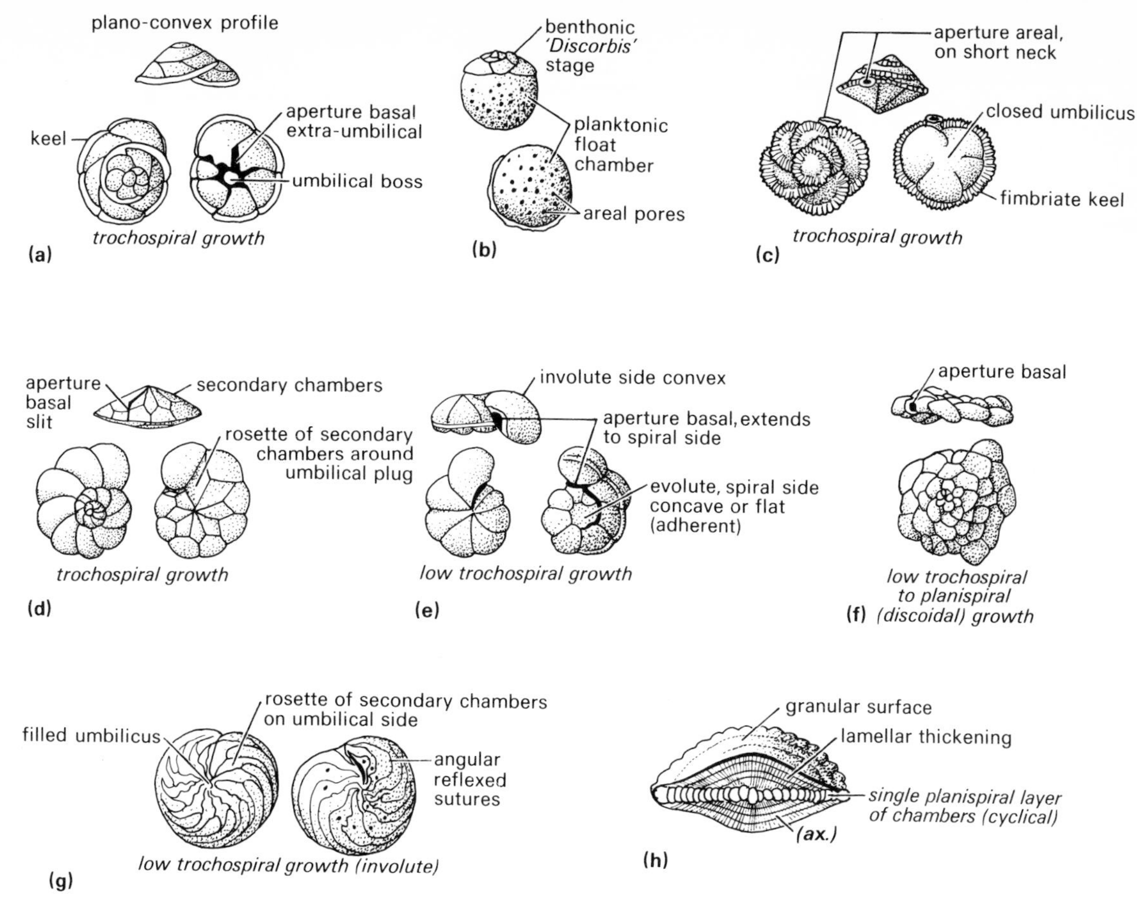

Figure 13.23 Suborder Rotaliina, Superfamily Discorbacea. (a) *Discorbis* ×57; (b) *Tretomphalus* ×34·5; (c) *Siphonina* ×31; (d) *Asterigerina* ×20; (e) *Cibicides* ×22·5; (f) *Planorbulina* ×15; (g) *Amphistegina* ×15; (h) *Linderina* ×47·5. ((a) after Loeblich & Tappan 1964 from Pokorný; (c) after Loeblich & Tappan 1964 from Reuss; (d) after Pokorný 1963 from d'Orbigny; (e) after Morley Davies 1971 from Macfadyen; (f) after Morley Davies 1971; (g) after Morley Davies 1971 from H. B. Brady; (h) modified from Morley Davies 1971 after Nuttall)

septum and numerous transverse septulae.

SUPERFAMILY DUOSTOMINACEA. The Duostominacea are an extinct group that may be intermediate in development between certain Endothyracea and most Rotaliina (Fig. 13.32). This is suggested by the wall structure which consists of both optically radial and microgranular calcite. In *Duostomina* (M. Trias., Fig. 13.25c) the low trochospiral test has a basal aperture divided into two by a flap.

SUPERFAMILY ROBERTINACEA. The Robertinacea have optically radial bilamellar walls composed of aragonite instead of calcite, although this may revert to the latter mineral with time in the fossil state. The aperture is typically a basal slit extending up the face of the last chamber. In *Robertina* (L. Eoc.–Rec., Fig. 13.26a) the test is high trochospiral, each elongate chamber subdivided by transverse partitions. *Ceratobulimina* (U. Cret.–Rec., Fig. 13.26b) has a moderately low trochospiral test whilst that of *Hoeglundina* (M. Jur.–Rec., Fig. 13.26c) is provided with a keel and peripheral slits marking the primary and relict (supplementary) apertures.

SUPERFAMILY GLOBIGERINACEA. The planktonic Globigerinacea typically have trochospirally coiled shells with inflated, coarsely perforate chambers

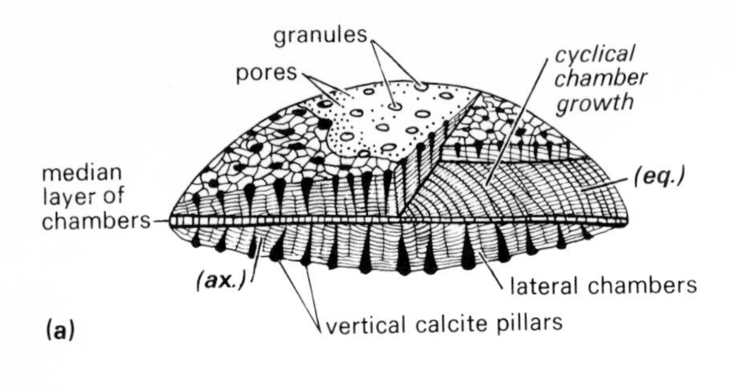

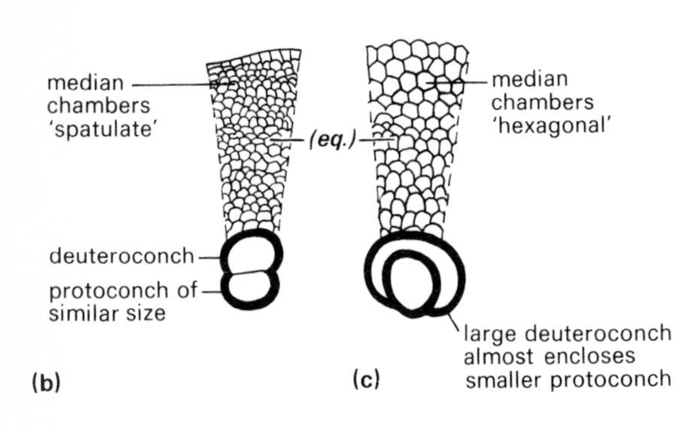

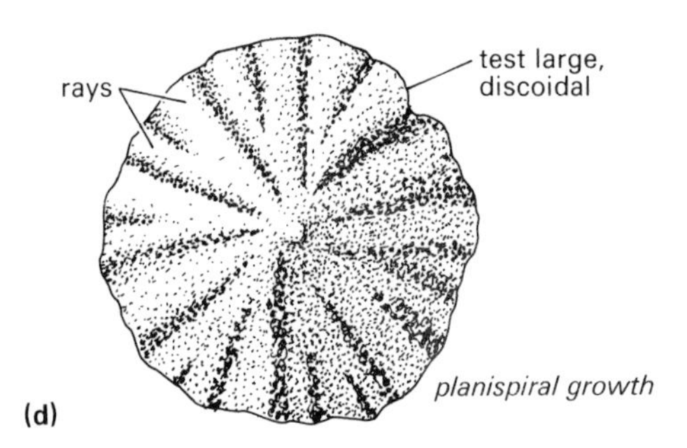

Figure 13.24 Suborder Rotaliina, Superfamily Orbitoidacea. (a) *Discocyclina* (*Discocyclina*) ×3; (b) *Lepidocyclina* (*Lepidocyclina*) ×13; (c) *Lepidocyclina* (*Eulepidina*) ×27; (d) *Discocyclina* (*Aktinocyclina*) ×7. ((a) and (d) after Loeblich & Tappan 1964 from Neumann)

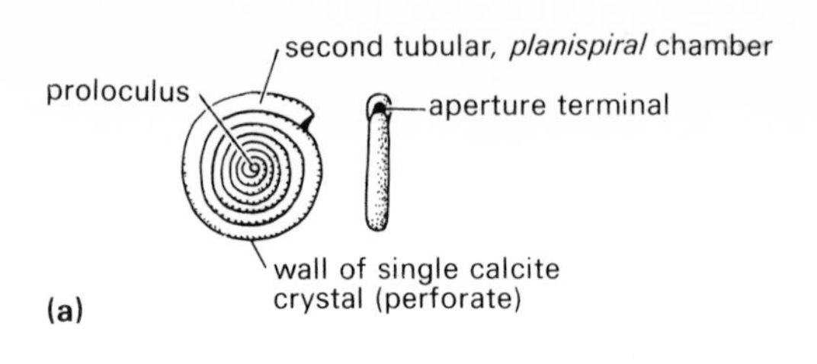

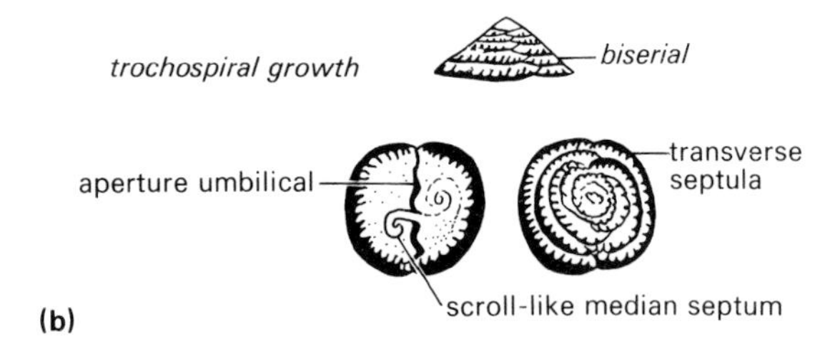

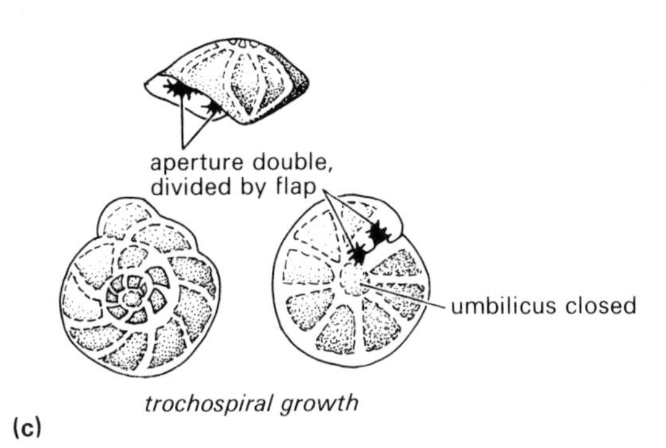

Figure 13.25 Suborder Rotaliina, Superfamily Spirillinacea. (a) *Spirillina* ×50; (b) *Patellina* ×33. Superfamily Duostominacea. (c) *Duostomina* ×60. ((b) after Loeblich & Tappan 1964; (c) after Loeblich & Tappan 1964 from Kristan-Tollmann)

bearing fine spines during life (e.g. *Hastigerinella*, Rec., Fig. 13.27g). These spines support a frothy ectoplasm, the pseudopodia being connected to the endoplasm through the coarse perforations. Inflated chambers, spines and frothy ectoplasm are all adaptations for greater buoyancy.

The wall of the Globigerinacea is composed of optically radial and bilamellar, low magnesian calcite (Fig. 13.27h). Although the primary aperture is usually basal, it may be modified through evolution to areal or terminal. Secondary sutural or areal apertures are also found. These apertures may be partially covered by one or several flaps called bullae. Although inflated chambers are characteristic, some genera have curious club-shaped (clavate) chambers (e.g. *Hastigerinoides*, L.–U. Cret., Fig. 13.27b), while others, including the Rotaliporidae, Globotruncanidae and Globorotaliidae have keeled outer margins (e.g. *Globotruncana*, U. Cret., Fig. 13.27c; *Globorotalia*, Palaeoc.–

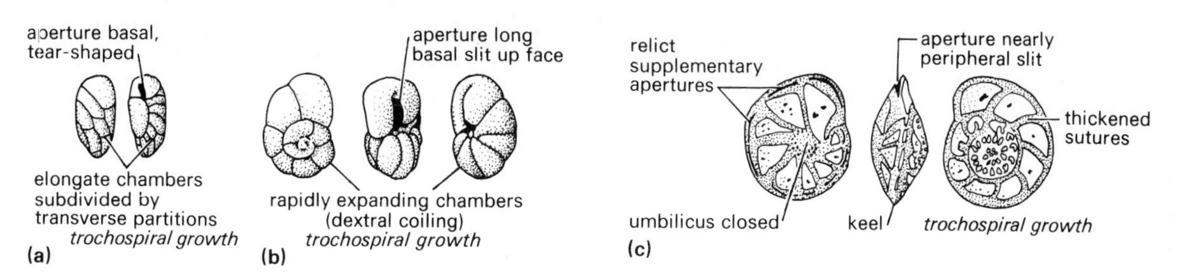

Figure 13.26 Suborder Rotaliina, Superfamily Robertinacea. (a) *Robertina* ×18·5; (b) *Ceratobulimina* ×30; (c) *Hoeglundina* ×10. ((a) after Loeblich & Tappan 1964 from Höglund; (b) & (c) after Loeblich & Tappan 1964)

Rec., Fig. 13.27d). Ornament is not prominent but the tests often have a rugose or pustulose surface, and rarely, longitudinal costae. Forms which are not trochospiral include the ancestral Heterohelicidae (high trochospiral–biserial–uniserial, e.g. *Heterohelix*, U. Cret., Fig. 13.27a) and the planispiral Hantkeninidae (e.g. *Hastigerinoides*, Fig. 13.27b).

Widely used for correlation are species of *Globotruncana*, *Globorotalia*, *Globigerina* (Palaeoc.–Rec., Fig. 13.27e) and *Orbulina* (Mioc.–Rec., Fig. 13.27f). The final, spherical **orbuline**, chamber of *Orbulina* completely envelopes the earlier, globigerine coil. This orbuline trend has occurred in several lineages and represents one of the most efficient adaptations for the maintenance of buoyancy.

SUPERFAMILY ROTALIACEA. Tests of rotaliaceans are built of optically radial bilamellar calcite. They are distinguished by the presence of rotaliid septal flaps and canals (Fig. 13.5d). Although primary apertures may be absent, basal foramina form by secondary resorption of the chamber wall. Generally, growth is planispiral or trochospiral, with a biconvex, lenticular test profile. In the commonly brackish-water genus *Ammonia* (Mioc.–Rec., Fig. 13.28a) the umbilicus is partly filled by small calcite pillars. *Elphidium* (L. Eoc.–Rec., Fig. 13.28b) is another common genus, with an involute planispiral test. A sutural canal system opens at the surface through sutural pores, the latter defined by backward projecting rods called **retral processes**. *Calcarina* (Rec., Fig. 13.28c) is a tropical genus in which the trochospiral test bears robust spines from a thick outer wall.

'Nummulites' are rotaliacean larger foraminifera widely used in correlating Eocene rocks from around the old-world Tethys Ocean but their descendants are still found today in the Indo-Pacific seas. Their tests are radial hyaline and perforate with rotaliid septa. Coiling is biconvex planispiral. Involute forms reveal V-shaped cavities in axial sections, and lateral extensions of these cavities that are called **alar prolongations** (e.g. *Nummulites*, Palaeoc.–Rec., Fig. 13.29b). Although distinctive, these alar prolongations are no more than an earlier or later chamber extended into the plane of section by the great curvature of the involute planispiral coil. Alar prolongations are not present in the evolute forms. Chambers may be simple or differentiated into **median** (i.e. equatorial) and **lateral** layers. They can also be subdivided into chamberlets (e.g. *Spiroclypeus*, Eoc.–L. Mioc., Fig. 13.29c). The course of the septa is indicated on the outer surface of the test by markings called **septal filaments** (Fig. 13.29a). These are the sutures between sinuously curved chambers. In some late Eocene and Oligocene nummulites, the sinuosity of the chambers is so great that successive chambers and their sutures overlap to give a distinctive **reticulate** appearance to the filaments. **Granules** are the surface representation of radiating pillars of calcite (Fig. 13.29c). Microspheric forms were often several times the size of megalospheric forms of the same species (Fig. 13.29d). Unfortunately, this has resulted in many species having two or more names, of which only the first one given remains valid.

SUPERFAMILY CASSIDULINACEA. The Cassidulinacea comprise small benthic foraminifera with optically granular, cryptolamellar calcite walls and slit-, tear- or loop-shaped apertures, generally areal or ter-

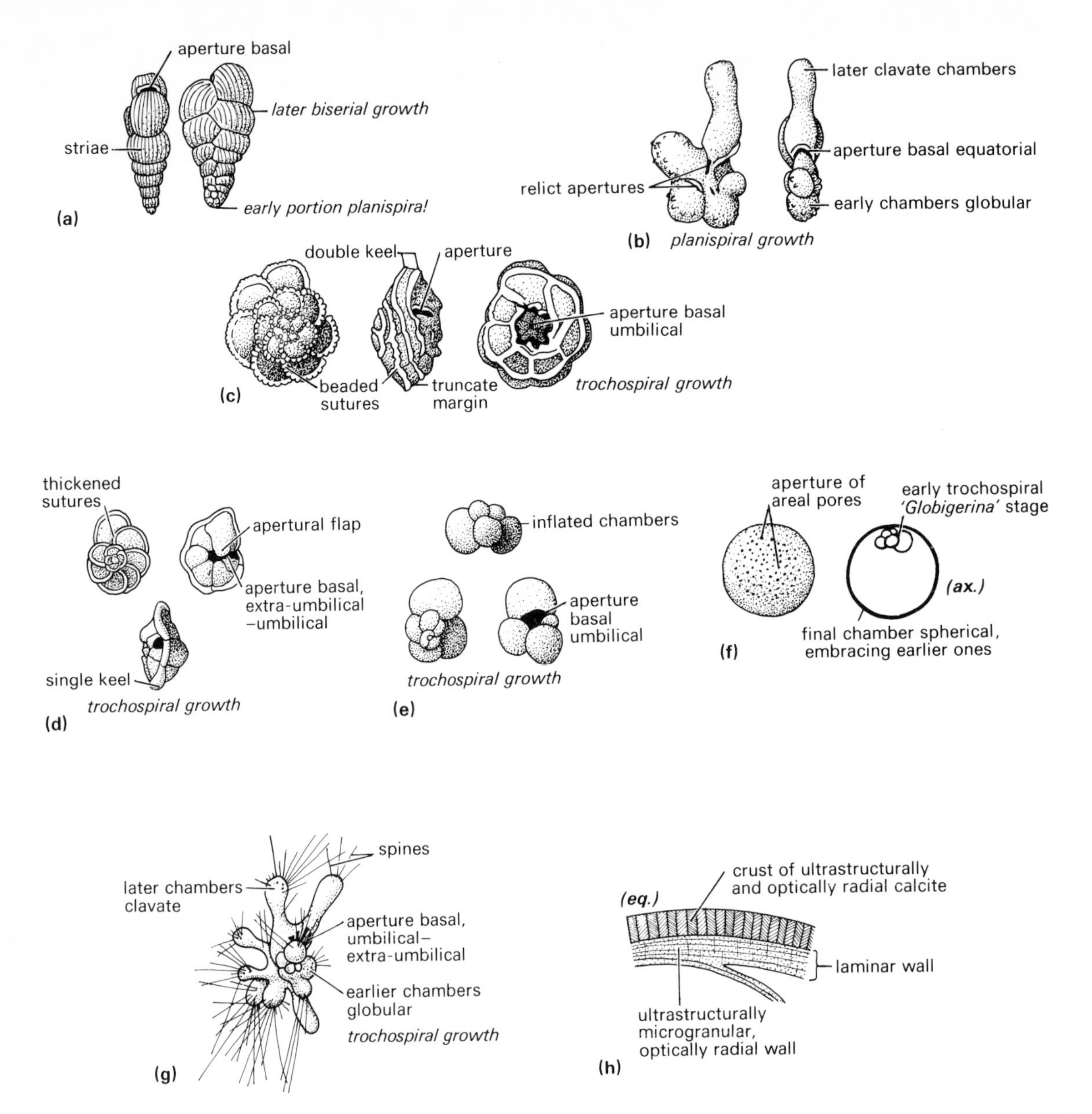

Figure 13.27 Suborder Rotaliina, Superfamily Globigerinacea. (a) *Heterohelix* ×97; (b) *Hastigerinoides* ×65·5; (c) *Globotruncana* ×36·5; (d) *Globorotalia* ×15·5; (e) *Globigerina* ×18·5; (f) *Orbulina* ×20; (g) *Hastigerinella* ×5·7; (h) outer wall structure of deep-water *Globorotalia* (diagrammatic). ((a) after Loeblich & Tappan 1964 from Loeblich; (b) after Loeblich & Tappan 1964; (c) after Glaessner 1945; (d) after Loeblich & Tappan 1964 from Bolli, Loeblich & Tappan; (e) after Morley Davies 1971 from H. B. Brady; (g) after Morley Davies 1971 from Rhumbler; (h) after Pessagno & Miyano 1968)

minal. In *Cassidulina* (Eoc.–Rec., Fig. 13.30c) the test is lenticular, consisting of biserially arranged chambers coiled in a plane spiral. Straight biserial followed by uniserial growth is seen in *Loxostomum* (U. Cret.–Palaeoc., Fig. 13.30b) and *Virgulinella* (Mioc.–Plioc., Fig. 13.30d), the latter with supplementary sutural apertures. *Pleurostomella* (L. Cret.–Rec., Fig. 13.30a) is uniserial throughout, with a terminal aperture and two 'teeth'.

SUPERFAMILY NONIONACEA. The wall structure of nonionacean tests is of optically granular, cryptolamellar or bilamellar calcite. The aperture is generally a basal slit. The involute planispiral tests of the genera *Nonion* (Palaeoc.–Rec., Fig. 13.31a) and *Melonis* (Palaeoc.–Rec., Fig. 13.31b) differ largely in the degree of chamber inflation. In *Osangularia* (L. Cret.–Rec., Fig. 13.31c), the test is trochospiral with a keel and a closed umbilicus.

SUPERFAMILY CARTERINACEA. The Carterinacea are represented by the single genus *Carterina* (Rec., Fig. 13.31d). In this unusual foraminiferid the wall is composed of monocrystalline calcite spicules arranged parallel within a calcite matrix. Unfortunately, the tests disintegrate after death and are not known as fossils. The aperture of *Carterina* is large and umbilical in position and the chambers are divided into chamberlets by septulae. It seems likely that this genus is most closely related to the discorbaceans.

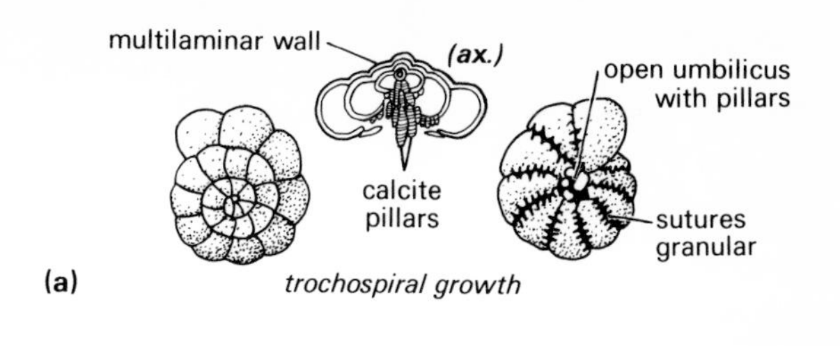

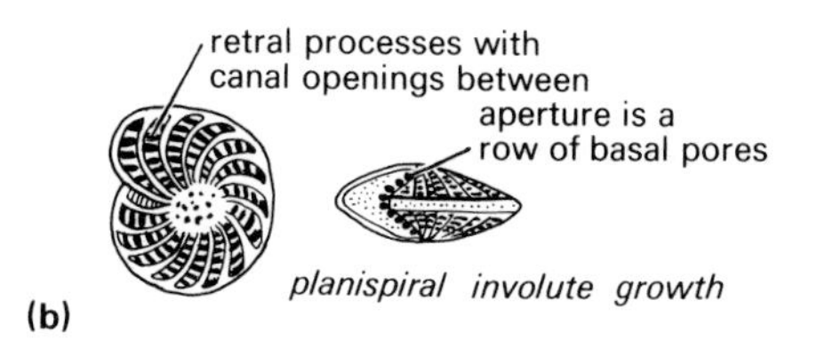

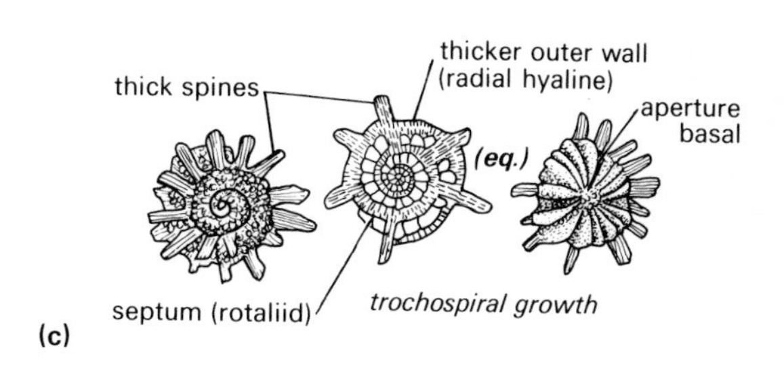

Figure 13.28 Suborder Rotaliina Superfamily Rotaliacea. (a) *Ammonia* ×22·5; (b) *Elphidium* ×32; (c) *Calcarina* ×6. ((a) after Banner & Williams 1973 and Morley Davies 1971 from Macfadyen; (c) after Loeblich & Tappan 1964 from Cushman, Todd & Post)

General history of foraminifera

Principal steps in the evolution of the foraminifera are illustrated in Figure 13.32. Although acritarchs are known from upper Precambrian rocks, there are no convincing examples of foraminifera of this age. Ammodiscaceans resembling *Bathysiphon* and *Tolypammina* have been reported from early Cambrian rocks (see Glaessner, pp. 9–24 *in* von Koenigswald *et al.* 1963) but records of calcareous forms of similar age may be misinterpretations of fossil algae (Brasier 1977). It is generally held that the unilocular condition was the most primitive and it seems likely that the agglutinated stocks of the early Palaeozoic had their origins in simple allogromiine ancestors in the late Precambrian, although evidence for this is wanting (see Glaessner, *op. cit.*). Conkin and Conkin (pp. 49–59 *in* Swain 1977) appear to dismiss the pre-Ordovician records but they note the appearance of simple ammodiscaceans in the Ordovician and Silurian and of lituolaceans in the late Devonian.

Foraminifera with hard tests were certainly scarce until the Devonian, during which period the Fusulinina began to flourish, culminating in the complexly constructed tests of the Fusulinacea in late Carboniferous and Permian times. This suborder died out at the end of the Palaeozoic. Miliolina first appeared in the early Carboniferous, arising probably from the agglutinated Ammodiscacea.

Important Mesozoic events include the appearance and radiation of the Rotaliina (largely from endothyracean stock), Miliolina and complex Textulariina in the Jurassic, soon followed by the appearance of the first unquestionably planktonic foraminifera. Cretaceous tropical regions witnessed a flowering of larger miliolines and rotaliines while the widespread chalk seas and newly opened Atlantic Ocean favoured a thriving planktonic

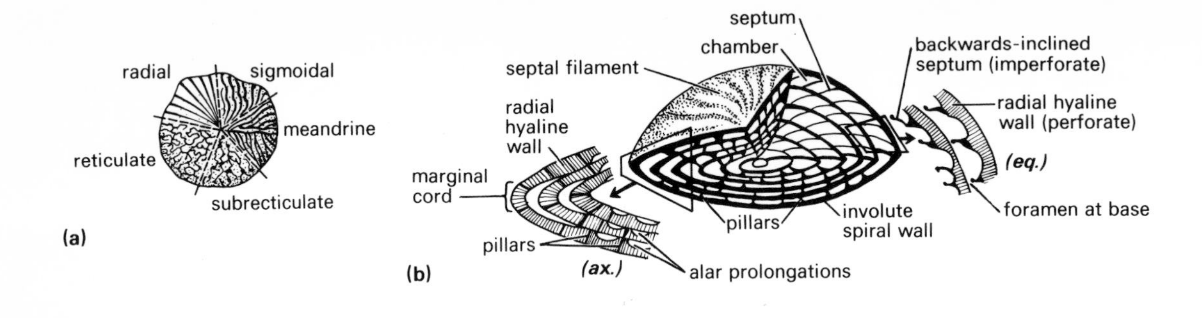

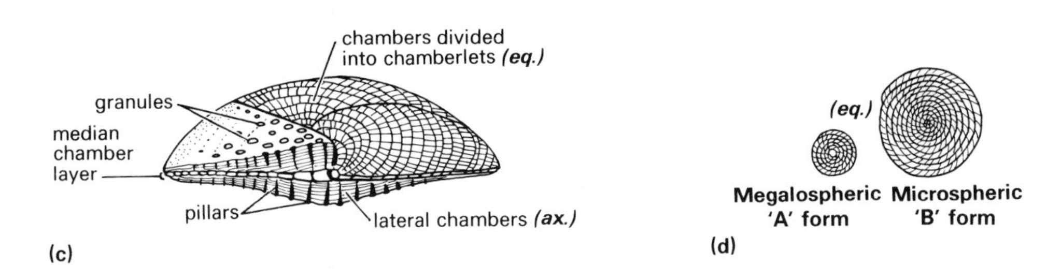

Figure 13.29 Suborder Rotaliina, Superfamily Rotaliacea. (a) Five main types of septal filament on *Nummulites* tests; (b) centre, *Nummulites* ×3·5 approx.; left, detail of axial section ×7; right, detail of spiral section ×10; (c) *Spiroclypeus* ×7 approx.; (d) megalospheric and microspheric forms of *Nummulites obesus* ×0·67. ((b) partly after Morley Davies 1971; (b) & (c) after Loeblich & Tappan 1964 from van der Vlerk & Umbgrove)

population. The planktonic Globotruncanidae became extinct at the end of the Cretaceous along with the spirocyclines and loftusiines.

A relatively rapid radiation followed in the Palaeocene with the appearance of the planktonic Globigerinidae and Globorotalidae and in the Eocene with the development of nummulites and soritids in the Old World and orbitoids in the New World, although they eventually became almost worldwide. Orbitoids died out in the Miocene, since which time larger foraminiferal stocks have progressively dwindled in distribution and diversity, mostly because of the climatic deterioration. Planktonics have also diminished in diversity since late Cretaceous times (Fig. 13.33), the 'boom' periods of the Eocene and Miocene corresponding with warmer climatic phases.

Applications of foraminifera

Foraminifera are in many respects ideal zonal indices for marine rocks, being small, abundant, widely distributed and often extremely diverse. Many also have an intricate morphology in which evolutionary changes can be readily traced. Planktonic foraminifera provide the basis of important schemes for intercontinental correlation of Mesozoic (especially upper Cretaceous) and Cainozoic rocks (see Blow, pp. 199–422 *in* Brönnimann & Renz 1969, vol. 1, Postuma 1971, and authors *in* Ramsay 1977 and Swain 1977). Some of the principles used to establish these zones are discussed by D. G. Jenkins (1965, 1970). Of the many examples of their employment we may note their value to deep-sea stratigraphy (Kennet 1973) as well as to on-land stratigraphy (Olsson 1970). Benthic foraminifera tend to be more restricted in distribution but provide useful schemes for local correlation (see Gradstein, pp. 557–67 *in* Schafer & Pelletier 1976) and sometimes for intercontinental correlation (see Barr 1970).

Environmental interpretations that use fossil foraminifera are founded mainly on comparisons

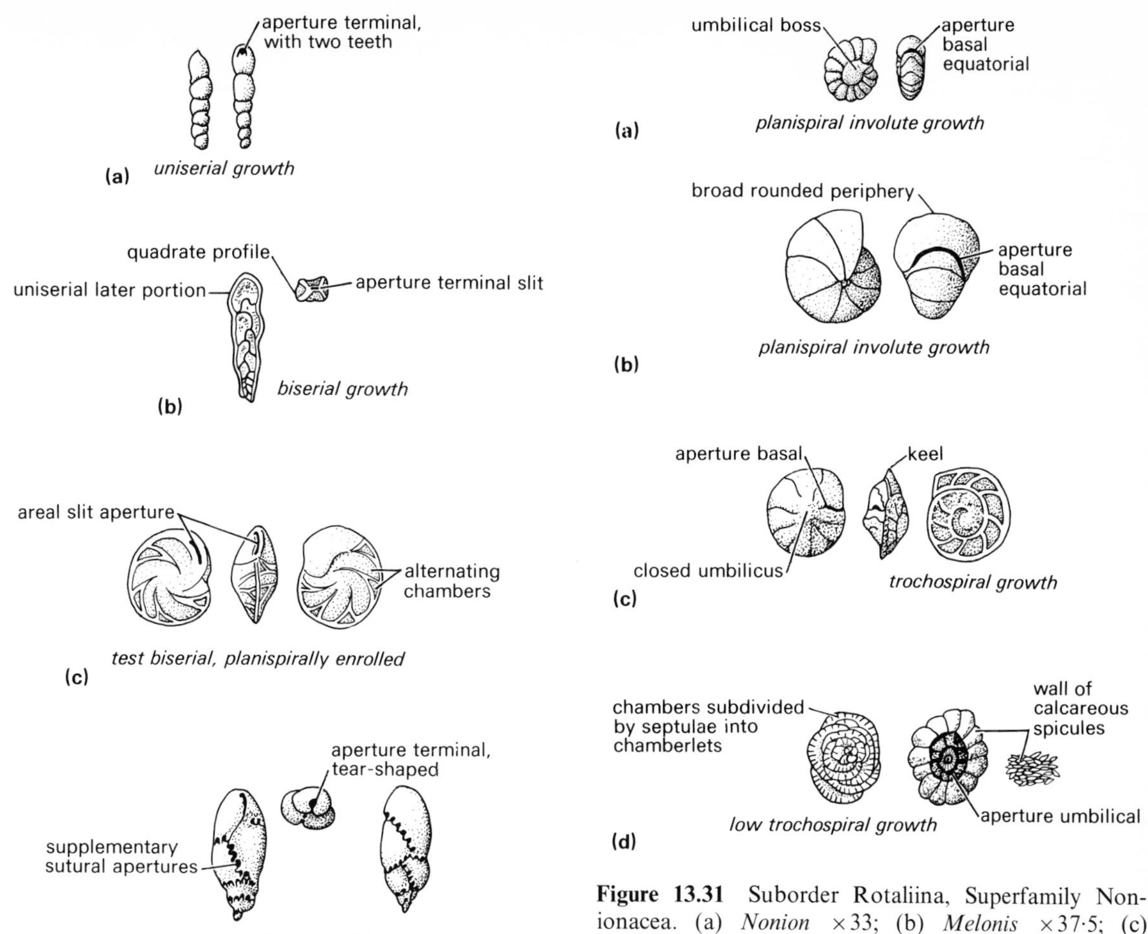

Figure 13.30 Suborder Rotaliina, Superfamily Cassidulinacea. (a) *Pleurostomella* × 16; (b) *Loxostomum* ×34·5; (c) *Cassidulina* ×26; (d) *Virgulinella* ×21·5. ((a), (b) & (d) after Loeblich & Tappan 1964; (c) after Loeblich & Tappan 1964 from Montanaro Gallitelli)

Figure 13.31 Suborder Rotaliina, Superfamily Nonionacea. (a) *Nonion* ×33; (b) *Melonis* ×37·5; (c) *Osangularia* ×37. Superfamily Carterinacea. (d) *Carterina* ×14. ((a) after Loeblich & Tappan 1964 from Voloshinova; (b) after Morley Davies 1971 from H. B. Brady; (c) & (d) after Loeblich & Tappan 1964)

with the numerous studies of Recent ecology, aspects of which are brought together by Phleger (1960), Murray (1973) and Boltovskoy and Wright (1976). For example, dramatic changes in depth, salinity and climate can be traced in late glacial and postglacial raised beaches and beach deposits from studies of their foraminifera (see Feyling-Hanssen *et al.* 1971, Scott & Medioli 1978). For studies of palaeogeography and palaeosalinity at more remote periods, Murray and Wright (1974) used the ratios between Textulariina, Miliolina and Rotaliina, as well as diversity studies of fossil populations.

The value of foraminifera as indicators of the depth of deposition has been outlined by Funnell (1967), comparison in most cases being made with the depth ranges of Recent genera and species. Evidence for vertical movements in young oceanic crust has been computed from such studies in Cainozoic rocks (see Hooper & Jones, pp. 481–5 *in* Schafer & Pelletier 1976, Resig 1976). Benthic depth-related assemblages can also be recognised in Cretaceous sediments (Sliter & Baker 1972, Olsson, pp. 205–30 *in* Swain 1977). Another method utilises the fact that planktonic life assemblages are depth-

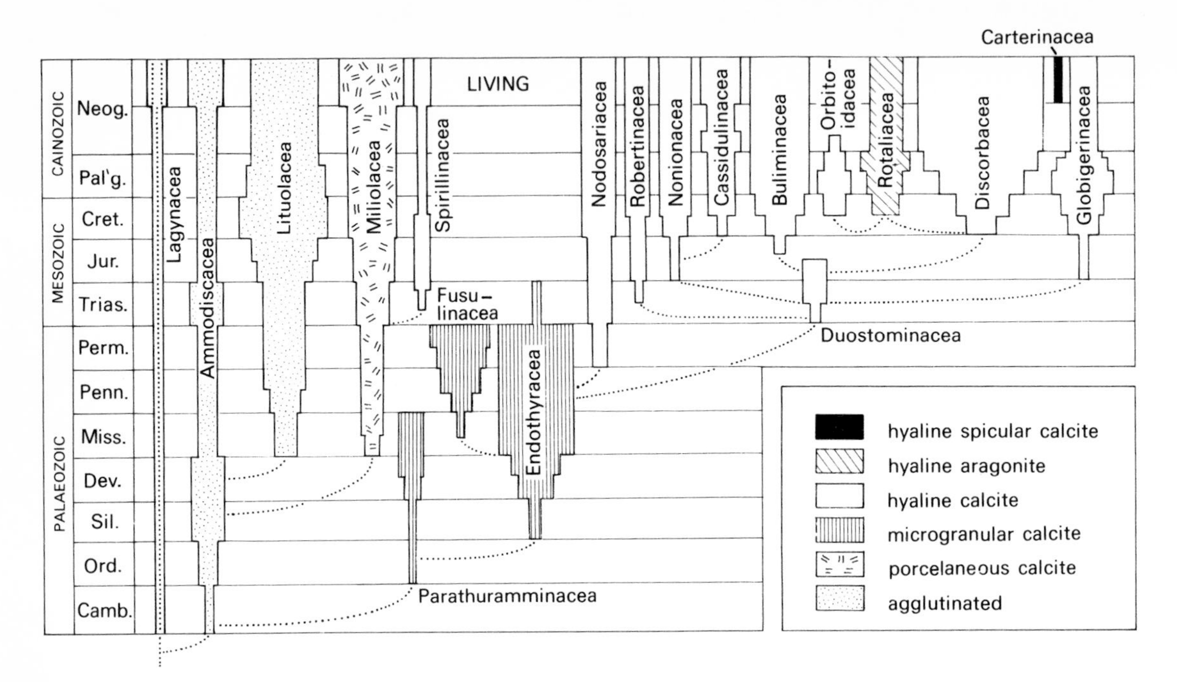

Figure 13.32 The phylogenetic history of the Foraminiferida. Width of the bars corresponds to the number of included families. (Modified from Tappan *in* Schafer & Pelletier 1976)

stratified and so give rise to higher-diversity death assemblages in deeper waters than in shallower waters (Kafescioglu 1971). A further useful guide is the ratio between planktonic and benthic individuals in a sample, the former increasing generally as the water becomes deeper. This ratio can be used for Jurassic and younger rocks (see Stehli & Creath 1964, Hart & Carter 1975).

The narrow temperature ranges of Recent planktonic species have become useful tools in palaeoclimatology, especially of Quaternary sediments (see Barash, pp. 433–42 *in* Funnell & Riedel 1971, Boltovskoy 1973). Plots of the changing proportions of warm- to cold-water species, of selected indicator species, or of coiling directions through a cored interval may allow the construction of palaeotemperature curves, as for example in some shallow-water Recent sediments (Schnitker, pp. 385–92 *in* Schafer & Pelletier 1976), the Quaternary of the deep sea (Rogl & Bolli 1973) or even for the whole Neogene (Bandy, pp. 46–57 *in* Brönnimann & Renz 1969, vol. 1, Ingle 1973). Studies on the oxygen isotope ratios of calcareous foraminiferid shells can also yield information on palaeotemperatures, such as the long Cainozoic history of climatic cooling and glaciations shown from Antarctic waters by Shackleton and Kennet (1975). The isotope method is reviewed by Emiliani (1966) and Hecht (pp. 1–44 *in* Hedley & Adams 1976).

Living planktonic foraminifera may be used, like drift bottles, as markers of current circulation and of distinct water masses (see Boltovskoy & Wright 1976, pp. 378–84). Cretaceous current patterns can even be reconstructed from the distribution of planktonic foraminifera (Sliter 1972).

Estimates of the relative rates of sedimentation have been calculated by comparing the proportion of living to dead (or living to total) foraminifera in benthic populations (Walton 1955, Phleger 1960). Calculations of the absolute rate of sedimentation, involving knowledge of the length of foraminiferid life cycles, have been suggested by Uchio (1960) but have not yet proved practical.

Little use has been made of the relationship between foraminiferid test morphology, habitat and environment such as that employed in studies of ostracods and many groups of macrofossils, although there are some preliminary studies relating to depth (Bandy 1964), substrate stability (Brasier 1975) and general environmental factors (Chamney, pp. 585–624 *in* Schafer & Pelletier 1976). This approach, for example, helped Brasier (1975a) to

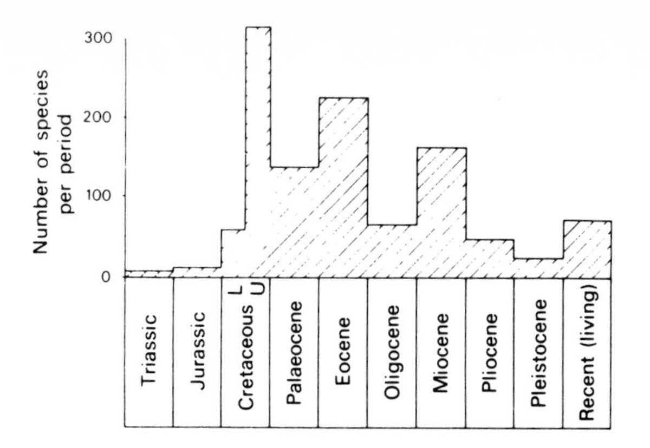

Figure 13.33 Changes in the specific diversity of planktonic foraminifera through time. (Based on Tappan & Loeblich 1973)

trace the gradual dispersal of sea-grass communities from Cretaceous to Recent times.

Foraminifera are not only useful to geologists; living benthic species can also be viable indicators of environmental pollution (see Seiglie 1975).

Further reading

Information on Recent biology and ecology is contained in the books by Phleger (1960), Murray (1973) and Boltovskoy and Wright (1976) and on Recent ecology and palaeoecology in Schafer and Pelletier (1976). Evolutionary trends are reviewed in von Koenigswald *et al.* (1963). Specimen identification may be assisted by reference to the *Treatise* (Loeblich & Tappan 1964) for genera, and to the *Catalogue* (Ellis & Messina 1940 to date) for species, as well as to the following general textbooks: Galloway (1933), Glaessner (1945), Cushman (1948), Jones (1956) and Pokorný (1963). Postuma (1971) is a useful guide to fossil planktonic foraminifera and their zones, Wagner (1964) to the fossil larger foraminifera, and Lewis (1970) to the Recent foraminifera.

Hints for collection and study

To collect living specimens of foraminifera, gather samples of relatively fibrous seaweed from marine or estuarine rock pools and tidal flats or scrape up the top 5 mm of mud from intertidal mudflats. The weed samples should then be placed in a bucket of nearby water and shaken vigorously to detach the foraminifera. Remove the seaweed and strain the water and sediment through a 200 mesh sieve. The mud samples should likewise be washed through a 200 mesh sieve. The sieved residues are then flushed into a container with more seawater for later examination in a petri dish with transmitted light. Living foraminifera can generally be distinguished from dead ones by their dark, protoplasm-filled chambers and by adherent food debris. Patient observation at high magnifications with condensed light should also reveal pseudopodia, locomotion and feeding habits. Arnold (pp. 154–206 *in* Hedley & Adams 1974) gives some useful tips for the collection and culture of living foraminifera.

Fossil foraminifera can be obtained from almost any post-Triassic marine sediment which has not undergone much leaching or become acid. Recent beach sands and lagoonal and estuarine muds are all excellent sources of material, although muds are generally richer in individuals and species than sands. To extract foraminifera from partially indurated argillaceous and marly rocks, methods C to E (especially D) are generally satisfactory (see Appendix). Good assemblages can also be coaxed out of chalks and other limestones by method B, but the hardest limestones will have to be thin sectioned or peeled (see method N). Disaggregated sediments can then be washed, dry sieved, concentrated and mounted by methods G, I, J and O. Most smaller foraminifera are studied in reflected light, sometimes stained with a solution of malachite green or a food dye to bring out the surface structures more clearly. Wall ultrastructure and growth plan are, however, better seen if the specimen is wetted and viewed with transmitted light. Larger foraminifera (and some smaller forms within indurated limestones) are generally studied in thin sections, preferably through both the equatorial plane and the growth axis. Thin sections of isolated specimens can also be prepared by embedding them in polyester resin: pour a little resin into the bottom of one cup from a polystyrene egg box or a plastic ice cube cup; scatter a dozen or so specimens over the resin and then cover with a further layer of resin. Bubbles can be discouraged if the cup is then placed in a vacuum. When dry, remove the block and prepare standard thin sections of the foraminiferid-rich portion. Further ideas on foraminiferid techniques are given by Todd *et al.* (pp. 14–20) and Douglass (pp. 20–5) *in* Kummel and Raup (1965).

14 Phylum Crustacea – Ostracods

The Class Ostracoda, sometimes known as 'seed shrimps', are small laterally compressed Crustacea enclosed within a protective shell. This shell is formed by two chitinous or calcareous **valves** that hinge above the dorsal region of the body. These creatures have adapted to various niches in the ocean plankton, on the sea floor, in freshwater ponds and even in humid forest soils. They are commonest, however, as shallow marine benthos, where they may number only less than the foraminifera amongst the fossil microfauna. Ostracods are particularly useful for the biozonation of marine strata on a local or regional scale and second to none as indicators of ancient shorelines, salinities and relative sea-floor depths. They have a long and well documented fossil record from the early Cambrian to the present day.

Soft body structure

The body of an ostracod is divided into an anterior **head** (or cephalon) and a median **thorax** although the junctions between these are indistinct (Fig. 14.1a). As in other arthropods the soft parts are covered by a rigid, jointed exoskeleton of **chitin**. The head is large, accounting for about half the body size, and bears a ventrally placed mouth and a dorsal, usually single, eye. The anus is at the posterior end of the body. On either side of the head/thorax junction arise large flap-like outgrowths that totally enclose the rest of the animal. These outgrowths are also provided with a chitinous exoskeleton, termed the **carapace** (see below).

There are few exterior signs of the typical body segmentation of arthropods excepting, of course, the seven pairs of jointed limbs borne on the ventral side of the body (Fig. 14.1a). As in other Crustacea, these appendages are basically **biramous**, comprising two distinct branches: an outer **exopodite** and an inner **endopodite** (Fig. 14.1b). In many instances, however, the exopodite has become reduced or lost during evolution, resulting in a **uniramous** limb. These ostracod appendages bear fine chitinous bristles called **setae** (which usually arise from just below the joints) and terminate in claws.

Four pairs of limbs arise from the head. At the front are the **antennules**, which are long, tapering, uniramous limbs attached to the forehead. They are employed variously in walking, swimming and feeding. Behind these are the **antennae**, used for walking, swimming and climbing. Mastication of the food is assisted by a pair of biramous **mandibles** beside the mouth and also by the biramous **maxillae** which lie behind the mouth. These maxillae can also whip up water to provide filter-feeding currents, to improve water circulation around the animal or to take up O_2, like gills.

The thoracic appendages vary in use and development but primarily comprise three pairs. Often these are uniramous and serve as walking or digging organs. At the end of the thorax occur a pair of unsegmented, leg-like structures, each called a **furca**. Although the furcae are not true appendages, they may supplement or replace the thoracic limbs as locomotory organs in the Order Myodocopida.

Because of their very small size (usually < 1 mm long) the ostracods have no need of special gill structures, respiration taking place by natural diffusion through the thin body wall. Blood vessels and hearts are also lacking in all except the relatively large and planktonic Myodocopida. Some of the latter are further distinguished by their pair of lateral eyes rather than the single dorsal eye typical of benthic forms. Even this one eye may be lacking in deeper-water genera.

The soft parts of ostracods provide the basis for the biological classification of Recent forms, special

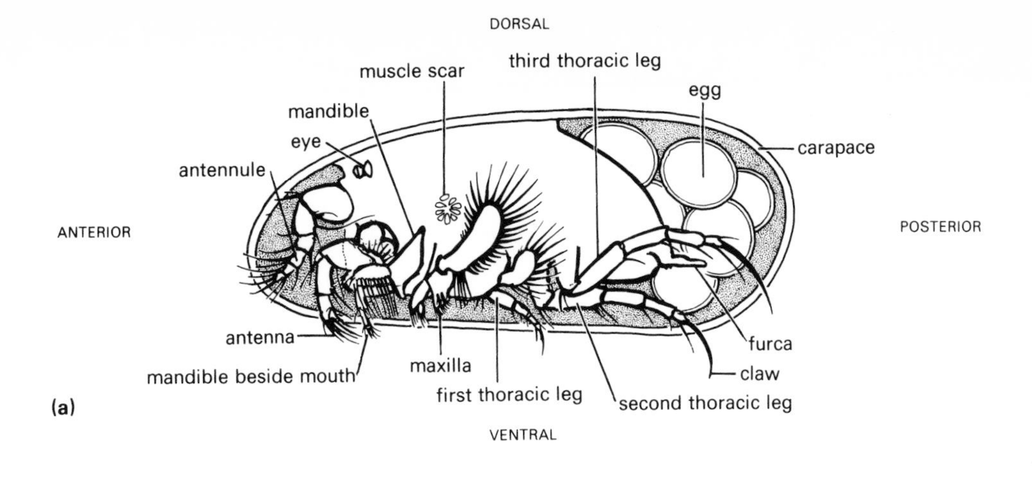

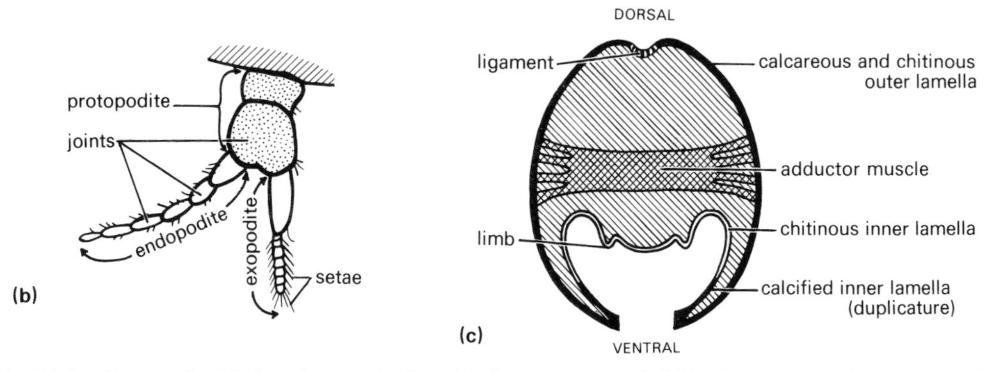

Figure 14.1 Living Ostracoda. (a) Female *Darwinula* with left valve removed; (b) basic structure of a crustacean limb; (c) schematic cross section through a living ostracod. ((a) modified from Benson *et al.* 1961 after Kesling and Sars; (b) modified from van Morkhoven 1962; (c) after Kornicker *in* Neale 1969)

note being made of the form and number of the appendages, the form of the furcae and of the reproductive organs. Unfortunately, these features are rarely preserved in fossils, although there are some spectacular exceptions (Bate 1972), so the palaeontologist is forced to base his taxonomic assessments on the nature of the preserved carapaces.

The ostracod carapace

The ostracod carapace is usually ovate, kidney-shaped or bean-shaped and from 0·3 mm to 300 mm long, although most adult carapaces measure only 0·5 to 3·0 mm long. The carapace consists of a right and a left valve, one of which is slightly larger and partially overlaps the other, with a hinge along the dorsal margin. Factors affecting carapace shape and the problems of distinguishing anterior from posterior, or dorsal from ventral, are discussed in subsequent sections.

The two lateral body flaps that form the carapace contain nervous tissues and a variety of glands. This tissue has a cover of epidermal cells which secrete a chitinous exoskeleton, that on the inner wall being called the **inner lamella** and that on the outer wall called the **outer lamella** (Fig. 14.1c). The latter is often a cryptocrystalline calcite shell with a chitinous framework in adult ostracods (see Bate & East 1972), but calcification is weaker in juvenile stages and in freshwater ostracods. In most of the planktonic Myodocopida the calcification is weak and amorphous, whilst in the extinct Archaeocopida the outer lamella appears to have been mainly of chitin that was secondarily calcified or phosphatised after

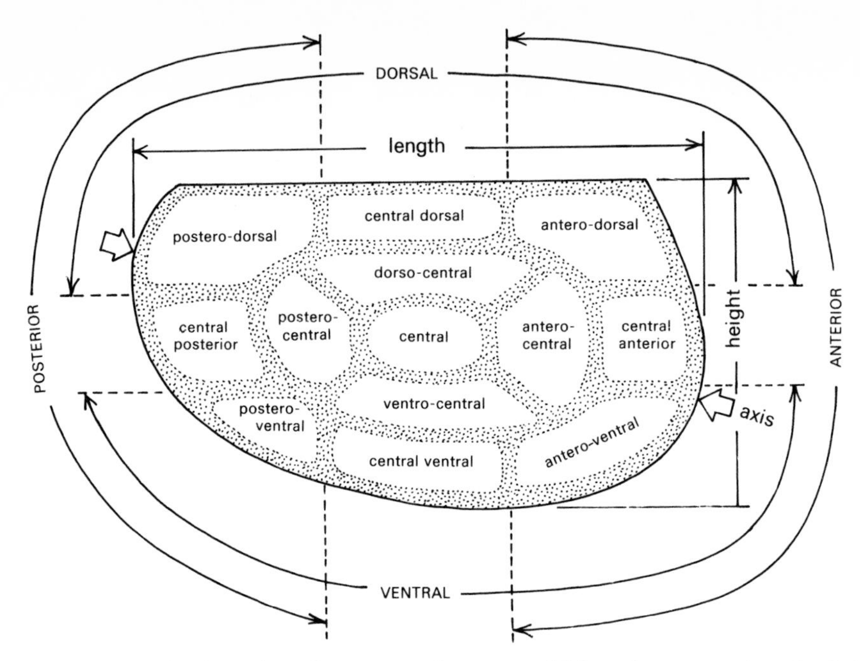

Figure 14.2 Nomenclature and orientation of an ostracod carapace (right valve, external surface). (Modified from Kesling 1951)

death. For this reason, juvenile or freshwater podocopids and marine myodocopids and archaeocopids are not common as fossils.

The margins of the thin chitinous inner lamella are calcified into a thicker **duplicature** in adult Podocopida and Myodocopida (Figs 14.1c, 14.3, 14.4). This serves to improve the interlocking of the free margins of the valves. When looking at the inside of the valve it is possible to see that the duplicature has an **outer margin** that is fused to the outer lamella, and an **inner margin** that may either be fused or free from the outer lamella (Fig. 14.4). In the latter case a space between these lamellae is called a **vestibulum**. The innermost line of contact between the fused lamellae is called the **line of concrescence**, and the area between this and the outer margin is known as the **marginal zone** (Figs 14.3, 14.4).

Articulation. Although the outer lamella appears to comprise two valves, it is in reality a single structure in which a dorsal **ligament** region remains uncalcified (Fig. 14.1c). The articulation of this dorsal margin is further stabilised by the development of teeth, sockets, ridges and grooves, collectively termed the **hinge**. Four basic kinds of dorsal hinge structure are found (Fig. 14.5). The **adont** hinge is the simplest, lacking teeth and sockets but often provided with a single groove along the margin of the larger valve and a corresponding ridge on the smaller valve. In the **merodont** hinge there are anterior, median and posterior hinge elements. The anterior and posterior ones are usually a tooth plus socket; the median element is a groove and ridge. The **entomodont** hinge differs in that the anterior portion of the median element is differentiated into coarse crenulations; the **amphidont** hinge has this portion developed into another tooth and socket. Slight modifications of these basic plans have received other names (see Sylvester-Bradley 1956).

Closure of the valves on the ventral side may be aided by ridges on the duplicature called **selvages** (Fig. 14.4).

As mentioned, the valves are united and opened by an elastic, chitinous portion of the outer lamella called the ligament, but they are closed by a series of transverse **adductor muscles** (Fig. 14.1c). These are generally attached to the inside of the valves in a region just anterior of the valve centre, leaving a cluster of adductor **muscle scars** on the inside of the

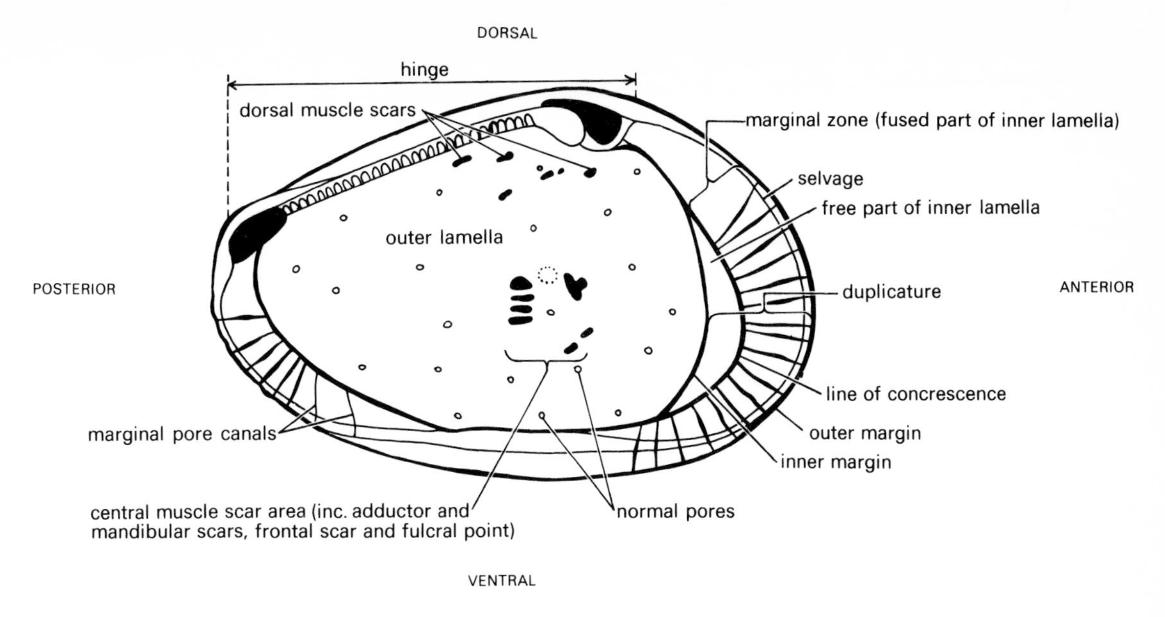

Figure 14.3 The main internal features of a podocopid carapace (left valve). (Modified from van Morkhoven 1962)

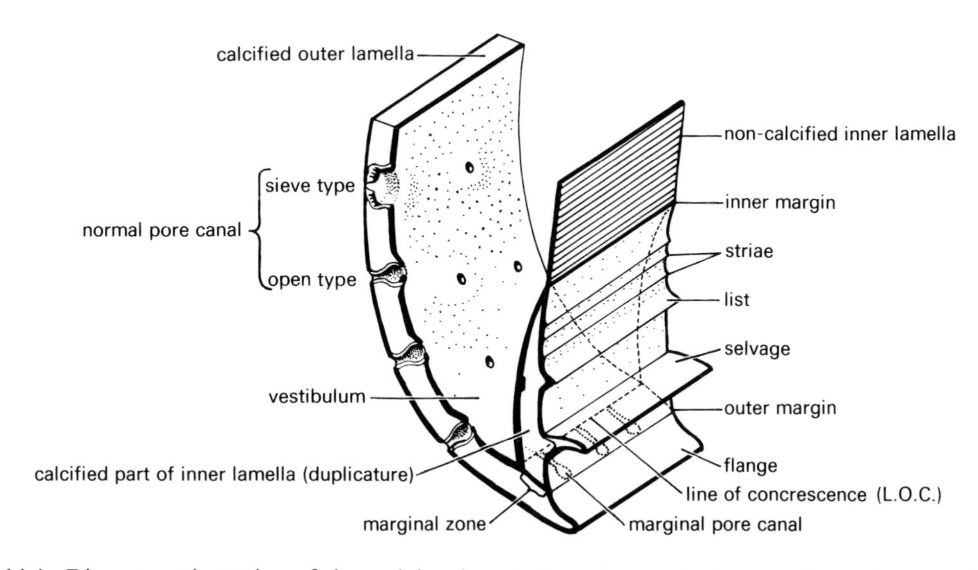

Figure 14.4 Diagrammatic section of the peripheral part of a podocopid ostracod valve, with outer lamella and duplicature. (Modified from Kesling 1951)

valve and either a **subcentral tubercle** or a **sulcus** on the outside (Fig. 14.3). The number and arrangement of scars is diagnostic for many of the higher ostracod taxa, but they are rarely seen in Archaeocopida and Palaeocopida. In the latter, the position of the muscles is marked by a prominent infold to the valve, the median sulcus, running from dorsum to venter (Fig. 14.9).

Other muscles, such as those operating the antennae and mandibles, may also leave small scars on the inside of ostracod valves. These, of course, lie anterior to the adductor muscle scars.

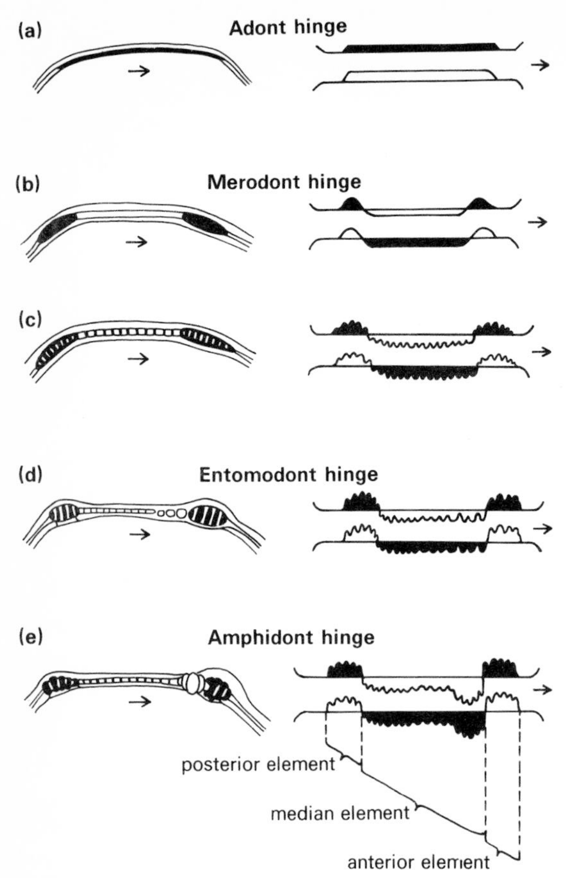

Figure 14.5 Some ostracod hinge types, seen in lateral view of the left valve and from above. (Modified from van Morkhoven 1962)

Tactile features. The ostracod is kept in touch with its surroundings by tactile bristles (setae) which penetrate the shell through slender **pore canals**. Those that penetrate the marginal zone are called **marginal** pore canals whilst those traversing the rest of the shell are simply termed **normal** pore canals (Fig. 14.4). Their form (e.g. branched or unbranched) and arrangement can be useful to taxonomy.

Response to external conditions may be further assisted by the development of clear **eye spots** or raised **eye tubercles** adjacent to the eyes, especially in shallow-water forms (see Fig. 14.8a, d).

Dimorphism. The male and female are distinct in the Ostracoda and often secrete carapaces of different size and shape. This sexual **dimorphism** is especially marked in the fossil Palaeocopida, where it has special taxonomic value (see Fig. 14.10). In this order the distinctive female forms are called **heteromorphs** and differ from the **tecnomorphs** (males and juveniles) in having a more inflated posterior region, pronounced ventral lobes, prominent hemispherical bulges called **brood pouches** (Fig. 14.10b), or wide **frills** extending beyond the free edges of the valves (Fig. 14.10e). Podocopid dimorphism is less obvious, but the males are generally longer and narrower and the females have greater posterior inflation (Fig. 14.12c). Occasionally, the impressions of paired ovaries or spirally wound testes have

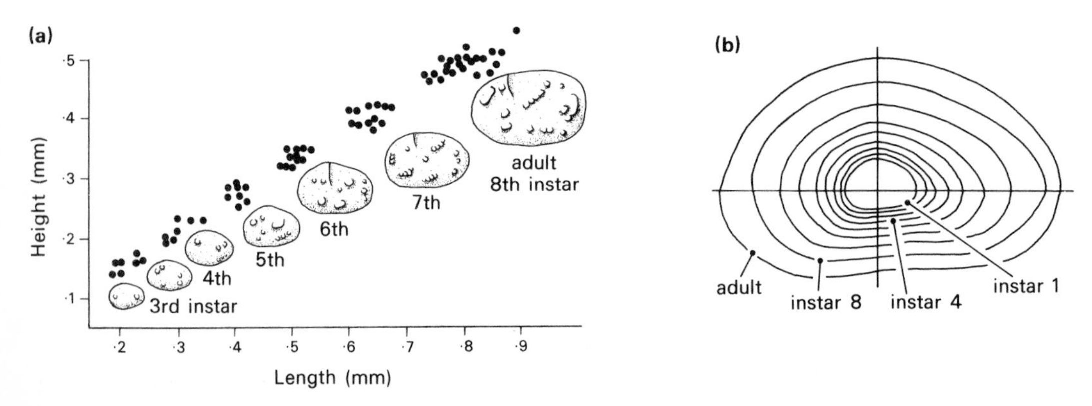

Figure 14.6 Ostracod growth. (a) Discontinuous size distribution and changes of shape of *Neocyprideis colwellensis* from the Eocene Lower Headon Beds; (b) Lateral outlines of successive instars of *Cypridopsis vidua*. ((a) modified from Keen 1977 and (b) from Kesling 1951)

been found in the posterior part of the inner surface of valves.

In certain freshwater species, females may comprise the whole population, reproduction taking place without the participation of males (**parthenogenesis**). Even in normal marine populations, the females may greatly outnumber the males, the ratio between them varying with environmental conditions.

Ostracod reproduction and ontogeny

Reproduction can take place at any time throughout the year. The testes of the male produce spermatozoa of unique proportions, being up to ten times the length of the male carapace. Copulation with the female results in fertilisation of the eggs (Fig. 14.1a) which are then either brooded in the carapace, shed into the water or laid singly amongst water weeds and stones. Most ostracod eggs are very resistant to dessication and cold, hence this stage can help survival through severe winters and prolonged droughts or even allow dispersal on the feet or feathers of birds.

The young ostracods grow in discontinuous stages called **instars** (Fig. 14.6). When the body of an instar has grown too large for its exoskeleton, the rigid chitinous and calcareous layers are shed. Rapid growth and development follow, together with the hardening of a new carapace. There are usually eight such instars between the egg and the adult stage.

The first instar possesses a thin bivalved carapace but the body lacks maxillae and thoracic legs. Maxillae generally develop at instar two, and the legs appear between instars four and six. By instar eight, all the limbs have developed, but the genitalia are rudimentary until the adult stage.

Naturally enough, the valves of ostracod instars increase progressively in size and become thicker and more heavily calcified. These changes are accompanied by modifications in shape and sculpture (Fig. 14.6), and in Podocopida by the increasing complexity of hinge, duplicature and marginal pore canals. Muscle scars are not usually seen before instar six, genital impressions before instar seven, and sexual dimorphism before the adult stage. Clearly, therefore, it is important to distinguish morphological variations that result from ontogeny and those that result from evolution or dimorphism. For these reasons most taxonomic work is based on adult specimens.

Ostracod distribution and ecology

Substrate and food. Ostracods today are predominantly benthic or pelagic throughout their life histories. The benthic ostracods occupy two main habitats: fresh water and sea water (Fig. 14.7). Freshwater ostracods belong almost entirely to the Superfamily Cypridacea (Order Podocopida) and most of them have smooth, thin, weakly calcified carapaces of a simple bean shape (e.g. *Cypris*, Fig. 14.14a). Many of these are filter feeders, consuming detritus or living organisms (e.g. diatoms, protists, bacteria) stirred up by the antennae or mandibles. *Cypridopsis* is a scavenger that holds dead plant and animal particles with its mandibles or antennae and tears at these with its maxillae. One species is known to attack the gastropod vector of the sickness 'Bilharzia' and hence has some medical value. Certain species of *Mesocypris* are even more remarkable in having left fresh water to plough through forest humus in search of food.

Whereas the freshwater ostracods may spend much time swimming several centimetres above the substrate, Recent marine benthic forms are heavier and tend to be either crawlers or burrowers. Again, they are filter feeding on detritus or on diatoms, foraminifera and small polychaete worms. Such ostracods thrive best in muddy sands and silts or on algae and sea grasses. They are scarcer in *Globigerina* oozes and scarcest in euxenic black muds, evaporites, well sorted quartz sands and calcareous sands.

It has often been observed that the size, shape and sculpture of benthic ostracods broadly reflects the stability, grain size and pore size of the substrate on, or in which, they live. For example, crawling forms dwelling on soft, relatively fine-grained substrates tend to have a flattened ventral surface perhaps with weight-distributing projections called **alae** (e.g. *Cytheropteron*, Fig. 14.18a) or frills, keels and lateral spines. Ostracoda dwelling on coarser substrates from the more turbulent, near-shore habitats are commonly thicker shelled with a coarse sculpture of ribs, reticulations or robust spines housing sensory setae (e.g. *Quadracythere*, Fig. 14.17a).

Interstitial ostracods, which live within or burrow through the pore spaces of sandy substrates, tend to be small, smooth and robust (e.g. *Polycope*, Fig. 14.20b); those that burrow through silts and muds need more streamlined carapaces and are usually smooth and elongated (e.g. *Krithe*, Fig. 14.18b). This burrowing is achieved with the assistance of short stout spines on the antennules. The Paradoxostomatidae (Order Podocopida, Suborder Podocopina) contain many forms that have developed probes for sucking the juices of sea grasses and algae. These **phytal** ostracods are generally smooth with slim and elongate valves (e.g. *Paradoxostoma*, Fig. 14.17b).

The Order Myodocopida contains some ostracods that spend their lives swimming in the oceans. This they do by means of two pairs of long and powerful antennae provided with long setae. Food particles in the water are moved towards the maxillae and thoracic limbs by water currents produced from the beating of epipodites on the modified first thoracic legs. In *Conchoecia* this filter feeding is supplemented by a carnivorous diet; *Gigantocypris* subsists largely on copepods, chaetognaths and small fish caught with its antennae.

As with other plankton, pelagic ostracods thrive in regions of current upwelling, rich in phosphates and nitrates. They sometimes grow very large, with *Gigantocypris*, for example, reaching lengths of up to 30 mm. Their carapaces are smooth, thin-shelled and ovate to subcircular in lateral profile (Figs 14.19, 14.20). The long and active antennules and antennae have in some cases led to the formation of **rostral incisures** and projecting **rostra** at the anterior end of the carapace (e.g. *Cypridina*, Fig. 14.20a).

A few freshwater ostracods, such as *Entocythere* (Order Podocopida), are **commensal**. These live attached to the appendages of larger Crustacea such as crayfish, isopods and amphipods, taking advantage of the feeding currents of their hosts. They are not common as fossils.

Salinity. Different species and genera of ostracods prefer to live under specific salinity ranges within the freshwater to hypersaline spectrum. As many Recent genera are relatively long ranging in time, they may be used as salinity indicators in the fossil record (Fig. 14.22). Even extinct taxa can be used to indicate palaeosalinities, because carapace morphology tends to vary with the salinity of the environment, as will be indicated below.

Three main salinity assemblages are distinguishable: freshwater ($<0.5^0/_{00}$), brackish-water (0.5–$30^0/_{00}$) and marine (30–$40^0/_{00}$). Hypersaline assemblages ($>40^0/_{00}$) mainly consist of marine and brackish-water forms (Fig. 14.7).

Freshwater ostracods thrive in lakes, rivers, ponds and ephemeral puddles. They may also be found in brackish-water estuaries, lagoons and seas, but the number of species drops sharply at salinities above $3^0/_{00}$. Nearly all Recent freshwater assemblages comprise species of the Family Cyprididae (Order Podocopida, Suborder Podocopina) with smooth, thin-shelled carapaces of simple morphology (e.g. *Cypris*, Fig. 14.14a). The hinge is generally adont (or rarely merodont) and eye spots or branched marginal pore canals are never present. Ostracod assemblages from isolated lakes around the world are sometimes remarkably similar, their distribution probably brought about by transport on the feet or feathers of migrant water birds, or in their faeces.

Various distinct ostracod assemblages thrive under different salinities within the brackish-water range. Their success under these conditions contrasts with that of most other invertebrate groups, such as the Foraminiferida, in which the brackish-water biofacies comprises only a few highly tolerant (**euryhaline**) marine species. None the less, brackish-water ostracod assemblages are of low diversity when compared with freshwater or marine assemblages, especially at salinities between 3 and $10^0/_{00}$. Above this, the number of **stenohaline** (less tolerant) species begins to increase and the number of individuals per unit area of sea floor begins to decrease. This is probably because both competition and predation are lessened in brackish waters, allowing very large populations of ostracods to survive. Most of these belong to the Family Cytheridae (Order Podocopida, Suborder Podocopina) and they tend to be thick-shelled, weakly ornamented forms with prominent normal pore canals and a merodont or amphidont hinge. Euryhaline marine species may react to lowered salinities by developing hollow tubercles on the valve sides (e.g. *Cyprideis*, Fig. 14.16c). As salinities become lower, these tubercles become more evident, appearing first in the juvenile instars and even developing in adults at salinities of $5^0/_{00}$ or less. Because such tubercles develop with environmental

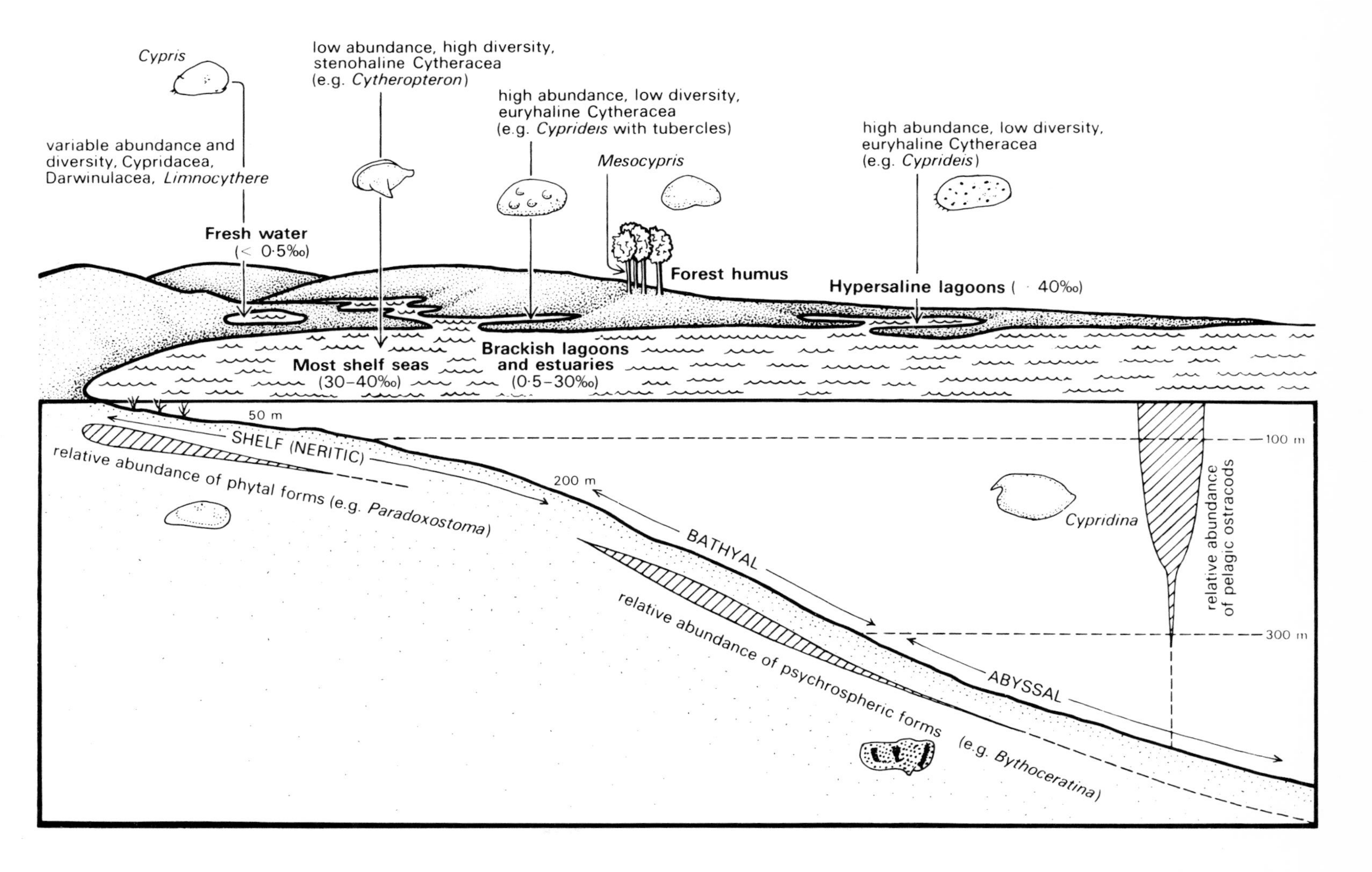

Figure 14.7 Diagram illustrating the ecological distribution of Recent Ostracoda, with some typical forms represented.

changes and the character is not transferred to the offspring, they are referred to as **phenotypic** characters.

Recent marine ostracod assemblages are much the most varied. The majority of species are adapted to salinities of around 35‰ (i.e. stenohaline), but euryhaline forms may also be present. The pelagic Myodocopida, with their thin-shelled and rounded carapaces, are restricted to these marine conditions. Benthic forms with heavy sculpture, amphidont hinges, branching marginal pore canals and well developed eye spots are found here and belong largely to the Cytheridae.

Hypersaline seas, lagoons and salt marshes tend to be colonised by euryhaline, brackish-water forms (e.g. *Cyprideis*, Fig. 14.16c) but the hollow tubercles are not developed here. As in brackish waters, the lack of competition and predation favours high numbers of a few tolerant species. Distinction between brackish and hypersaline faunas in the fossil record may therefore depend on the nature of the associated biota and sediments.

Depth. Although depth in itself is not thought to affect ostracod distributions, it does control the variation of some important ecological factors. For example, both water density and hydrostatic pressure increase directly with depth, light is reduced, and substrates tend to become finer grained. Salinity and oxygen levels also vary slightly in the present oceans, but their stratification was probably much greater in earlier oceans without polar ice caps.

As most freshwater bodies are relatively shallow, freshwater ostracods reveal little variation with depth. In the seas and oceans, however, there are distinct depth assemblages in both benthic and pelagic Ostracoda. In the latter a surface assemblage (< 250 m) of rich diversity may overlie an impoverished layer at 300–400 m, with richer assemblages at 450–625 m and at 720 m downwards (Angel 1969). These daytime zonations with their distinct species are partly disrupted by upward migrations at night time, but in general appear to correspond with different water masses found at different depths.

Benthic ostracod depth assemblages may be categorised broadly as inner-shelf, outer-shelf and bathyal–abyssal. The shelf (or neritic) assemblages live between 0 and 200 m depth, and include many of the marginal marine forms mentioned above. Whereas the densest populations are found in the marginal areas, however, the highest diversities tend to occur in shallow-shelf seas. Of these forms, the paradoxostomatids with their plant-dwelling habit are mainly restricted to waters shallower than 100 m, although they can feed on transported plant debris at greater depths. The presence of thick valves with eye spots, strong sculpture, amphidont hinges and conspicuously branched pore canals are features common in Recent shallow water ostracods from coarse-grained substrates (e.g. *Quadracythere*, Fig. 14.17a). Deeper-water neritic substrates, which tend also to be finer grained, support forms with smooth, thin, often translucent carapaces with relatively weak hinges and no eyes or eye spots (e.g. *Krithe*, Fig. 14.18b; *Argilloecia*, Fig. 14.14e). Bathyal and abyssal assemblages are neither abundant nor diverse, but they are distinctive. Often referred to as the **psychrospheric** fauna, they occur mostly at depths of 1000–1500 m and at temperatures of 4–6°C. Psychrospheric ostracods have adapted to conditions of darkness, constant salinity and temperature, and fine grain size. As these conditions stretch more or less uniformly throughout the deep ocean floors of the world, these ostracods have a cosmopolitan distribution. However, they inhabit shallower waters at the Poles than at the Equator. Most of them are blind forms with relatively large carapaces (> 1 mm long) and thin, highly sculptured walls (e.g. *Bythoceratina*, Fig. 14.18c).

Temperature. As well as the vertical temperature controls outlined above, there is a latitudinal temperature control of shallow-water forms, giving rise to numerous localised (**endemic**) assemblages from the poles (at temperatures below 0°C) to the subtropics and tropics (where they may reach 51°C). This endemism is no doubt heightened in benthic ostracods by the lack of a planktonic larval stage for dispersal. As with most groups, tropical assemblages tend to be more diverse than those in higher latitudes. Some of the latter are, however, of relatively large body size, explained by their slower metabolism and the longer time it takes them to reach maturity. As well as affecting the metabolic rate, maturation and food supply, temperature can also control the breeding season and, in some freshwater forms, the incidence of parthenogenesis.

Ostracods and sedimentology

Like other microfossils, ostracods are readily transported after death. Because the valves tend to spring open when the adductor muscles decay, there is a likelihood that turbulence, scavengers or deposit feeders will disarticulate them. This factor can be turned to advantage in palaeoecology by studying the ratio of valves to whole carapaces. According to Oertli (pp. 137–51, *in* Oertli 1971), whole carapaces predominate where the rate of sedimentation is high and disarticulated valves where it is low.

Transport and mixing of freshwater, brackish-water and marine faunas is particularly evident in estuaries and deltas (see Kilyeni, pp. 251–67 *in* Neale 1969). The freshwater ostracods here drift out in a layer of less dense water, overlying a salt wedge in which tidal currents bring marine forms landwards.

Ostracods are rarely so abundant as to make up the bulk of the sediment. Where they do, these 'ostracodites' are usually of freshwater or brackish-water origin.

Classification

Kingdom ANIMALIA
Phylum CRUSTACEA
Class OSTRACODA

The Class Ostracoda have the organisation of the Superphylum Arthropoda and of the largely aquatic Phylum Crustacea (see Manton 1977 for classification). They form a distinct class within the latter phylum because of the laterally compressed body, the undifferentiated head, the seven-or-less head and thoracic limbs, the pair of furcae and, of course, the bivalved perforate carapace lacking in growth lines. Biologists subdivide the still-living members of the group on differences in their soft parts, particularly their appendages. Generally, these taxa correspond with the carapace-based taxa of palaeontologists. This biological approach, however, cannot be extended to the extinct Palaeozoic orders, which are diagnosed entirely on carapace features.

Van Morkhoven (1962–3) has discussed the

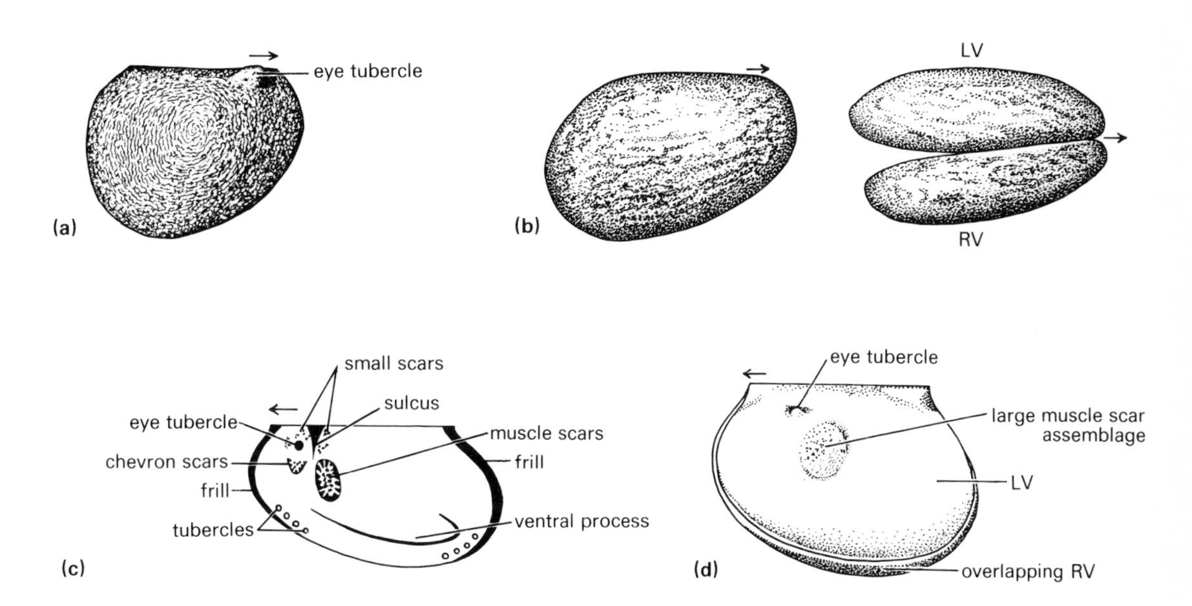

Figure 14.8 Orders Archaeocopida and Leperditicopida. (a) *Bradoria* exterior RV ×8; (b) *Indiana*, exterior RV ×7; dorsal view ×7; (c) common structural features of leperditicopids; (d) *Leperditia*, exterior LV ×1·3. ((a) & (b) after Benson *et al.* 1961 from Ulrich & Bassler; (c) after Abushik 1971; (d) after Benson *et al.* 1961 from Triebel)

In this and the following figures in this chapter, RV = right valve, LV = left valve, ♂ = male, ♀ = female, and the arrows point anteriorly from the dorsal margin.

criteria utilised for taxonomic groupings within the post-Palaeozoic ostracods, and these and the criteria for Palaeozoic forms are discussed by Scott (pp. 74–92 *in* Benson *et al.* 1961). In summary, we may note the value of the following carapace features in approximate order of importance: (a) basic carapace shape; (b) muscle scar position and arrangement; (c) degree of development and fusion of the duplicature with the outer lamella; (d) structure, shape, size and arrangement of normal and marginal pore canals; (e) nature, location and degree of valve overlap; (f) hinge elements; (g) nature of sexual dimorphism, if present; (h) nature of surface sculpture, and presence of eye spots; (i) nature of marginal zone; and (j) form of selvages and flanges.

Obviously it is essential to be able to distinguish dorsal from ventral and posterior from anterior in fossil ostracod valves. In the still-living Orders Podocopida and Myodocopida, orientation presents no great problem, but in the extinct Archaeocopida, Leperditicopida and Palaeocopida the correct orientation is less certain. Guidelines for orientation, following the currently accepted practices, are therefore included below.

Order Archaeocopida. The case for ostracod affinities of this ancient arthropod stock have been outlined by Sylvester-Bradley (pp. 100–102 *in* Benson *et al.* 1961). Appearing early in the Cambrian but dying out in Ordovician times, the archaeocopids may have been the ancestral stock of the younger Leperditicopida, sharing a similar shape and orientation (see below and Fig. 14.8). The carapace was thin, flexible, porous and of chitinous, phosphatic or weakly calcified material. They had a straight dorsal margin lacking in hinge elements and a ventral margin that was convex and highest at the posterior end. It lacked a duplicature. No muscle scars are found, although subcentral tubercles may indicate their original positions. Neither is there evidence for sexual dimorphism in this ancient stock. Carapace morphology suggests that many archaeocopids swam about in well illuminated, probably shallow, waters because the ventral margin is not flattened and eye spots are found. This pelagic existence is further suggested by their occurrence in Cambrian pelagic limestones along with agnostid trilobites and paraconodonts.

Bradoria (L.–M. Camb., Fig. 14.8a) has a smooth or wrinkled, subquadrate carapace with prominent anterodorsal eye tubercles. *Indiana* (L. Camb., Fig. 14.8b) was smooth, lacked eye tubercles and had a more elongate carapace.

Order Leperditicopida. The Leperditicopida probably developed from archaeocopid stock in late Cambrian or early Ordovician times, differing from them in their thick calcified shell, large and distinctive muscle scars and their relatively large size (5–30 mm long). This was a very successful Ordovician to Devonian group in which both the eye spots and the smooth valves with convex ventral margins indicate a shallow-water, pelagic mode of life. Their large size suggests they had a carnivorous diet, as do the large Myodocopida. Recognition and orientation of the valves should be helped by the guidelines below and in Figure 14.8c.

(a) The dorsal margin is long and straight, often ending in distinct corners called **cardinal angles**. The hinge is adont. Eye spots and eye tubercles, where present, occur in an anterodorsal position.
(b) The ventral margin is convex and undifferentiated, except in certain genera with flattened rim-like borders, or with ventral tubercles on the inner margins.
(c) Viewed laterally with the hinge horizontal, the narrowest end is anterior (often with a more acute cardinal angle) and the highest end is posterior (often with a more obtuse cardinal angle).
(d) Muscle scars are large, comprising a compact cluster of numerous (> 200) rounded scars placed anterior of the valve centre. A subcentral tubercle or sulcus may mark their position on the outer surface.
(e) The valves are strongly unequal or subequal in size, the right one overlapping the margins of the left.

Leperditia (L. Sil.–U. Dev., Fig. 14.8d) was a widespread genus with a purse-shaped, smooth or punctate carapace bearing distinct eye tubercles and adductor muscle scars.

Order Palaeocopida The Palaeocopida were an important Palaeozoic group (range L. Ord.–M. Perm.) that can be recognised by their long straight

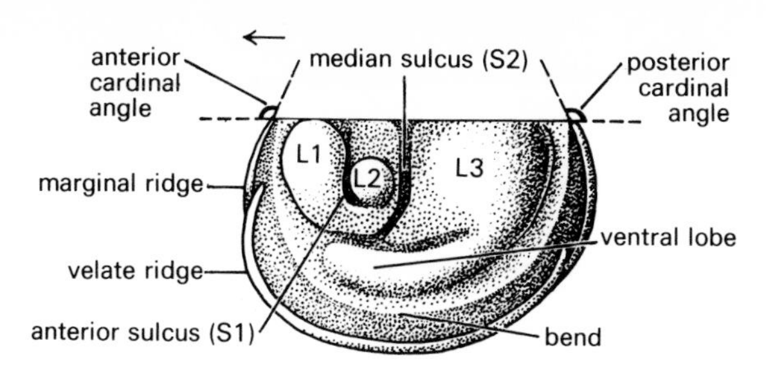

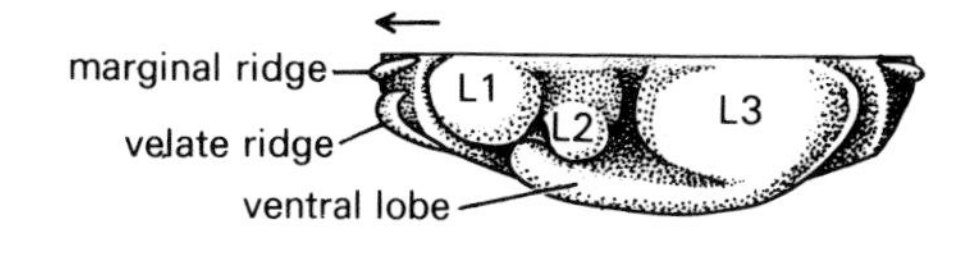

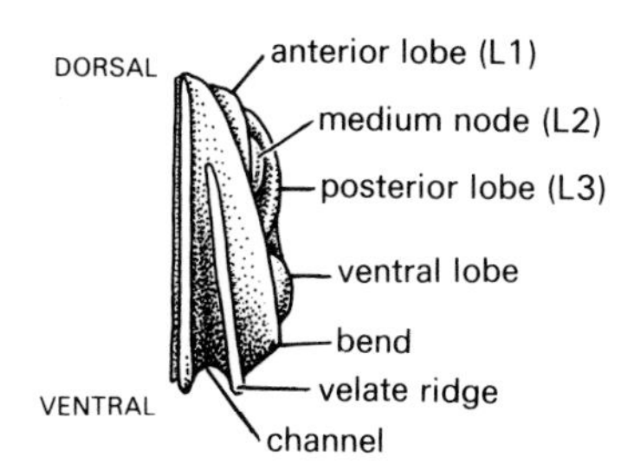

Figure 14.9 Order Palaeocopida, Suborder Beyrichicopina. Morphology of a palaeocopid valve in lateral, anterior and dorsal view, with nomenclature. (After Kesling 1951)

hinge line, lobate and sulcate sculpture, and often by the distinctive sexual dimorphism. Orientation of the carapace should be helped by Figure 14.9 and the guidelines below.

(a) The dorsal margin is long and straight, often ending in distinct cardinal angles. Eye spots and eye tubercles, if present, occur in an anterodorsal position. The **sulci** and **lobes** are more sharply defined towards the dorsal margin.
(b) The ventral margin is often convex and may be provided with frills, flanges, brood pouches or spines, especially in the heteromorphs. Commonly a **ventral lobe** runs parallel to the ventral margin (Fig. 14.9).
(c) Viewed laterally with the hinge horizontal, the narrowest end is posterior (often with an acute cardinal angle) and the highest end is anterior (often with a more obtuse cardinal angle).
(d) Viewed from above, the greatest width is usually posterior. However, the wide brood pouches of heteromorphs in the Superfamily Beyrichiacea are anteroventral in position (Fig. 14.10c).
(e) The median sulcus, if present, approximates to the position of the numerous adductor muscles on the inside, but muscle scars are rarely seen. Both features are generally anterior of the valve centre.
(f) Major spines and alae tend to be directed backwards.

This diverse group is usually subdivided on the basis of general shape, the nature of dimorphism (if any), the form of lobes and sulci and on superficial sculpture (e.g. spines, striae and reticulation). There are two suborders: the Beyrichicopina and the Kloedenellocopina.

The Beyrichicopina (L. Ord.–M. Perm.) have sub-equal valves, the right one being slightly larger Cardinal angles, lobes, sulci and sexual dimorphism are usually well developed. The latter often takes the form, in heteromorphs, of either anteroventral brood pouches or broad striate frills that parallel the free margins (**vela**). The hinge elements are adont. Morphological features of the group suggest they were not burrowers or crawlers but swam in shallow waters, perhaps near to the sea floor, and filter fed on detritus, benthos and plankton.

Beyrichia (L. Sil.–M. Dev., Fig. 14.10a, b) was a widespread genus with three distinct lobes and a granular or pitted surface. The heteromorph has globular brood pouches. In *Aechmina* (M. Ord.–L. Carb., Fig. 14.10d) the dorsal margin sprouts a remarkably large spine, and the ventral margin bears short spines. Dimorphism is unknown in this genus. *Hollinella* (M. Dev.–M. Perm., Fig. 14.10c) has essentially four lobes, but the anterior and posterior ones (L_1 and L_4) are united with a prominent ventral lobe. The marginal frill of the heteromorph is broader than that of the tecnomorph. *Oepikium* (L.–U. Ord., Fig. 14.10e) has three lobes, the anterior and posterior ones (L_1 and L_3) being very broad and the median one (L_2)

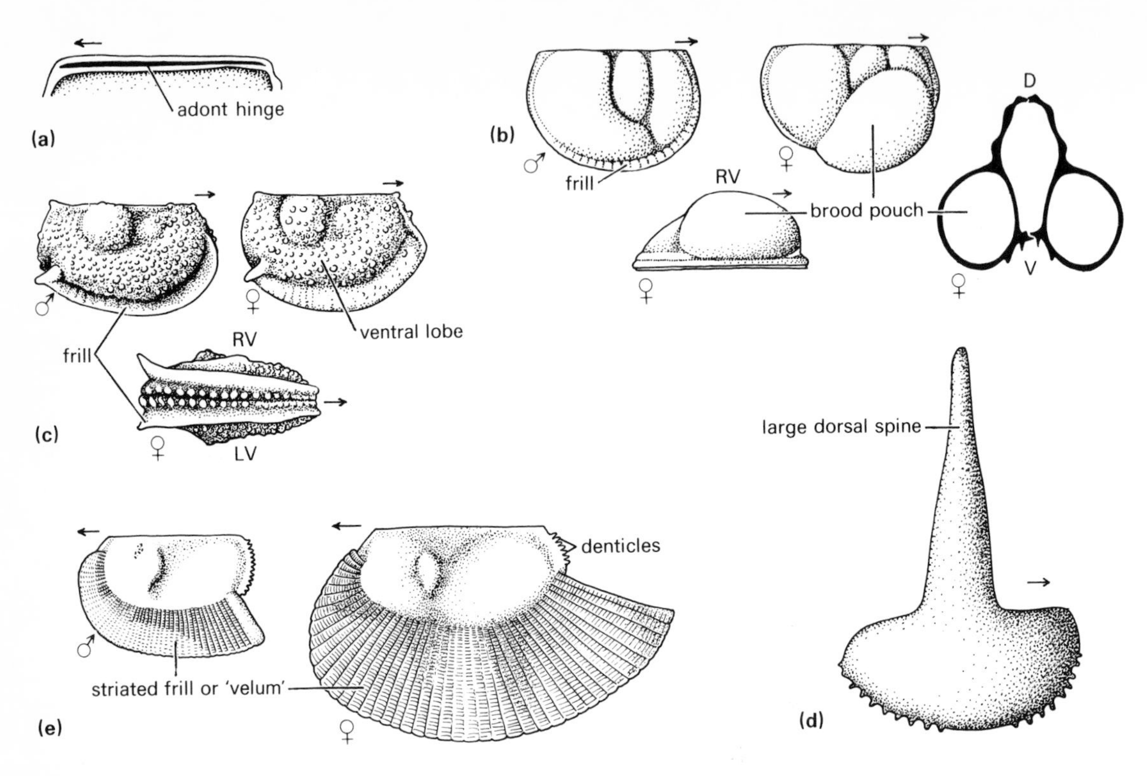

Figure 14.10 Order Palaeocopida, Suborder Beyrichicopina. (a) *Beyrichia*, adont hinge of RV. (b) *Beyrichia*, exterior RV of ♂ ×17·5, exterior RV of ♀ ×17·5, ventral view of RV of ♀ ×17·5, transverse section of ♀ ×16; (c) *Hollinella*, exterior RV ♂ ×20, exterior RV of ♀ ×20, ventral view of ♀ ×20; (d) *Aechmina*, exterior RV ×40; (e) *Oepikium*, exterior LV of ♂ ×14, exterior LV of ♀ ×20. ((b), (c) & (e) after Benson *et al.* 1961 from Kesling; (d) after Benson *et al.* 1961 from Bouček)

virtually united with L_1. Both the heteromorphs and the tecnomorphs possess radially striated frills, that of the former being amongst the largest known. *Aparchites* (L.–M. Ord., Fig. 14.11a) has an ovate, non-sulcate carapace in which the frill is reduced to a low ridge, often provided with small spines. Dimorphism is not known here.

The Suborder Kloedenellocopina (L. Ord.–Perm.) comprises ostracods with subovate valves of unequal size, the larger one (usually the right valve) overlapping the smaller one along all, or part, of the free margin. The cardinal angles are invariably rounded off, the lobes and sulci are less numerous or less prominent than in most Beyrichicopina, and sexual dimorphism is either less conspicuous (e.g. it may be manifest as a greater posterior swelling of the heteromorph carapace) or absent. The ventral margin is also rather less convex and can be straight or concave. It is thought that many of the Kloedenellocopina were crawlers and burrowers in sand and mud.

In *Kloedenella* (Sil.–Dev., Fig. 14.11b) the left valve overlaps the right one and both bear two prominent anterodorsal sulci. The heteromorph has a higher posterior region than the tecnomorph. *Glyptopleura* (L. Carb.–M. Perm., Fig. 14.11c) has a subovate, unusually costate carapace in which the right valve overlaps the free margins of the left. The median sulcus (S_2) is represented by a sub-central pit. *Cryptophyllus* (M. Ord.–U. Dev., Fig. 14.11d) is atypical in that the earlier moult stages are retained, each new outer lamella being added to the inside of the preceding one. Hence the external surface reveals up to six concentric ridges.

Order Podocopida. The Podocopida comprise the bulk of the Mesozoic and Cainozoic fossil ostracods, although they have a longer history (L. Ord.–Rec.). Living forms are largely diagnosed from their soft parts, but fossil taxa have been

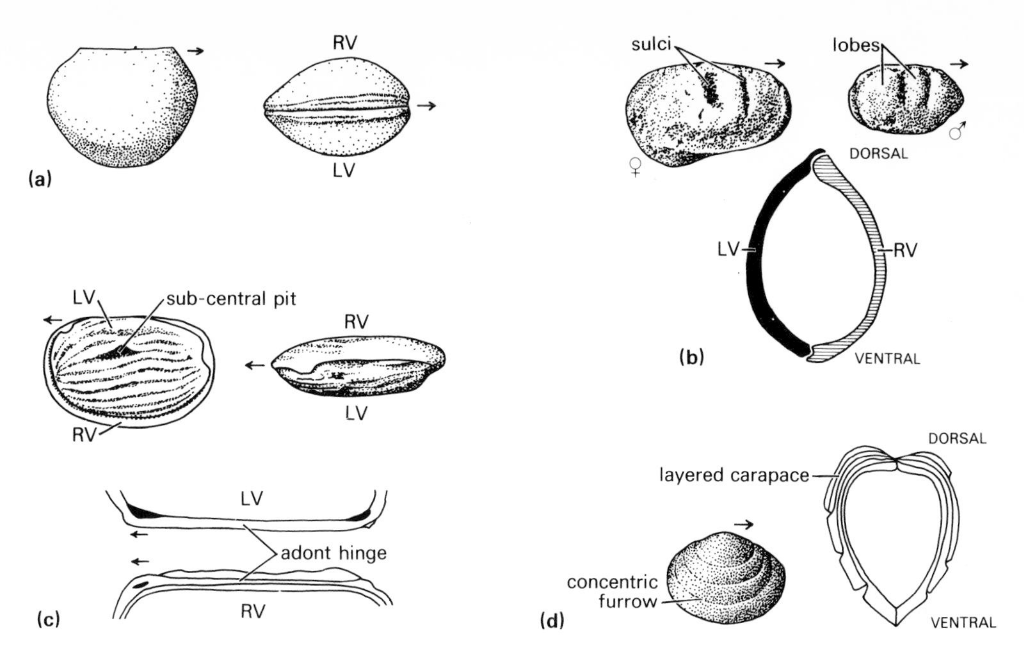

Figure 14.11 Order Palaeocopida, Suborders Beyrichicopina and Kloedenellocopina. (a) *Aparchites*, exterior RV ×7, ventral view ×7; (b) *Kloedenella*, exterior RV of ♀ ×20, exterior RV of ♂ ×20, transverse section ×45; (c) *Glyptopleura*, exterior LV ×18, dorsal view ×18, hinge ×40; (d) *Cryptophyllus*, exterior RV ×33, transverse section ×53. (All after Benson *et al.* 1961; (a) from Jones, (b) from Swartz & Whitmore and from Sohn, (c) from Scott and (d) from Levinson)

erected on carapace morphology. Podocopid valves commonly have a convex dorsal margin and a weakly convex, straight or concave ventral margin. Lobes and sulci are uncommon and muscle scars and duplicature are prominent. Podocopid valves may be orientated using the following guidelines.

(a) The dorsal margin is convex or straight but less than the total length of the carapace. It bears adont, merodont, entomodont or amphidont hinge elements. Eye spots and eye tubercles, if present, occur in an antero-dorsal position.
(b) The ventral margin is often concave but may be straight or convex. The duplicature, where present, is narrow in the Suborders Metacopina and Platycopina and wider in the Podocopina, with marginal pore canals in the marginal zone. The ventral region may also be provided with prominent spines, frills, flanges and wing-like alae.
(c) Adductor muscle scars are variable in number and arrangement but are invariably situated just anterior of the valve centre. Their position may be marked on the outer surface by a subcentral tubercle.
(d) Viewed laterally, the more pointed end is generally posterior whilst the higher, blunter end is anterior.
(e) In dorsal or ventral view, the broadest region occurs near the posterior end in adults and is often more swollen in female carapaces.
(f) The more complex portions of entomodont and amphidont hinges (i.e. crenulation or lobation of the median element) are developed towards the anterior end of the hinge (Fig. 14.5).
(g) Major spines, tubercles and alae generally point backwards.
(h) The marginal area of the Suborder Podocopina tends to be broader and with more marginal pore canals at the anterior end.

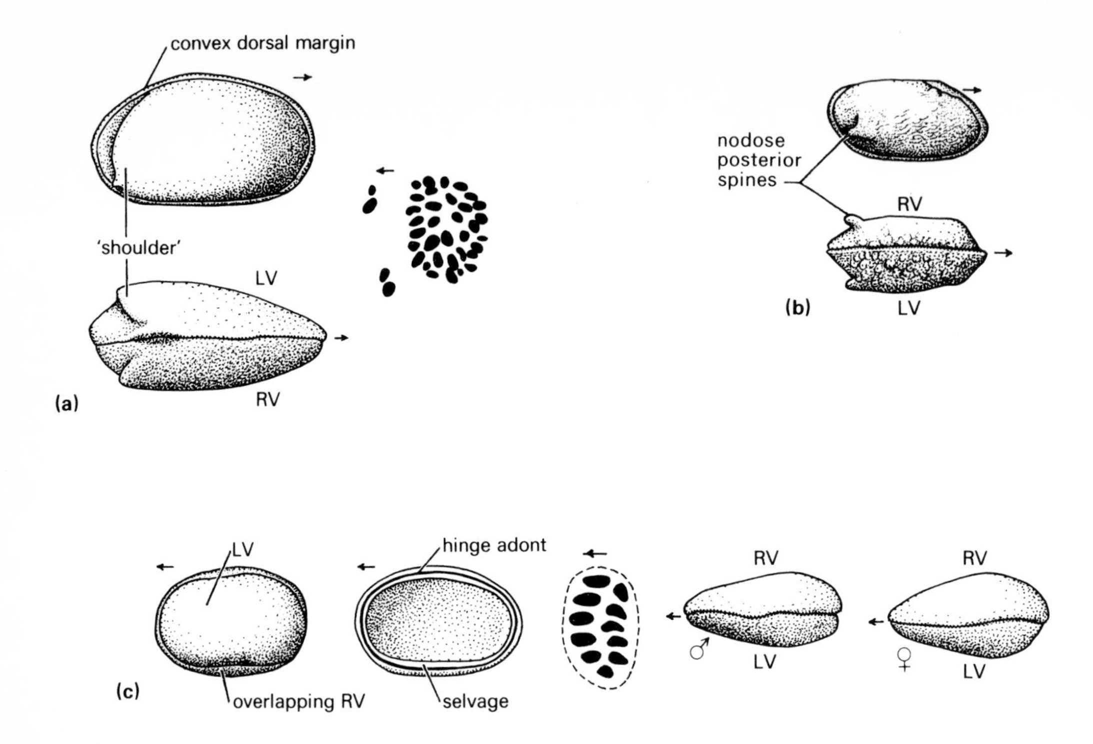

Figure 14.12 Order Podocopida, Suborders Metacopina and Platycopina. (a) *Healdia*, exterior RV of ♂ ×67, muscle scars × 133, dorsal view of ♀ ×67; (b) *Quasillites*, exterior RV ×20, ventral view ×20; (c) *Cytherella*, exterior LV ×27, interior RV ×27, detail of adductor muscle scar, dorsal view of ♂ ×27, dorsal view of ♀ ×27. ((a) after Shaver and (b) after Sohn, both in Benson *et al.* 1961; (c) after Andreev 1971)

The majority of Podocopida have adapted to crawling and burrowing niches in marine sediments or on seaweeds. However, this order also includes the terrestrial and freshwater Cyprididae and fresh- and brackish-water genera of the Cytheridae. Three suborders are recognised by Benson *et al.* (1961): the Metacopina, Platycopina and Podocopina, of which the Podocopina are the most diverse.

The Suborder Metacopina were marine and are known only from fossil carapaces (?L. Ord., M. Ord.–L. Cret.). They were ancestral to the Platycopina and probably to certain Podocopina. The muscle scars are numerous (> 25) and assembled in a compact group (Fig. 14.12a). The hinge elements are either simple and adont or differentiated into merodont form; the duplicature is indistinct and of narrow width. Typically the left valve is slightly larger, overlapping the free margins of the right valve (i.e. the converse of the Platycopina). For example, *Healdia* (Dev.–Perm., Fig. 14.12a) has a smooth, rounded carapace with backwardly directed 'shoulders' near the posterior end. The dorsal hinge is a simple crenulated bar and groove (i.e. adont). *Quasillites* (Dev.–L. Carb., ?U. Carb., Fig. 14.12b) has a carapace with a straight dorsal margin, merodont hinge and an external sculpture of ridges, grooves and nodose posterior spines.

The Suborder Platycopina are a still-living marine group which arose from the Metacopina in the Triassic period. They differ from that group in having a larger right valve, in the entirely adont hinge, and in the adductor muscle scar pattern. The latter comprises 10–18 or so elongate scars arranged in two slightly curved rows (Fig. 14.12c). However, platycopine and metacopine carapaces share the same ovate shape, an inequality of valve size, weakly developed duplicature and a prominent selvage with contact grooves around the free margins. Recent Platycopina bear biramous antennae (c.f. Podocopina) and three pairs of thoracic appendages that serve as maxillae, although the third pair are rudimentary in females. There are

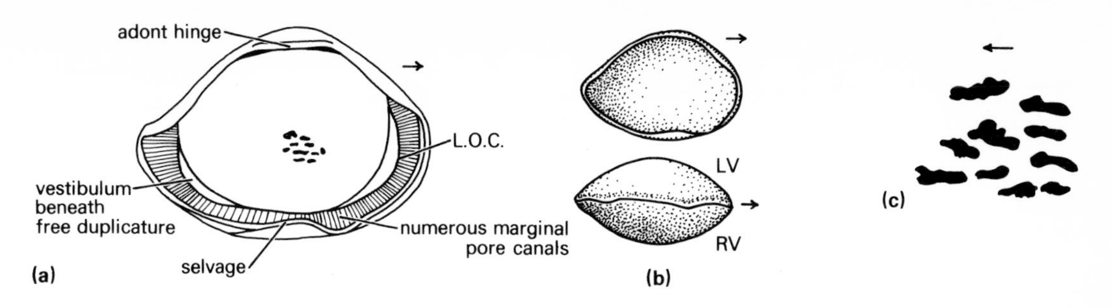

Figure 14.13 Superfamily Bairdiacea (Suborder Podocopina, Order Podocopida). (a) *Bairdia*, interior LV, about ×40; (b) *Bairdia*, exterior RV and dorsal view ×43; (c) detail of adductor muscle scars. L.O.C. = line of concrescence. ((a) after van Morkhoven 1963; (b) after Andreev 1971; (c) after van Morkhoven 1962)

only a few genera, of which *Cytherella* (Jur.–Rec., Fig. 14.12c) is the commonest. This has a smooth ovate carapace with the rear end more inflated in dorsal view, especially in the larger female specimens.

The Suborder Podocopina (L. Ord.–Rec.) bear carapace features which may be polyphyletic, arising from both palaeocopid and metacopine ancestors. None the less, they do have similar soft-part morphology, with almost uniramous antennae and three pairs of thoracic limbs employed for walking and grasping. The dorsal margin is usually convex with adont, merodont, entomodont or amphidont dentition. The ventral margin is concave with a prominent, wide duplicature. Muscle scar elements are fewer than in the Metacopina and their arrangement serves to distinguish four major superfamilies: the Bairdiacea, Cypridacea, Darwinulacea and Cytheracea.

The Bairdiacea (L. Ord.–Rec.) are an ancient marine stock that were common in Palaeozoic seas. Their carapaces are commonly thick and smooth with a 'bairdioid' shape, i.e. with a strongly convex dorsal margin, a blunt anterior end and a pointed posterior end (Fig. 14.13). The duplicature is wide with a prominent vestibulum, and the adductor scars consist of 6–15 elongate elements arranged radially, irregularly or aligned (Fig. 14.13c). The articulation is weak and adont or merodont. *Bairdia* (Ord.–Rec., Fig. 14.13) is in fact the longest-ranging ostracod genus. Its large left valve partly overlaps the margins of the right and the hinge is adont, with a simple ridge and groove articulation.

The Cypridacea (Dev.–Rec.) include most of the freshwater ostracods and a few marine forms. Because of low salinities they secrete smooth, thin, chitinous or weakly calcified shells, often with a rather low preservation potential. The hinge is adont or rarely merodont. The adductor scar pattern consists generally of one large dorsal element, three anterior elements and two posterior elements, all elongated and more or less aligned (Fig. 14.14b). The duplicature is incompletely fused to the outer lamella, leaving prominent vestibula and a relatively narrow marginal zone. Recent cypridaceans are distinguished by their appendages, the more so because their carapaces are very similar. The palaeontologist may therefore face problems with the taxonomy of fossils specimens, making necessary the accurate measurement of all carapace features. *Cypris* (?Jur., Pleist.–Rec.; Fig. 14.14a) thrives in freshwater ponds. It has a smooth, subtriangular carapace of relatively large size (< 2·5 mm long). *Carbonita* (?L. Carb., U. Carb.–Perm.; Fig. 14.14c) was typical of fresh or slightly brackish-water facies around coal swamps. It has a smooth, elongate carapace with a somewhat larger right valve. *Cypridea* (U. Jur.–L. Cret., Fig. 14.14d) is another fossil form typical of fresh to slightly brackish waters. Both dorsal and ventral margins are relatively straight and the anteroventral margin is provided with a beak and notch. The hinge is merodont and the surface is usually pitted and pustulose. *Argilloecia* (Cret.–Rec., Fig. 14.14e) has adapted to outer shelf and bathyal marine conditions, particularly to *Globigerina* oozes. The cara-

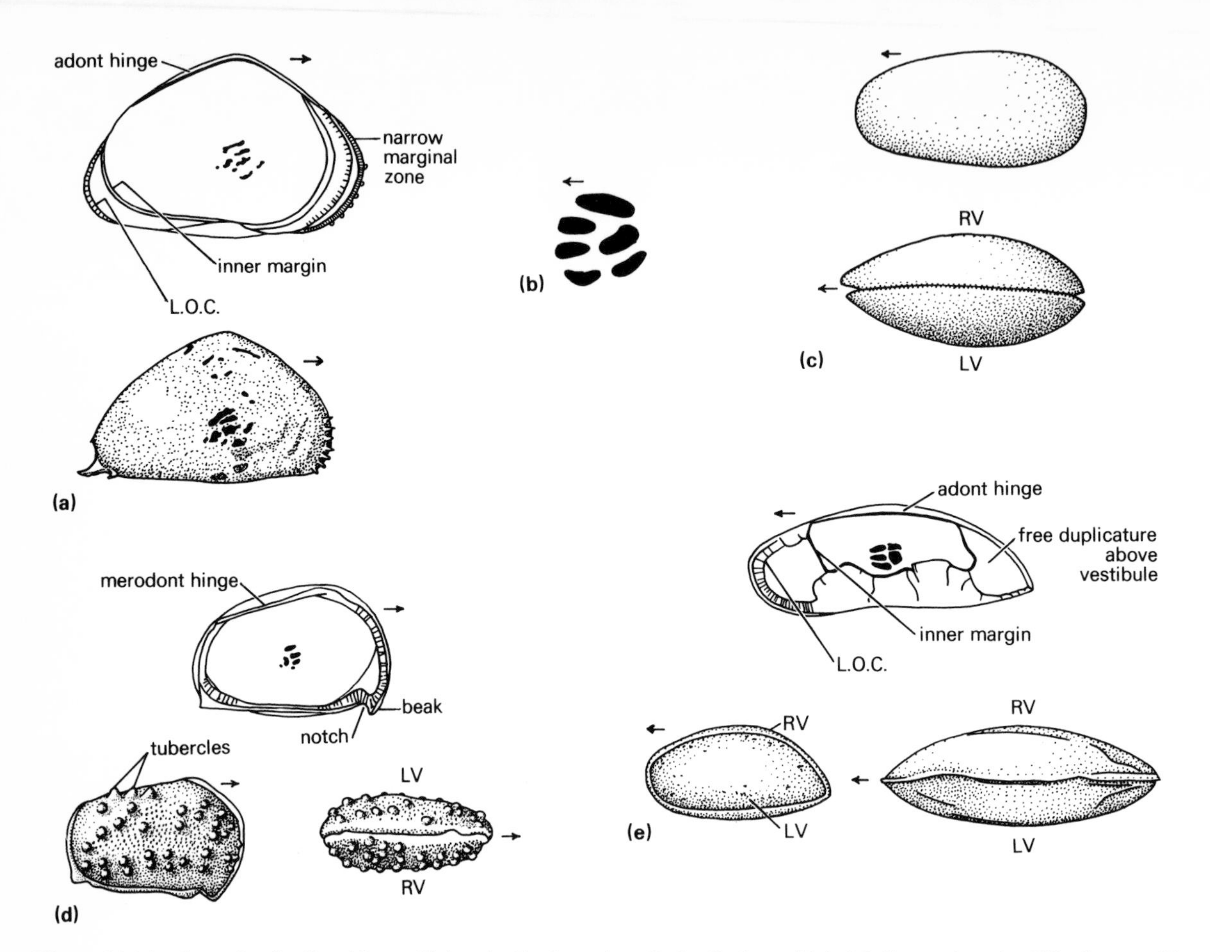

Figure 14.14 Superfamily Cypridacea (Suborder Podocopina, Order Podocopida). (a) *Cypris*, interior LV, about × 16, exterior RV × 13; (b) detail of cypridacean (*Paracypris*) adductor muscle scars; (c) *Carbonita*, exterior LV × 27, dorsal view × 27; (d) *Cypridea*, interior LV × 30, exterior RV × 20, dorsal view × 20; (e) *Argilloecia*, interior RV × 72, exterior LV × 63, dorsal view × 72. ((a) and (d) (part) after van Morkhoven 1963 from Sylvester-Bradley; (b) after van Morkhoven 1962 from Sars; (c) after Benson *et al.* 1961 from Jones; (d) (part) after Benson *et al.* 1961 from Kesling; (e) after van Morkhoven 1963 from Mueller and after Pokorný 1958 from Alexander)

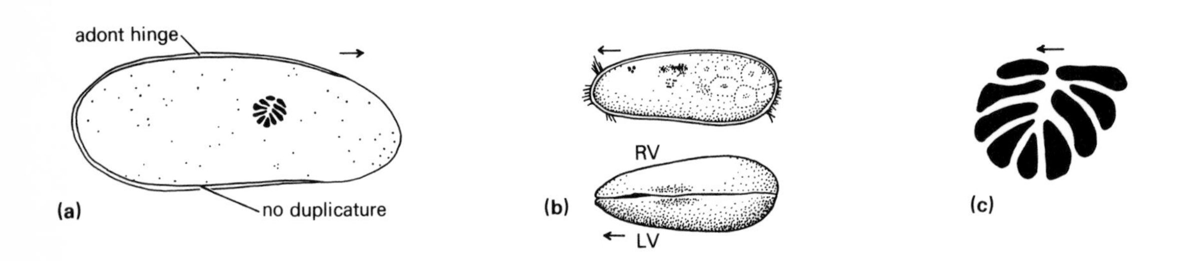

Figure 14.15 Superfamily Darwinulacea (Suborder Podocopina, Order Podocopida). (a) *Darwinula* interior LV × 63; (b) exterior LV and dorsal view × 38; (c) detail of muscle scar. ((a) after van Morkhoven 1963 from Wagner; (b) after van Morkhoven 1963 from Sars; (c) after van Morkhoven 1962)

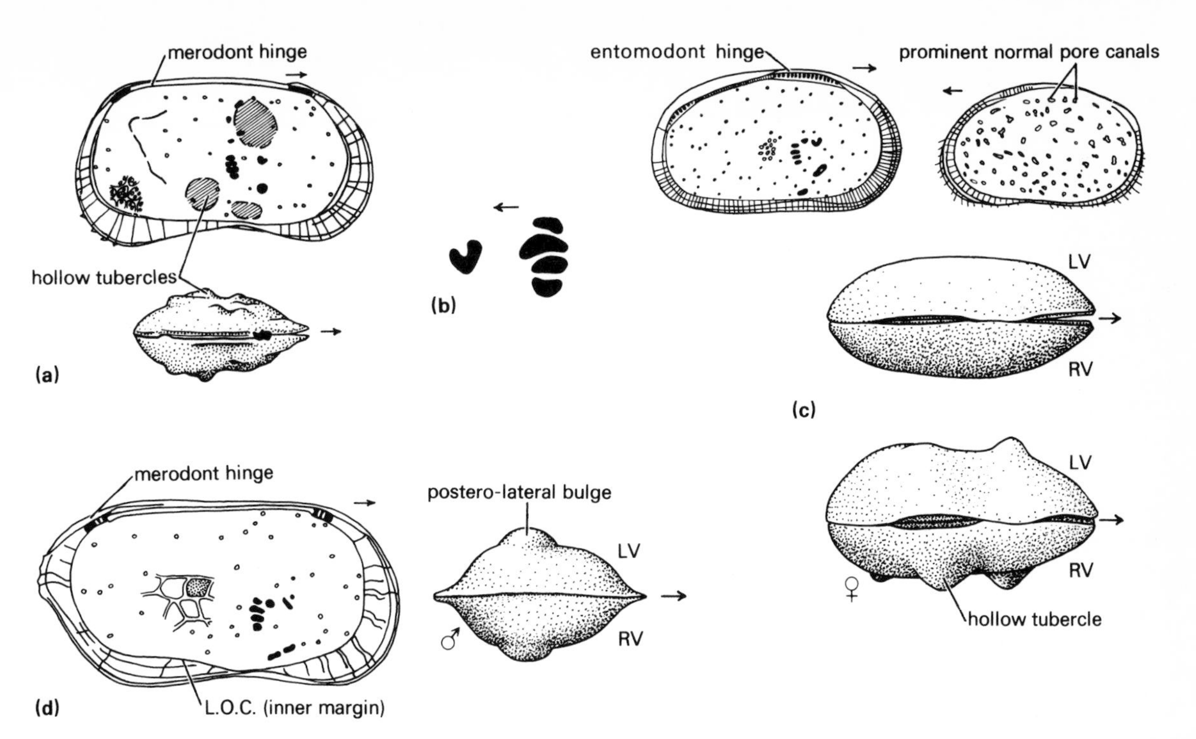

Figure 14.16 Superfamily Cytheracea (Suborder Podocopina, Order Podocopida). Fresh- and brackish-water genera. (a) *Limnocythere*, interior LV ×67, dorsal view ×38·5; (b) detail of a cytheracean muscle scar; (c) *Cyprideis*, interior LV ×50, exterior LV ×27, dorsal view of ♂ (above) and ♀ with hollow tubercles (below); (d) *Cytherura*, interior LV ×87·5, dorsal view ×55·5. ((a) after van Morkhoven 1963 from Wagner and Mueller; (b) after van Morkhoven 1962; (c) after van Morkhoven 1963 from Wagner and Klie and after Benson *et al.* 1961 from Goerlich; (d) after Benson *et al.* 1961 from Wagner)

pace is smooth and elongate with a blunt anterior and a pointed posterior end, the right valve slightly larger than the left one.

The Superfamily Darwinulacea (?Ord., U. Carb.–Rec.) consists of a single genus of freshwater ostracods sporting a distinctive muscle scar pattern, i.e. an almost symmetrical rosette of 9–12 elongate scars (Fig. 14.15c). The carapace of *Darwinula* (Fig. 14.15) is smooth, thin-shelled, elongate and ovate, provided with an adont hinge but lacks a duplicature.

The Superfamily Cytheracea (M. Ord.–Rec.) are morphologically the most varied of this order. They have distinctive adductor muscle scars of four or five elements aligned in a vertical row, anterior of which are found several discrete mandibular and frontal muscle scars (Fig. 14.16b). The hinge of cytheraceans is usually merodont or amphidont and the duplicature with its marginal zone is prominent, often with branched marginal pore canals. Ecologically the Cytheracea are a varied group, as the following examples will show.

Limnocythere (Jur.–Rec., Fig. 14.16a) is one of the few freshwater cytheraceans. It has a thin, chitinous carapace with a merodont hinge and marginal zone bearing many straight marginal pore canals. The hollow tubercles may be phenotypic, induced by the low salinities, as in the genus *Cyprideis* (Mioc.–Rec., Fig. 14.16c) which lives mostly in brackish or hypersaline waters. It has a subovate, essentially smooth carapace with an entomodont hinge. Immature valves from brackish waters may bear phenotypic hollow tubercles. Species of *Cytherura* (Cret.–Rec., Fig. 14.16d) often

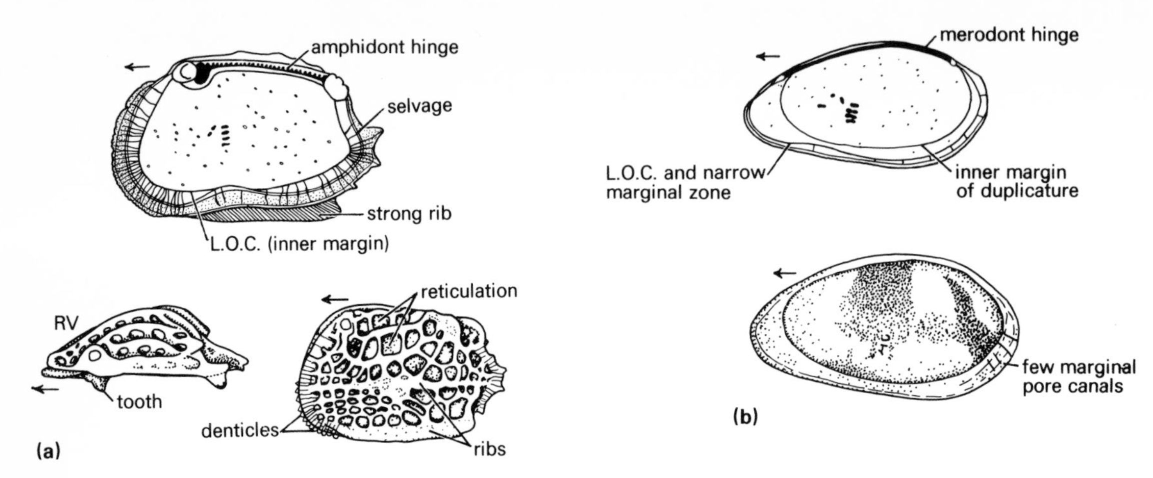

Figure 14.17 Superfamily Cytheracea (Suborder Podocopina, Order Podocopida). Shallow marine genera. (a) *Quadracythere*, interior RV ×50, dorsal RV ×40, exterior LV ×40; (b) *Paradoxostoma*, interior RV ×50, exterior LV ×62. ((a) after van Morkhoven 1963 partly from Hornibrook; (b) after van Morkhoven 1963 from Wagner and after Pokorný 1958 from Sars)

thrive in brackish or very shallow marine waters. In these the carapace is smooth and oblong, the males more elongate and the females provided with postero-lateral bulges. The hinge is merodont and the duplicature narrow, without vestibula.

Quadracythere (?U. Cret., Palaeoc.–Rec., Fig. 14.17a) is a predominantly shallow-water marine genus that likes to crawl over coarse sediment substrates. Its carapace is thick and subquadrate in shape, with a heavy reticulated and ribbed sculpture, adorned along the anterior margin with denticles. The hinge is strong and amphidont and the duplicature is of an even width, completely fused with the outer lamella. *Paradoxostoma* (?Cret., Eoc.–Rec., Fig. 14.17b) and its relatives thrive in rock pools along the intertidal zone and on subtidal seaweeds. The genus has a thin, elongate carapace which is very narrow in dorsal view and more pointed at the anterior end. Its hinge is merodont whilst the marginal zone is narrow with a few simple pore canals.

Species of *Cytheropteron* (L. Jur.–Rec., Fig. 14.18a) live at various depths but being, blind, particularly on soft fine-grained sediments in deeper marine waters. In these, the carapace is ovate or subrhomboidal in outline with distinctive alae as well as an upturned **caudal process** at the posterior end. The hinge is merodont. *Krithe* (U. Cret.–Rec., Fig. 14.18b) is another blind mud-dweller from the outer shelf and bathyal habitats. Its smooth, thin, elongate carapace has a weak adont hinge and a duplicature of varying width that bears a broader anterior and a narrower posterior vestibulum. *Bythoceratina* (U. Cret.–Rec., Fig. 14.18c) is a typical psychrospheric ostracod, thriving best at depths between 2000 m and 3000 m. Its carapace is subquadrate with a straight dorsal margin bearing a merodont hinge. The ventrolateral margins have developed pointed alae whilst the posterior end sports a short caudal process. The outer surface of *Bythoceratina* is commonly reticulated, but it may be smooth or spinose.

Order Myodocopida. The Myodocopida (Ord.–Rec.) include most of the pelagic ostracods. Their soft parts are distinct from those of other living groups in having a biramous antenna with a large basal segment, the whole organ modified for swimming. The furcae are also large and used for locomotion. Muscle scars of the Myodocopida are variable in number, shape and arrangement and of less taxonomic value than those of the Podocopida. Recognition and orientation of myodocopid carapaces may be helped by the following guidelines.

(a) The dorsal margin is commonly convex (but may be straight) and provided with weak, adont hinge elements. In the Superfamily

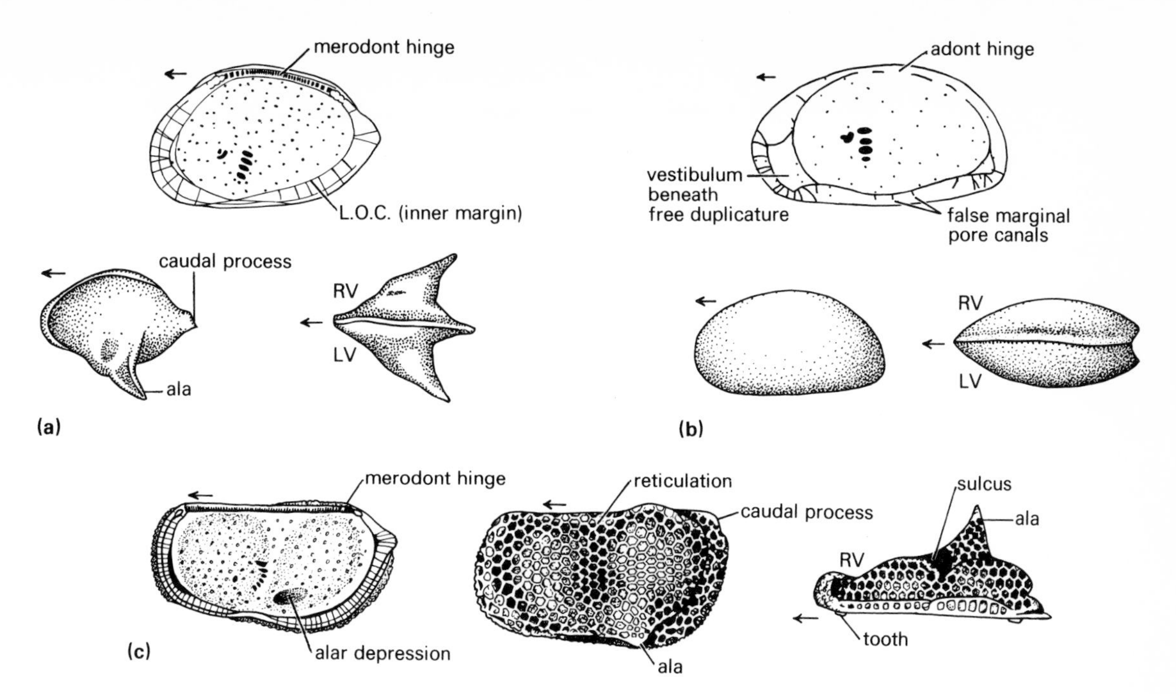

Figure 14.18 Superfamily Cytheracea (Suborder Podocopina, Order Podocopida). Deeper water forms. (a) *Cytheropteron*, interior RV ×47, exterior LV ×20, dorsal view ×20; (b) *Krithe*, interior RV ×50, exterior LV ×40, dorsal view ×40; (c) *Bythoceratina*, interior RV ×50, exterior LV ×50, dorsal view RV ×50. ((a) after van Morkhoven 1963 from Sars and after Pokorný 1958 from G. S. Brady; (d) after Benson *et al.* 1961 from G. S. Brady, Crosskey & Robertson; (c) after Benson *et al.* 1961 from Hornibrook)

Cypridinacea there is a prominent anterior beak (rostrum) overhanging a rostral incisure, the former pointing anteroventrally and the latter dorsally (Fig. 14.20a).

(b) The ventral margin is convex, occasionally furnished with a pronounced ventral spine or with ventral swellings. The duplicature is narrow.

(c) The anterior end of the Cypridinacea bears the rostrum and is higher than the more pointed posterior end. The valve of the Superfamily Entomozoacea bears a median, C-shaped **nuchal furrow** whose convex side points posteriorly (Fig. 14.19a). An anterodorsal swelling is present in some of these.

(d) Viewed from above, the broader end is posterior in many genera.

Although known from offshore facies in Palaeozoic rocks, Myodocopida are rare in Mesozoic and Cainozoic deposits, probably because later forms secreted thinner, amorphously calcified or chitinous carapaces. Two main suborders are distinguished by Benson *et al.* (1961): the Myodocopina and the Cladocopina.

The Suborder Myodocopina (Ord.–Rec.) are subdivided by Benson *et al.* (1961) into several living and extinct superfamilies: the Entomozoacea, Entomoconchacea, Thaumatocypridacea and the Cypridinacea. The Entomozoacea are a Palaeozoic group exemplified by the genus *Richteria* (Sil.–Perm., Fig. 14.19a). In this the oblong carapace bears concentric and longitudinal striae, transected by a median nuchal furrow.

The Entomoconchacea of the Devonian and Carboniferous were large subglobular ostracods, as for example in *Entomoconchus* (L. Carb., Fig. 14.19b), which has a **siphonal gape** at the posterior end and a distinctive set of muscle scars.

Although Recent species of *Thaumatocypris* (Thaumatocypridacea, M. Jur.–Rec., Fig. 14.19c) are thin, weakly calcified and pelagic, the fossil forms are thicker with a heavy ornament (Fig. 14.19d) and were presumably benthic, swimming for short distances only. Both kinds bear the characteristic anterior spines. Recent Cypridinacea

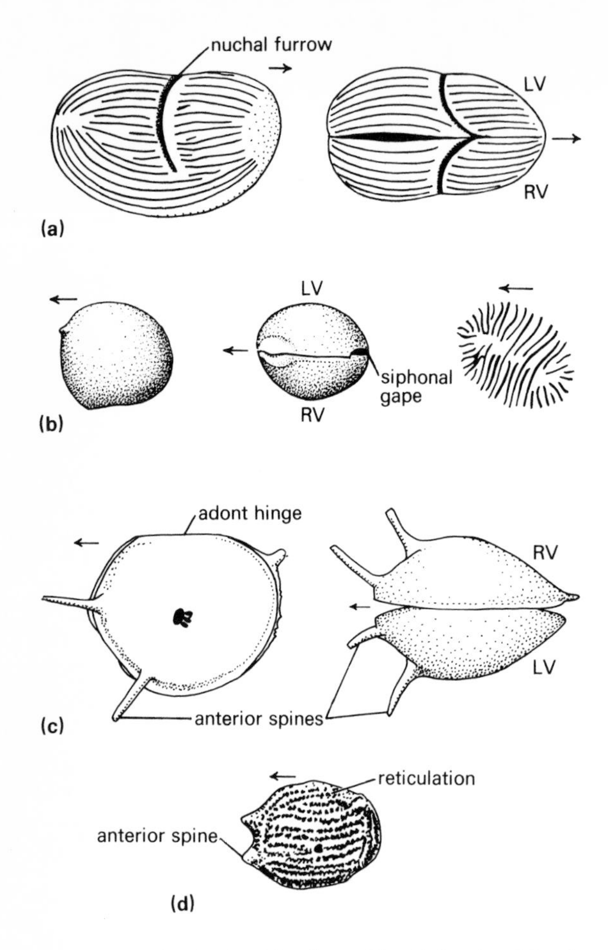

Figure 14.19 Order Myodocopida, Suborder Myodocopina. (a) *Richteria*, exterior RV × 10, dorsal view × 10; (b) *Entomoconchus*, exterior LV × 1, ventral view × 1, detail of muscle scars ×9; (c) *Thaumatocypris*, exterior LV of Recent species ×20, dorsal view of Recent species ×20; (d) exterior LV of Jurassic species × 16·5. ((a) after Benson *et al.* 1961 from Canavari; (b) after Benson *et al.* 1961 from Sylvester-Bradley; (c) after Benson *et al.* 1961 from Mueller; (d) after Pokorný 1958 from Triebel)

are pelagic filter feeders and carnivores, provided with two stalked, compound eyes and a median simple eye. Their carapaces can be recognised from Silurian times onwards by the prominent anterior rostrum and rostral incisure (e.g. *Cypridina*, U. Cret.–Rec., Fig. 14.20a).

The Suborder Cladocopina differ from the above in lacking eyes, heart and the second and third thoracic limbs. *Polycope* (?Dev., Jur.–Rec.; Fig. 14.20b) is the principle genus and typical of the group. Its carapace is very well rounded and lacks a rostrum whilst the inner surface bears the distinct cladocopine adductor muscle scars (three spots in the centre of each valve). *Polycope* is a weak swimmer and prefers to live interstitially in the substrate.

General history of ostracods

Ostracods of archaeocopid kind are among the first known fossil arthropods, having appeared alongside the trilobites in the early Cambrian. Unfortunately, like other invertebrates of this age, their origins are obscure. Their swimming, filter-feeding niche appears to have been taken over in the late Cambrian by the Leperditicopida. Early Ordovician times were marked by a major radiation of ostracods (Fig. 14.21), probably associated with a worldwide transgression and the expansion of available niches. At this time appeared the first Palaeocopida (both Beyrichicopina and Kloedenellocopina), the first Podocopida (i.e. Bairdiacea) and perhaps the first Myodocopida (i.e. Entomozoacea). The Ordovician proved in fact to be the heyday of the Palaeocopida, their generic diversity tending to dwindle from then until their apparent extinction in the Permian. Curiously, though, several genera of Recent Palaeocopid-like ostracods have been reported in dredge samples from the South Pacific (Hornibrook 1949).

Of the Palaeozoic ostracod fauna, that of the Devonian and early Carboniferous was the most varied, comprising at that time almost as many podocopid genera as in the Jurassic period and more fossil myodocopids than at any other time. It was also at this time that the first freshwater ostracods (i.e. the Cypridacea) appear to have evolved. The late Devonian, however, witnessed the extinction of the leperditicopids and numerous other early Palaeozoic genera, and although new taxa appeared after this, impoverishment continued until the Jurassic. By early Triassic times all palaeocopids appear to have become extinct and only a few podocopids (e.g. *Bairdia* and *Darwinula*) survived the Permo-Triassic boundary.

Triassic times saw the beginning of podocopid dominance amongst benthic assemblages, with the Podocopina now greatly exceeding the Metacopina in diversity. The Platycopina evolved from the dwindling metacopine stock at about this time. Many of the Jurassic Cytheracea are of restricted

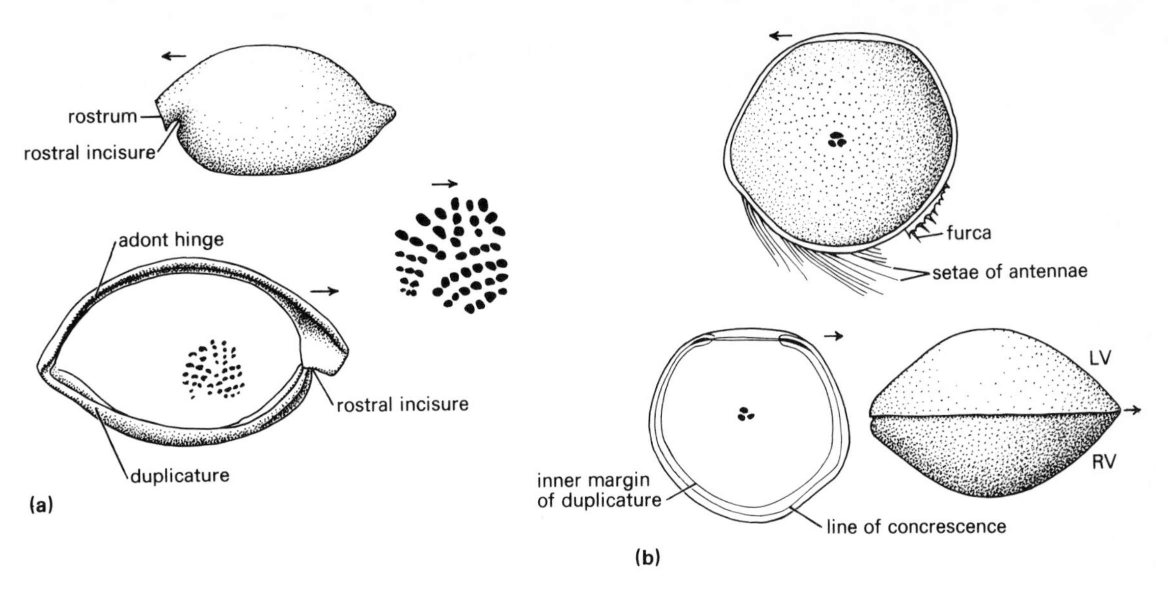

Figure 14.20 Suborders Myodocopina and Cladocopina (Order Myodocopida). (a) *Cypridina*, exterior LV ×20, interior LV ×33, detail of muscle scars ×73·5; (b) *Polycope*, exterior LV of living specimen ×47, interior LV ×47, dorsal view ×47. ((a) after Benson *et al.* 1961 from Mueller and after van Morkhoven 1963 from Keij; (b) after Pokorný 1958 and Benson *et al.* 1961 from Sars and from Sylvester-Bradley in Benson *et al.* 1961)

time range and useful for biostratigraphy (e.g. *Camptocythere*). It is also from these times that the value of ostracods as palaeoecological indicators becomes marked.

A diverse Cretaceous fauna of cytheraceans (many of them with amphidont hinges by late Cretaceous times) suffered a slight setback during the major facies changes that ushered in the Cainozoic Era. Since the Palaeocene, diversity of ostracod assemblages has tended to increase, although the very high numbers of Pleistocene to Recent genera (Fig. 14.21) also reflects the contribution of poorly calcified groups (e.g. myodocopids and cypridaceans) and the active interest of zoologists. The present deep-sea ostracod fauna is thought to have developed in late Cretaceous times (Benson &

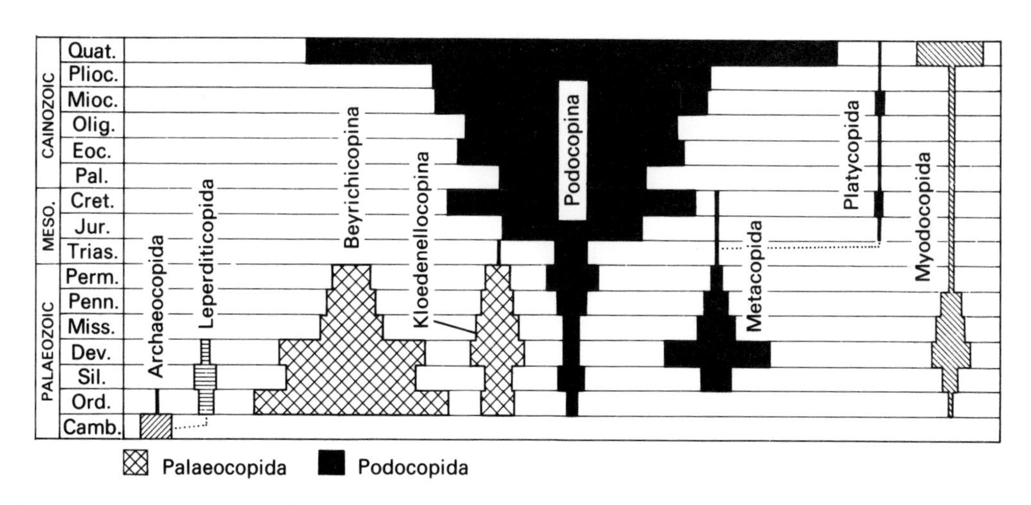

Figure 14.21 Diversity of ostracod taxa through time. Width of bars corresponds to the number of genera. (Data mainly from Benson *et al.* 1961)

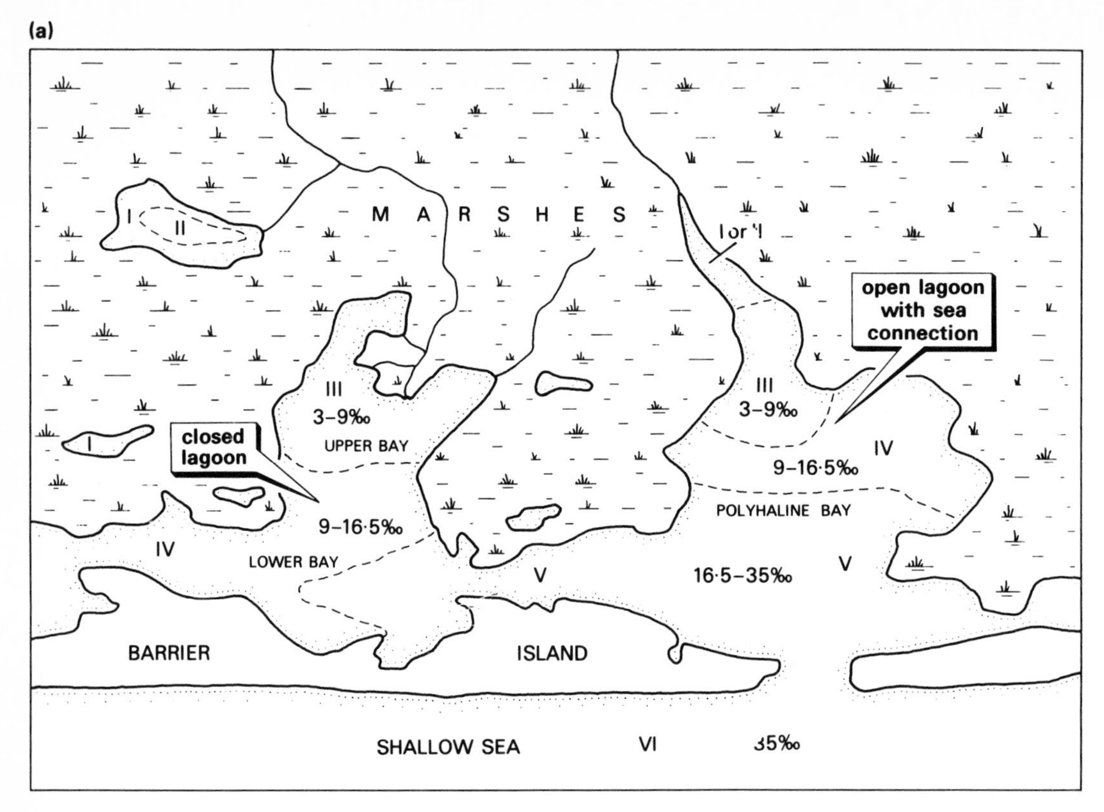

Figure 14.22 Ostracod palaeoecology in the Late Eocene of the Hampshire basin, England. (a) The environments as reconstructed from the ostracod fauna.

Sylvester-Bradley, pp. 63–91 *in* Oertli 1971) coinciding with a major phase of sea-floor spreading and the development of the Atlantic Ocean.

Applications of ostracods

The Ostracoda have considerable biostratigraphic utility, especially in Jurassic to Pleistocene rocks of shallow marine, brackish or freshwater origin. Their value here derives from their small size and relative abundance. Ostracod species, when properly defined, can also be seen to display negligible variation in form within a population and to have a relatively restricted vertical (time) range. Some examples of ostracod evolution can be traced, as for example those given by Bettenstaedt (1958) and Benson (1972). Such factors render ostracods useful as the basis for various kinds of biozones. Unfortunately, however, the restricted and often benthic niche of these creatures means that they are not geographically widespread and are therefore less suitable for long-distance correlations. In the Jurassic system, where many other microfossil assemblages are poorly developed, ostracods have proved particularly useful for correlation (e.g. Bate & Robinson 1978). Their value on a regional scale is well demonstrated in younger rocks by Glintzboeckel and Magné (1959) and by Keen (1972).

Because the majority of Recent ostracod genera are found in Miocene rocks and many have close relatives in Mesozoic assemblages, the principle of uniformity (uniformitarianism) can be used in palaeoecology. Even where the biological affinities of a fossil are uncertain, palaeoecology can be inferred from carapace morphology. Both lines of reasoning are best balanced against the evidence from studies of preservation (e.g. the valve to whole-carapace ratio), sedimentology and the associated flora and fauna. Fortunately, there are many excellent studies of Recent ostracod ecology and distribution. Particularly instructive here are the papers by Benson (1959), Neale (pp. 247–307 *in* Puri 1964), Puri *et al.* (pp. 87–199 *in* Puri 1964) and

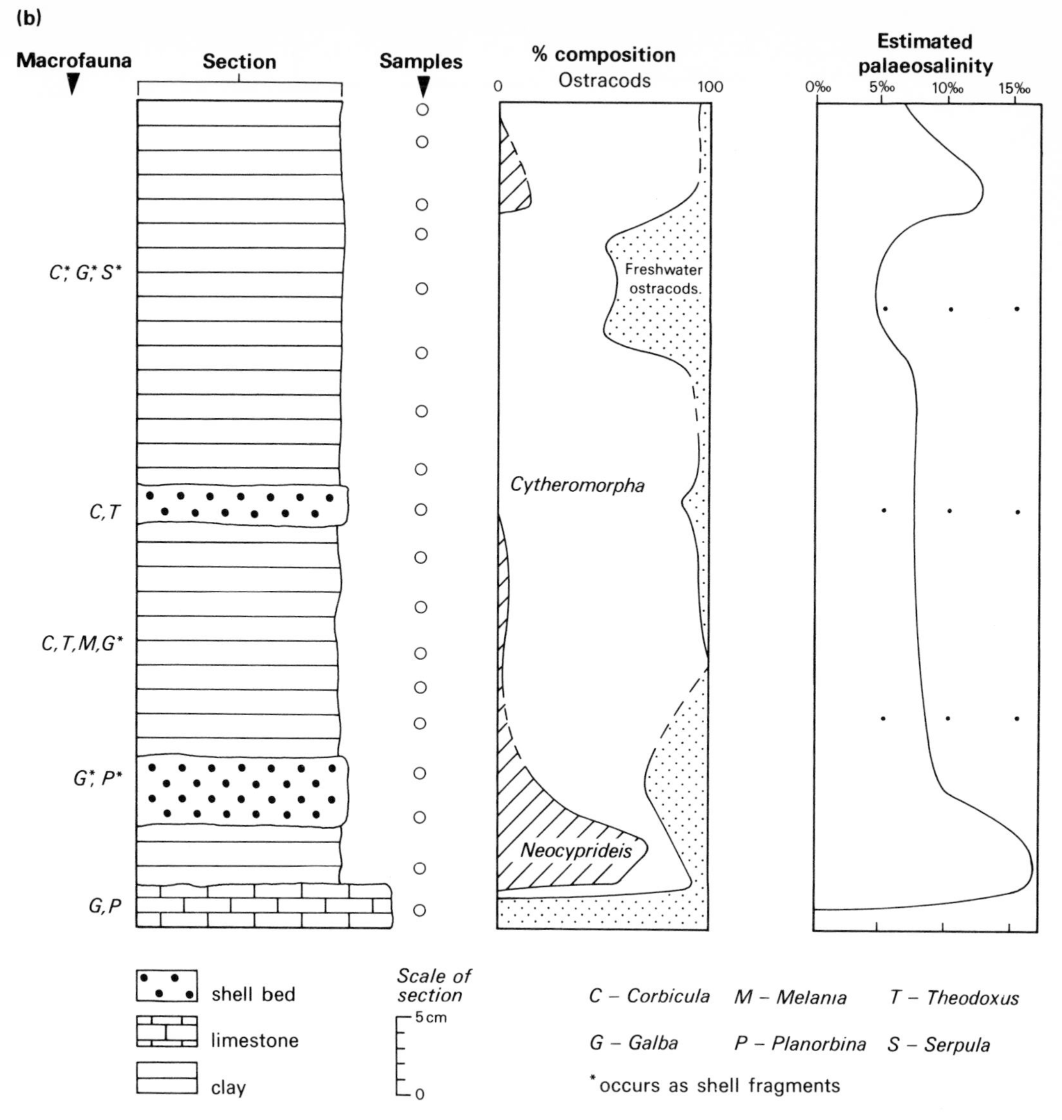

Figure 14.22 (b) Changes in the proportions of fresh- and brackish-water Ostracoda with inferred salinity changes, from part of the Lower Headon Beds. (Modified from Keen 1977)

Kilenyi (pp. 251–67 *in* Neale 1969). There are also reviews by Puri *in* Funnell and Riedel (1971, pp. 163–9 and 353–8).

Where the ecological parameters of living species are precisely known, the history of changes in rainfall, temperature, salinity and alkalinity recorded in Quaternary lake sediments, for example, can be charted (see Delorme, pp. 341–7 *in* Oertli 1971, and Cameron *et al.*, pp. 335–52 *in* Löffler & Danielpol 1977). Even the palaeoclimatology of Miocene through to Recent times can be studied in some marine sediments with the aid of ostracods (see Hazel, pp. 361–75 *in* Oertli 1971).

Ostracods are especially useful for outlining the nature of palaeosalinities and their fluctuations in marginal marine sequences such as those of the late Carboniferous (Pollard 1966), middle Jurassic (Bate 1965), late Jurassic to early Cretaceous (Anderson 1963, Kilenyi and Allen 1968) and the Cainozoic (Keen 1977; see Fig. 14.22). The remarkable history of the deep ocean basins, including the Mediterranean, is also clearly outlined by studies of ostracods, particularly the psychrospheric forms (see Benson & Sylvester-Bradley, pp. 63–91 *in* Oertli 1971; and Benson 1975).

Ducasse & Moyes (pp. 489–514 *in* Oertli 1971) demonstrate how ostracods can be employed to plot the changing position of shorelines in the Tertiary rocks of Aquitaine. The same symposium volume contains numerous fine examples of their value to palaeoecology, as for example in the Devonian of the Eifel region where lagoon, backreef, reef-core, fore-reef and offshore ostracod assemblages can be recognised, controlled largely by water turbulence (Becker, pp. 801–16 *in* Oertli 1971). As with other fossils the biogeography of fossil ostracods can be used to test palaeogeographic reconstructions (Donze, pp. 441–50 *in* Löffler & Danielpol 1977).

Finally, we may note the value of ostracods to sedimentology. Krutak (1972), for example, suggests that the ostracod valve length can be used to estimate the original grain size in recrystallised sedimentary rocks. Oertli (pp. 137–51 *in* Oertli 1971) outlines how ostracod valves can be used to gauge sedimentation rate, current strength and compaction in sedimentary rocks. The sometimes difficult distinction between pelagic and turbiditic sediments can also be alleviated by ostracods because they are more abundant and better preserved in the former (Dieci *et al.*, pp. 409–32 *in* Oertli 1971).

Further reading

The *Treatise* on Ostracoda by Benson *et al.* (1961) gives a relatively balanced and detailed review of the whole group, with taxonomy down to genus level. The two volumes on post-Palaeozoic ostracods by van Morkhoven (1962–3) are confined mainly to the Podocopida, with some discussion of the Myodocopida. However, on the former group, the first volume provides invaluable information on morphology and reviews their ecology, and the second volume has a key to the families and a well organised systematic section on the genera. For specific identification, palaeontologists should consult the *Catalogue of Ostracoda* by Ellis and Messina (1952 to date) or the journal *A stereo atlas of ostracod shells*, published by the British Micropalaeontological Society (Llandudno).

Bate and Robinson (1978) have recently brought together a useful volume on some stratigraphically important ostracods, ranging from Ordovician to Pleistocene, and Swain (1977) contains eight useful reviews of the stratigraphic distribution of fossil ostracods around the Atlantic basin and borderlands. Hartmann (1976) contains articles on the evolution of post-Palaeozoic ostracods. For papers dealing with ostracod ecology the reader is referred to the symposium volumes edited by Puri (1964), Neale (1969), Oertli (1971) and Löffler and Danielpol (1977), and for palaeoecology to the latter two volumes; the biology and palaeobiology are discussed in Swain *et al.* (1975).

Hints for collection and study

Living ostracods can be collected from seaweeds and surface scrapes from mud flats along with foraminifera (*q.v.*) or from freshwater ponds. The latter kind are readily cultivated in a tank provided with pondweed and a little manure. To examine their general behaviour, study the washed muds or pond water in a glass petri dish using reflected light. The morphology and the limb movements are better seen if a specimen is placed with a blob of water under a cover slip on glass cavity slide and viewed with transmitted light.

To extract ostracods from argillaceous rocks and marls, employ methods C to E (especially D; see Appendix). Method B is useful for hard chalks and limestones and method F where the carapace is phosphatic or silicified. Wash and dry the disaggregated sample as in methods G and I and mount as in method O.

Isolated ostracod valves should be examined in reflected light on both the internal and external surfaces. To see the muscle scars, pore canals and duplicature more clearly, place the specimen on a glass slide, cover with a drop of water (or glycerine, immersion oil or Canada Balsam) and a cover slip and view with transmitted light. Further suggestions for collection, preparation and study are given by Sohn *et al.* (pp. 75–89 *in* Kummel & Raup 1965).

15 Group Chitinozoa

The Chitinozoa are flask- or bottle-shaped, hollow organic vesicles of uncertain affinity. Appearing first in the Ordovician (or ?late Precambrian), they lived throughout the Palaeozoic Era, but the majority became extinct at the end of the Devonian Period. Chitinozoan walls are unusually resistant to oxidation, thermal alteration, tectonism, and recrystallisation of a $CaCO_3$ matrix. Indeed, Chitinozoans may be the only fossils recognisable in some rocks (such as slates) and from this derives their particular value to biostratigraphy.

The vesicle

The chitinozoan **vesicle** ranges from 30–1500 μm, but most are 150–300 μm long. The vesicle has a longitudinal axis of symmetry, sections taken at right angles to this being radially symmetrical. The wall (Figs 15.1, 15.2) is two-layered and of a dark brown or black chitin-like substance (**pseudochitin**). It encloses an empty **body chamber** that once housed the organism. The **oral** end, which bears the **aperture**, is usually produced into a neck, whilst the **aboral** end is broader and closed off. The aperture is occluded by a separate **operculum**, whose form and position is of taxonomic value.

The outer wall of the vesicle may be smooth, striate, tuberculate, hispid (i.e. hairy), folded into hollow spines or extended into a tubular **sleeve**. The inner wall can also give rise to spines that penetrate through the outer wall. Many chitinozoans are found united in long chains or clusters, the vesicles welded together at the operculum (i.e. the **oral pole**) and at the base (i.e. the **aboral pole**, Fig. 15.1a). In certain genera the operculum is deeply recessed within the neck so that adhesion of the adjacent vesicle must be achieved by a basal, tubular appendage called a **copula** (Fig. 15.2c).

Distribution and ecology of chitinozoans

Chitinozoa are known only from marine sediments, They are common in rocks deposited under well-aerated, shallow-water conditions, especially in shales and siltstones, but they are also found in limestones, dolomites, graptolitic shales, slates and cherts. The wide facies tolerance and distribution tends to suggest that Chitinozoa were either planktonic or attached to floating objects for at least part of their life cycle. Laufeld (1974) considered they were zooplanktonic with evidence for blooms in 'late autumn'.

Classification

Group Chitinozoa

Because Chitinozoa are invariably opaque with their internal structure obscured, classification has concentrated largely on form genera defined on their outline shape (i.e. silhouette). However, Jansonius (1967, 1970) amongst others, has argued for higher taxonomic subdivisions based on the form and position of the operculum as seen in specimens clarified by infra-red filters and by infra-red photography. He has divided the Chitinozoa into two subgroups: those with simple opercula ('Simplexoperculati') and those with more complex, recessed opercula ('Complexoperculati'). Each subgroup contains two families.

The Simplexoperculati are the older stock, known perhaps from late Precambrian times and certainly from early Ordovician times. In the more primitive Desmochitinidae, the operculum was a simple external lid. *Desmochitina* (?U. Precamb., Ord.–Sil.; Fig. 15.1a, b) had a relatively small subspherical vesicle with short **lips** but no neck and

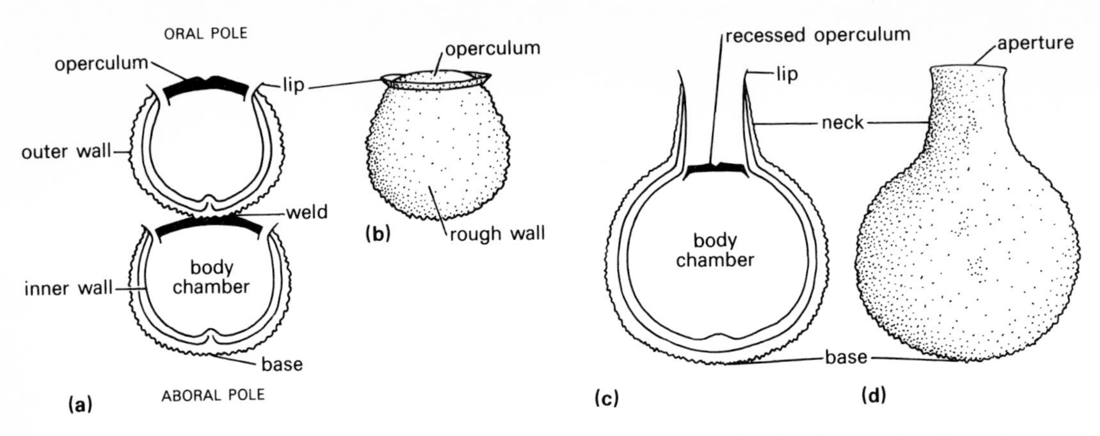

Figure 15.1 Simplexoperculate Chitinozoa (diagrammatic). (a) *Desmochitina*, longitudinal section through two welded vesicles; (b) *Desmochitina*, exterior view; (c) *Lagenochitina*, longitudinal section; (d) *Lagenochitina*, exterior view. (Based partly on Jansonius 1970)

was commonly united in chains. The operculum is recessed within the neck of the Conochitinidae, as in *Lagenochitina* (Ord.–Sil., Fig. 15.1c, d) which had a relatively large vesicle with a cylindrical neck.

The Complexoperculati bear a recessed operculum provided with a sleeve-like extension, the **flange**, together called the **prosome** (Fig. 15.2). This prosome is simple in the Sphaerochitinidae, whose vesicles lack aboral sleeves and copulae, as for example in *Ancyrochitina* (Ord.–Dev., Fig. 15.2a, b) which has a flask-shaped vesicle with a ring of spines around the base. The Tanuchitinidae, display elaborate differentiation at the aboral end and their vesicles are often tubular. *Velatachitina* (Ord.–Sil., Fig. 15.2c, d) is subcylindrical with a sleeve at either end, formed from the outer wall. The inner wall is produced aborally into a copula, whilst the prosome has an orally extended tube with ring-like markings (**annulations**).

Affinities of chitinozoans

The chitinoid wall of Chitinozoa suggests animal affinities but whether metazoan or protistan is still uncertain. Comparison may be made with the egg

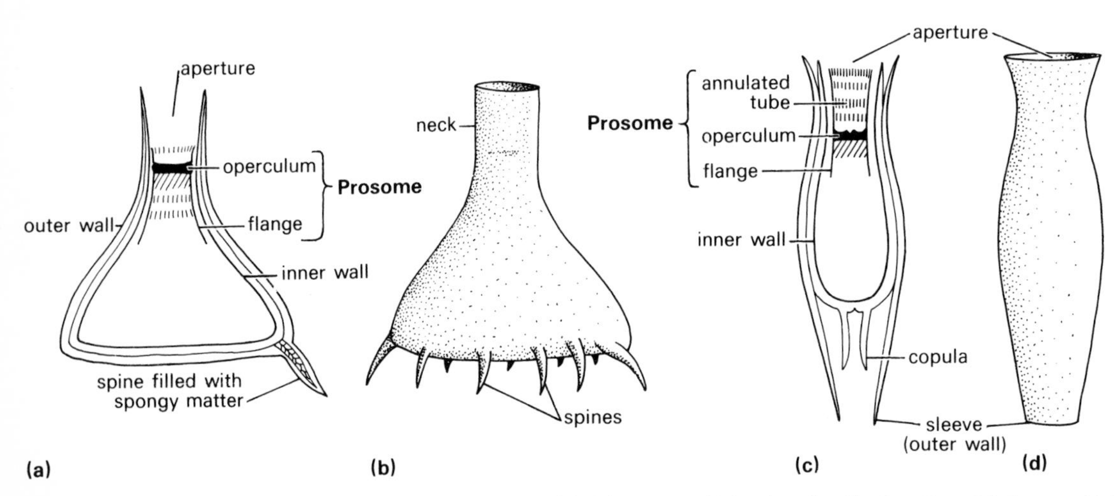

Figure 15.2 Complexoperculate Chitinozoa (diagrammatic). (a) *Ancyrochitina*, longitudinal section. (b) *Ancyrochitina*, exterior view. (c) *Velatachitina*, longitudinal section. (d) *Velatachitina*, exterior view. (Based partly on Jansonius 1970)

cases of worms and recent gastropods, an analogy that is further strengthened by discoveries of fossil cocoons filled with chitinozoan chains and clusters. It is unlikely that these represent the eggs of graptolites, as has been suggested (W. Jenkins 1970), because the geological ranges are dissimilar. Evidence of asexual reproduction (i.e. budding) and of secondary thickening of the wall (Cramer & Diez 1970) are more compatible with a protistan origin. Similar pseudochitinous shells are built by testacean and foraminiferid rhizopods and by ciliated protists like the tintinnids. The chains and clusters may also be compared with those formed by dinoflagellates and acritarchs. Laufeld (1974), however, believes they were the egg cases of some planktonic metazoan. Studies of growth and functional morphology should help to solve this problem of affinity.

General history of chitinozoans

The oldest chitinozoans may be the *Desmochitina*-like sacs from the upper Precambrian Chuar group of Arizona (*c.* 750 Ma BP, Bloeser *et al.* 1977), but this affinity has not yet been demonstrated conclusively. The group is unknown in Cambrian rocks but appears again in a major radiation of Ordovician (including Tremadocian) age. During the Ordovician, morphologically simple vesicles of desmochitinid and conochitinid type were characteristic, the latter stock gradually dwindling through the Silurian and dying out at the end of it. The more complex tanuchitinids also appeared in the Tremadocian, whilst the sphaerochitinids appeared in the late Ordovician and characterised the Silurian and Devonian assemblages. Stout basal horns tend to be typical of many of these Silurian and early Devonian species, although late Devonian forms are more often covered with short spines. Chains of Chitinozoa are also prominent in late Silurian and Devonian assemblages. Carboniferous chitinozoans are rare, following extinctions in middle and late Devonian times, but specimens have been reported from Permian sediments (Tasch 1973, p. 826).

Applications of chitinozoans

In the absence of other fossils, Chitinozoa can be useful for local stratigraphic correlations (see Taugourdeau & Jekhowsky 1960, Jodry & Campau 1961, Grignani 1967, Cramer 1973). Gradual evolutionary changes are detectable in some assemblages (W. Jenkins 1969) and may prove useful for wider correlation but there has been little use of international zonal schemes as yet. The occurrence of Chitinozoa in metamorphosed sediments may allow an approximation of the depositional age, such as the Ordovician age indicated for the Macduff Slates of the Scottish Highlands (Downie *et al.* 1971). Männil (1972) infers the existence of ecologically controlled assemblages and Zalusky (pp. 151–66 *in* Swain 1977) discusses the biogeographic provinces of fossil Chitinozoa, but these aspects require further study before they can be more widely applied.

Further reading

Useful introductions to the Chitinozoa may be found in Taugourdeau (1961), van Oyen and Calandra (1963), Pokorný (1963), Jansonius (1967, 1970), Evitt (pp. 468–71 *in* Tschudy & Scott 1969), Tasch (1973, pp. 821–33) and Jansonius and Jenkins (pp. 341–57 *in* Haq & Boersma 1978). Illustrations of most known species and information on their stratigraphic ranges are brought together by Taugourdeau (1967). Laufeld (1974) contains numerous excellent photographs of Silurian Chitinozoa.

Hints for collection and study

Fossil Chitinozoa may be extracted from Palaeozoic argillaceous rocks by the same techniques recommended for acritarchs and other organic walled microfossils (*q.v.*). Make permanent mounts from strews of the organic residue on glass slides. Mount with Caedax or Canada Balsam and view with transmitted light.

16 Group Conodontophorida

Conodonts are tiny phosphatic tooth-like structures that occur in marine rocks of late Precambrian to late Triassic age. Despite over 100 years of careful study, the affinity and function of these curious fossils has remained uncertain but their value to biostratigraphy continues to grow. This value derives from their small size, relative abundance, widespread distribution and their distinctive 'evolutionary' transformations. The durable phosphatic composition renders them fairly resistant to the rigours of diagenesis and tectonism and amenable to extraction from rocks with weak organic acids or by physical methods. Unfortunately, it also means that conodonts are readily reworked and transported into sediments of younger age.

The shape of single conodont elements

The **conodont element** may be from 0·1–5·0 mm long and of an amber, white, grey, brown or black colour. Four basic shapes of element can be distinguished: simple cones, bars, blades and platforms.

The **simple cone** comprises a curved, horn-like **cusp**, pointed at the oral end and expanded into a **base** at the **aboral** end (Fig. 16.1a). The **anterior** edge is convex whilst the **posterior** edge is concave. Careful examination of the aboral side reveals that the base bears a conical hollow termed the **basal cavity**. In a well preserved specimen, this cavity is lined by a second, inner cone called the **basal body**, itself provided with a **cone cavity**.

Bars and **blades**, together known as **ramiform** elements, are compound elements formed by the secretion of additional cusps, called **denticles**, borne on rounded bars or flattened blades upon either the posterior or anterior side of the **main cusp**, or on both sides (Fig. 16.1b). These posterior and anterior projections are termed **processes**, and they may be supplemented by lateral processes. Processes can be united aborally by a deep basal cavity or by grooves, and provided with a basal body. The main cusp of the oral surface is recognised by its large size and by its location above the deepest part of the basal cavity on the aboral surface.

In **platform** conodonts, the processes (especially the posterior one) are thickened into a basal ledge-like platform, sculptured on the oral side by ridges, tubercles and pits (Fig. 16.1c). A central row of

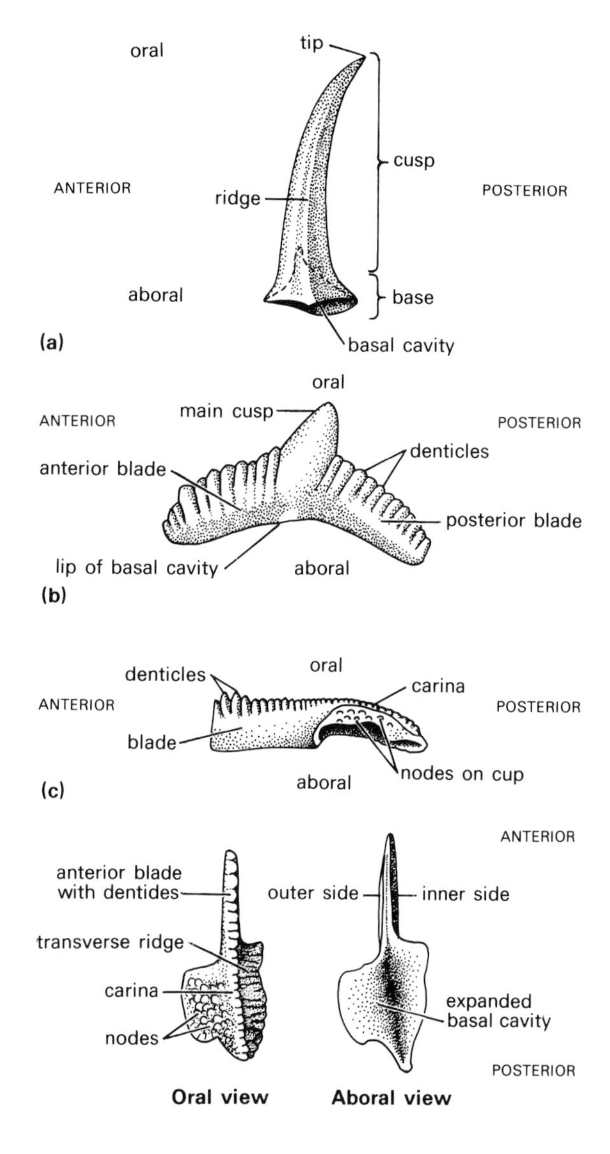

Figure 16.1 Terminology of the main kinds of conodont element. (a) *Distacodus*, a simple cone; (b) *Ozarkodina*, a compound blade; (c) *Gnathodus*, a platform element. (Based on Hass in Hass *et al.* 1962; (a) after Pander and (c) after Branson & Mehl)

denticles forms a keel-like **carina**, continuous with the denticles on the blade of the anterior process. Some platform elements may also bear a furrow on the oral side called a **trough** (e.g. Fig. 16.7f). The basal cavity on the aboral side may be large or almost non-existent; basal bodies, where present, can be relatively large (e.g. Fig. 16.7c).

Many conodonts are differently developed on either side of the denticles, with a convex **outer** side and a concave **inner** side (e.g. Fig. 16.7a). Recognition of 'right-hand' and 'left-hand' elements from this indicates that conodont elements were often formed in pairs.

Composition, ultrastructure and growth of elements

Conodont elements are built of tiny crystallites of calcium carbonate fluorapatite about 0·5 to 1·0 μm in diameter, with additional amounts of dark organic matter containing amino acids. The crystallites were laid down in layers called **lamellae** from 1–5 μm thick, with growth beginning at a **growth centre** and accreting by the addition of lamellae in various directions. Sometimes these lamellae did not quite touch over the growth axis, leaving **interlamellar spaces** (Fig. 16.2d).

Studies of conodont ultrastructure reveal three major modes of growth that have led Bengtson (1976) to define three basic groups: the protoconodonts, paraconodonts and euconodonts.

Protoconodonts are the most ancient group and comprise simple cones with deep basal cavities built of hyaline phosphates with abundant organic matter. Growth was initiated at the tip, with younger lamellae added by internal secretion in successive layers towards the base (Fig. 16.2a). There is no division into conodont proper and basal body.

Paraconodonts are mainly simple cones built of hyaline phosphates with somewhat less organic matter. A concentric growth centre just below the tip is added to by relatively few, thick lamellae, growing in a basal direction (Figs 16.2b, 16.5c). These lamellae may barely overlap on the oral and aboral edges, leaving a deep and often wide basal cavity. It is, however, possible that protoconodonts and paraconodonts are homologous with the basal bodies of euconodonts.

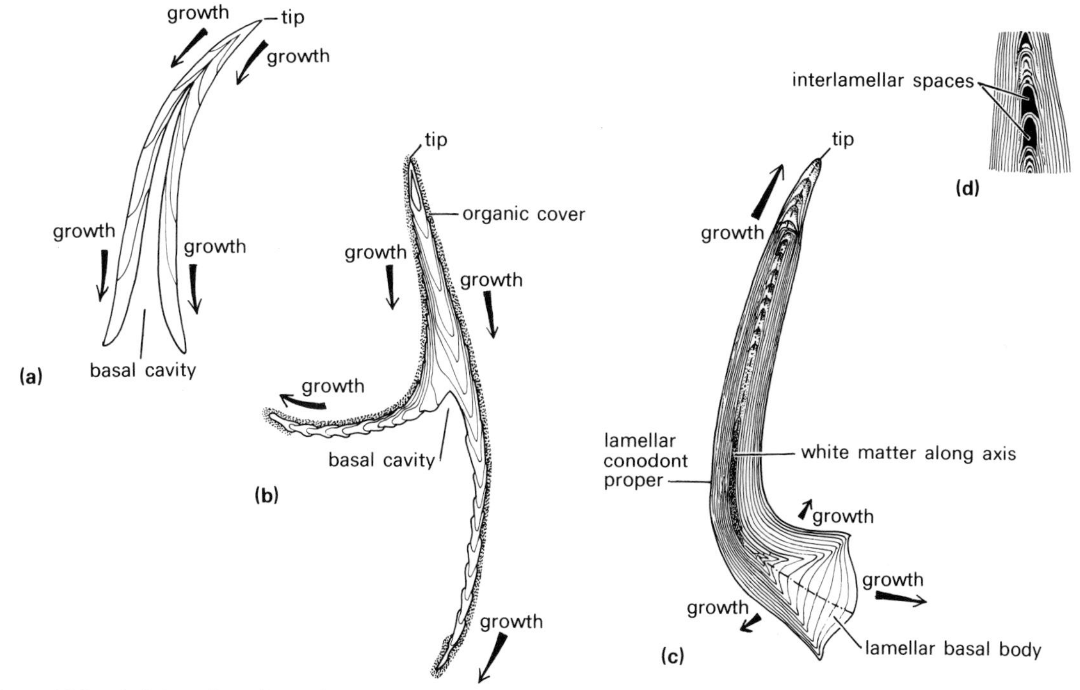

Figure 16.2 Axial sections through conodont elements showing lamellar growth and ultrastructure. (a) Protoconodont (*Hertzina*); (b) paraconodont (*Furnishina*); (c) euconodont (*Acontiodus*); (d) euconodont (*Polygnathus*). ((a) based on Bengtson 1976; (b), (c) and (d) on Müller & Nogami 1972)

Euconodonts consist of a conodont proper and a distinct basal body, although these may be fused together. Both are constructed of lamellae, but in the former they are added orally and laterally, and in the latter they are added aborally and laterally, the combined euconodont having overall concentric lamellar growth (Figs 16.2c, 16.7c). The lamellae of the basal body appear to have been layed down at the same time because they pass confluently into those of the conodont proper.

Although newly deposited lamellae of the euconodont proper and its basal body are clear and **hyaline**, older parts of the cusps and denticles may contain opaque **white matter** when viewed in reflected light (Fig. 16.2c). In ultrastructure this consists of tiny holes arranged in patches or rows that obliterate primary lamellae and hence are of secondary origin. Its presence in aborted and overgrown **germ denticles** indicates that white matter was, none the less, formed during life. This feature allows further subdivision of the euconodont elements, those with white matter being called **cancellate** and those with very little or without being called hyaline. A variety of hyaline conodonts, known as **fibrous** conodonts (or neurodonts) have fibrous, longitudinal fractures rather than clean transverse breaks as in the other kinds. This may be because the crystallites were never fused together, although they were certainly arranged in lamellae.

Symmetry transition series

Many conodonts show gradual morphological transitions from one plan to another. Members of these symmetry transition series, which may be either simple cones or ramiform elements, often receive different generic or specific names. Such transition series need have little to do with evolution, deriving instead from their varying position and function within the multi-element apparatus.

Conodont apparatuses and assemblages

Although most conodonts are found as discrete elements, numerous examples are also known of conodont **assemblages**, often comprising a combination of 12–22 elements from several form genera. A common and typical assemblage, for example, is one up to 9 mm long with a basket of *Hindeodella* or *Lonchodus* elements comprising four or five pairs arranged about an axis with bilateral symmetry (e.g. '*Scottognathus*', Fig. 16.3c). To these may be added one or two pairs of arched blades (*cf. Ozarkodina*), pick-shaped blades (*cf. Neoprioniodus*) and platform elements arranged about the axis. From such assemblages (representing the remains of a portion of the original conodont **apparatus**) the various types of apparatus can be reconstructed. They may also be reconstructed from analysis of the stratigraphic, geographic and relative frequency distributions of discrete conodont elements (Jeppson 1971).

At least four basic types of apparatus have so far been recognised in Palaeozoic rocks. These generally comprise various combinations of homologous platform elements ('P elements', or 'I' if icriodontan elements), ozarkodinian elements (O), neoprioniodontan elements (N) and a transition series of ramiform elements of hindeodellan or other types (A_1–A_3 or B_1–B_3; see Klapper & Philip 1971).

The conodont animal

The concentric mode of growth in euconodonts suggest that they were surrounded by soft, secretory tissues, i.e. they were internal structures. This contrasts with the superficially similar jaws of annelids ('scolecodonts'), fish denticles and gastropod radulae, which are all borne on top of the secretory tissues. However, it is possible that the protoconodonts and the paraconodonts were borne in this way. The chemical composition is unlikely to be a reliable indicator of biological affinity as a wide range of organisms have phosphatic skeletons (e.g. chordates, inarticulate brachiopods, certain arthropods and annelids, and many problematical late Precambrian to early Palaeozoic invertebrates).

Early ideas on conodont affinities are reviewed by Rhodes (1954), Hass *et al.* (1962) and Lindström (1964), with the general conclusion that the conodontophorid was a soft-bodied, bilaterally symmetrical and nektonic organism. Since then two important discoveries have brought to light some possible conodont animals.

Melton and Scott (pp. 31–65 *in* Rhodes 1973) discovered a number of compressed cigar-shaped creatures, 70 mm long and 15 mm wide, in the fish-

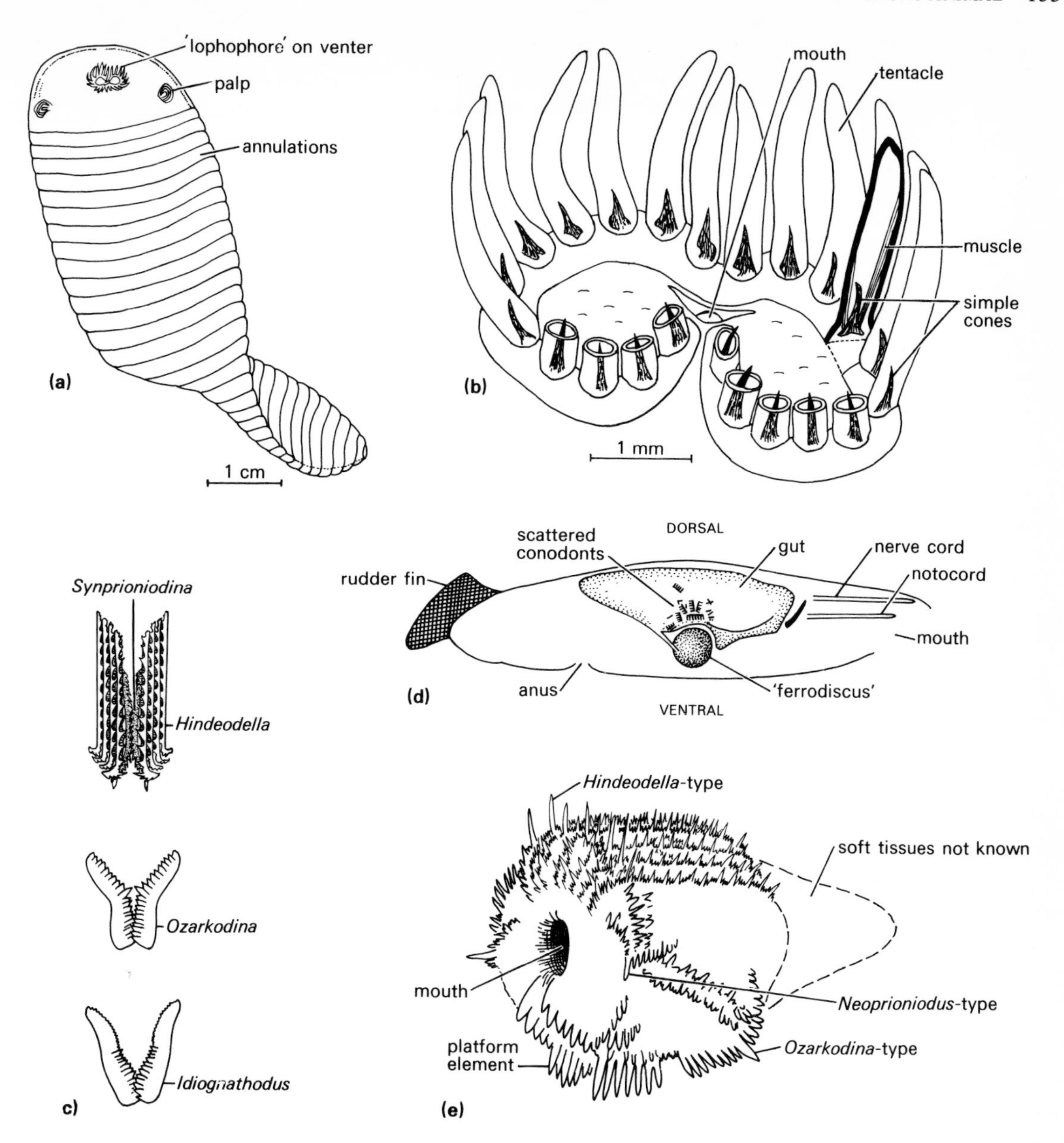

Figure 16.3 The conodont animal. (a) Reconstruction of *Odontogriphus omalus* and (b) reconstruction of its 'conodont'-supported lophophore; (c) natural assemblage of conodonts ('*Scottognathus*'); (d) conodont-bearing chordate animal ('*Lochreia*') from the Namurian Bear Gulch limestone; (e) hypothetical conodont animal. ((a) & (b) after Conway-Morris 1976; (c) after Hass in Hass *et al.* 1962 from Rhodes; (d) after Melton & Scott in Rhodes 1973; (e) after Lindström 1974)

bearing Bear Gulch limestone from the Namurian of Montana (Fig. 16.3d). Equipped with a mouth, gut, dorsal nerve cord, notocord, anus and rudder-fin, this bilaterally symmetrical animal seems to have had chordate affinities. Interest has centred on the presence of 'conodont assemblages' within the gut, interpreted by them as part of a food-filtering system. Unfortunately, however, these assemblages are incomplete, inconsistent and scattered, without any semblance of the familiar arrangement. Lindström (pp. 85–102 *in* Rhodes 1973 and Lindström 1974) therefore dismissed these fossils as conodont eaters, constructing instead a hypothetical conodont animal from studies of the natural assemblages (Fig. 16.3e). He concluded that conodonts were the internal supports of food-gathering tentacles, probably of a lophophore-bearing creature.

Lindström's reconstructions are perhaps enhanced by the discovery of *Odontogriphus*, a flattened worm-like animal, 60 mm long, from the middle Cambrian Burgess Shale of British Columbia (Fig. 16.3a, b; Conway-Morris 1976). The body was soft and gelatinous with annular constrictions and a distinct head, bearing a U-shaped structure interpreted as a lophophore. At the root of each of the 20–25 'tentacles' is a compressed 'simple cone' closely resembling some contemporaraneous conodonts in appearance (Fig. 16.3b), although the present composition does not appear to be phosphatic. *Odontogriphus* was apparently nektonic and suspension feeding, with the cones supporting the food-gathering tentacles of a lophophore like those of the Superphylum Lophophorata (comprising the Phyla Brachiopoda, Ectoprocta and Phoronidea). But while the latter can protect their apparatus by means of a shell or burrow, the nektonic *Odontogriphus* apparatus would have been vulnerable unless it was able to withdraw or introvert it within the head. Introversion best explains the compact, inward pointing elements seen in natural assemblages of younger compound euconodonts and there is an analogy here with the paired, tooth-like apparatus of hagfish (Priddle 1974), chaetognaths (Rietschel 1973) and annelids (Rhodes 1954). Folds of tissue protecting such a retracted apparatus may then have deposited lamellae on the oral side of the elements, while the basal body developed at the aboral attachment site (e.g. Bengtson 1976). Hitchings and Ramsay (1978) put forward a similar model in which euconodont elements filtered out all but the microplanktonic particles.

The altogether different ultrastructure and the claw-like assemblages of the protoconodonts prompted Landing (1977) to suggest that they were grasping rather than filtering organs.

Conodont distribution and ecology

The conodontophore was certainly marine, but whether nektonic, benthic or nekto-benthic has been the subject of some debate (e.g. Barnes & Fåhraeus 1975). They can be found in black shales deficient in benthic fossils, in cephalopod and agnostid limestones of pelagic origin and in phosphorites formed by upwelling currents and plankton blooms. Conodonts are most abundant as fossils in conditions such as the above where terrigenous supply was very slight, leading to condensed sequences with up to 23000 elements per kilogramme of rock. Although found with most kinds of Palaeozoic marine fossils, conodonts are especially associated with graptolites, radiolarians, fish remains, brachiopods, cephalopods, trilobites and palaeocopid ostracods. They are less frequent in coral, stromatoporoid, sponge, crinoid and algal beds and rarely found in fusulinid limestones.

Assemblages found in deeper-water sediments are somewhat less diverse than nearshore ones and contain a variety of platform elements that show little evidence of transition series. Such conodonts are rarely robust and often have small basal cavities and relatively constant proportions of white matter. Nearshore assemblages tend to consist of larger and more robust elements with bigger basal cavities, variable amounts of white matter and distinct symmetry transition series. Accessory denticles and processes may also occur in these elements.

Numerous facies-related conodont assemblages have been reported (see authors *in* Barnes 1976), especially the contrast between nearshore forms such as *Icriodus*, *Polygnathus*, *Cavusgnathus* and deeper-water forms such as *Palmatolepis*, *Ancyrodella* and *Idiognathodus*. It has been suggested that these depth-related biofacies are the product of nektonic, depth-stratified conodont communities (Druce, pp. 191–238 *in* Rhodes 1973, and Seddon & Sweet 1971; see Fig. 16.4). Many authors, however, note a relationship with lithofacies and a tendency for lateral displacement of one assemblage by another. Such features are more suggestive of benthic or nekto-benthic communities (Barnes & Fåhraeus 1975; Le Fevre *et al.* pp. 69–89 *in* Barnes 1976). Jeppson (pp. 105–18 *in* Barnes 1976) even produces evidence from which he infers that migrations of conodontophores took place during growth. It seems that the majority of conodont animals were mobile and benthic but capable of swimming, whilst the simple-cone-bearing creatures were probably nektonic.

Ideas on the feeding mode of the conodont

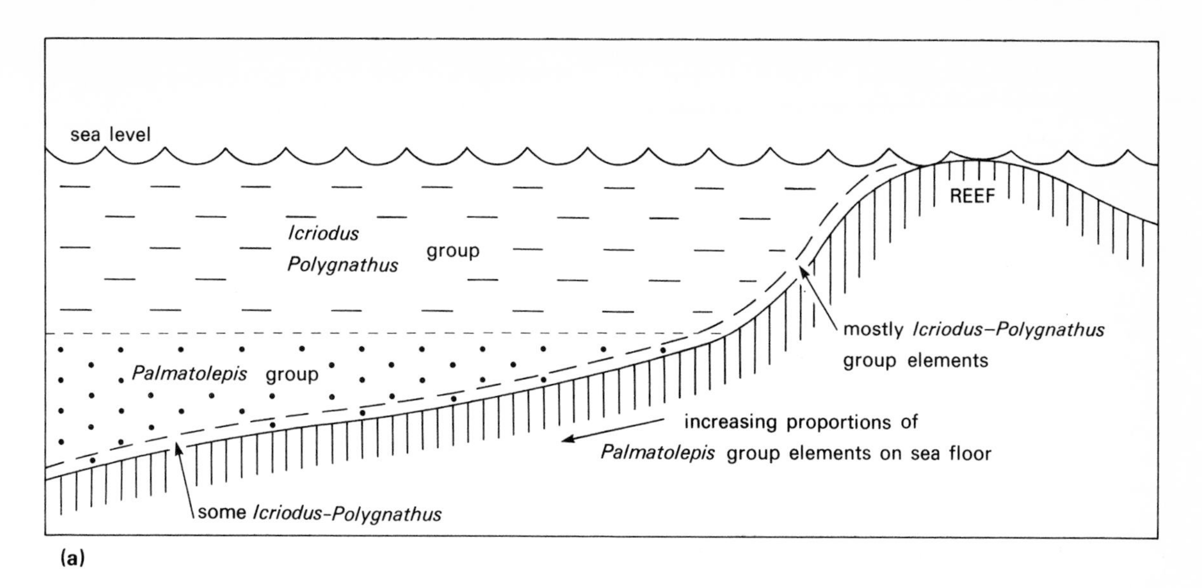

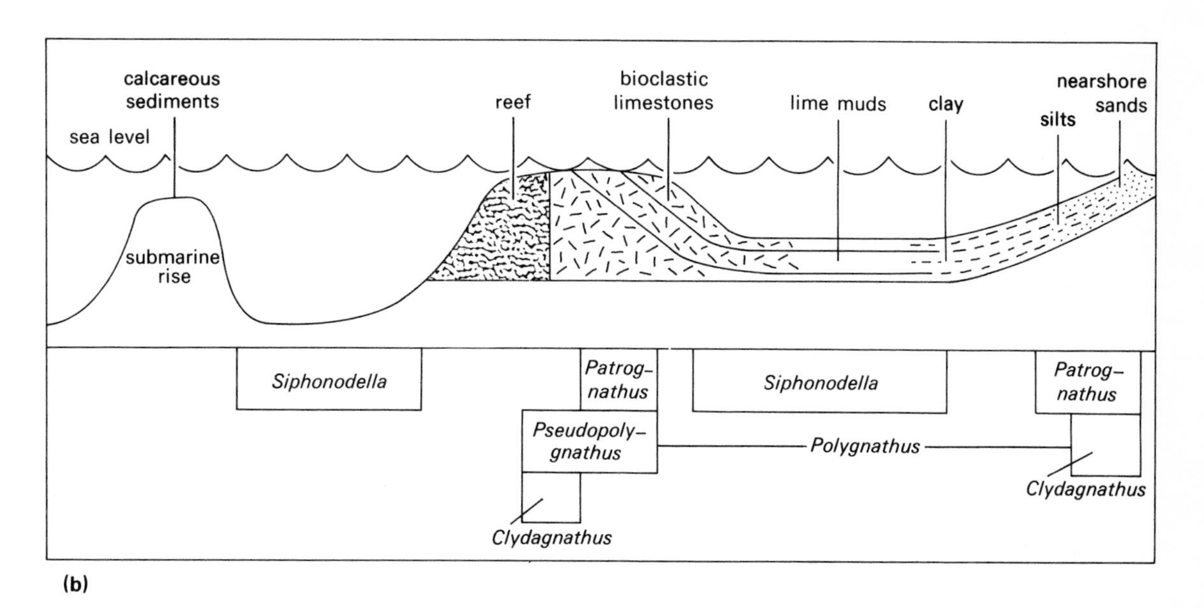

Figure 16.4 Depth and substrate-related distribution of platform conodonts in the late Palaeozoic seas. (a) Late Devonian (mainly after Seddon & Sweet 1971); (b) early Carboniferous (after Austin *in* Barnes 1976)

animal vary with the affinities preferred, from suspension feeding to predatory. Even so, the conodontophore was clearly no match for the similarly sized conodont-eating animal of Melton and Scott (pp. 31–65 *in* Rhodes 1973).

Classification

Group Conodontophorida

The classification of conodonts also poses some

problems. Each conodont element has been accorded a Linnaean binomial name on the basis of external shape. Unfortunately, the validity of many of these names is challenged by the evidence that different form genera and species can occur within one apparatus, as witnessed by the fossil assemblages and the fact that the same conodont form genus or species can occur in different assemblages. As the assemblage represents the true biological unit, it can only have one name, which must be that of the first described conodont element that it contains. All the other elements in that assemblage, no matter how distinct in appearance, must take that name. Unfortunately it was also customary to give these natural assemblages yet another, often invalid, new name, resulting in a 'dual nomenclature' – one utilitarian (for elements) and one biological (for assemblages). Some of these problems are discussed by Hass *et al.* (1962) and Lindström (1964, 1970).

The higher taxonomic units used by Hass *et al.* (1962) were based on external morphological features and geological ranges of single elements, so they are entirely artificial and biologically invalid. Those of Lindström (1970) include some data on internal structure, composition and natural assemblages, but have not yet been widely adopted. Druce *et al.*(1974) outline some familial and generic characters amongst Carboniferous euconodont assemblages that incorporate evolutionary information, and theirs therefore approaches a natural classification. The ultrastructure and growth studies of Barnes, Sass and Monroe (pp. 1–30 *in* Rhodes 1973), Müller and Nogami (1972) and Bengtson (1976) should eventually provide a basis for a supra-familial classification.

Protoconodonts. The protoconodonts (U. Precamb.–L. Ord.) are narrow, slightly curved simple cones with a deep basal cavity and aborally–internally deposited lamellae. For example, *Hertzina* (?U. Precamb., M.–U. Camb., Figs 16.2a, 16.5a) has a very thin wall with a flattened area on the posterior edge, bounded by two longitudinal carinae.

Paraconodonts. The paraconodonts (M. Camb.–Ord.) are dark brown to black elements in which the lamellae are added aborally, with little overlap. Most are pixie-cap shaped with a deep basal cavity, as in *Furnishina* (U. Camb., Figs 16.2b, 16.5b) which has a broad flat area on the anterior side of the cusp and a very flared base. *Westergaardodina* (M. Camb.–L. Ord., Fig. 16.5c) is unusual in having a small central cusp flanked by two larger, lateral denticles forming a W- or U-shaped element which contains lateral cavities.

Euconodonts. Euconodonts (L. Ord.–U. Trias) consist of basal body and a conodont proper with aboral–lateral and oral–lateral directions of lamellar growth, respectively. They may be arbitrarily divided into simple cones, bars, blades and platforms, but it must be remembered that this is not a biological distinction. All the following examples are cancellate unless otherwise stated.

SIMPLE CONE ELEMENTS. Simple cones have a single cusp without accessory denticles or processes and with a deep basal cavity. Deriving from paraconodonts, these basic euconodonts dominated many late Cambrian and Ordovician assemblages and gave rise to the more complex elements of later times. The genera are distinguished on ultrastructure, cusp cross section and profile, base cross section and profile (especially the shape of the oral edge and the aboral margin). The angle which the oral edge makes with the cusp may also be diagnostic. *Oneotodus* (U. Camb.–U. Dev., Fig. 16.5d) has a slender cusp with a circular cross section, whereas *Drepanodus* (L. Ord.–U. Dev., Fig. 16.5f) is more laterally compressed with sharper anterior and posterior edges that give to the cusp a lenticular cross section. *Oistodus* (L. Ord., Fig. 16.5h) differs from the latter in having a strongly expanded and reclined base and also in having a hyaline rather than cancellate ultrastructure. In *Distacodus* (L. Ord.–M. Sil., Figs 16.1a, 16.5e) there are carinae on either the anterior or lateral sides of the cusp, but in *Acontiodus* (L.–U. Ord., Figs 16.2c, 16.5g) these ridges flank the posterior edge.

COMPOUND ELEMENTS – BARS AND BLADES. Compound elements developed from simple cones by the addition of denticles along anterior, posterior and lateral processes. These 'evolutionary' trends were repeated in the growth of each element, so that studies of the initial stages of growth can be used to indicate the apparent phylogenetic relationships. In

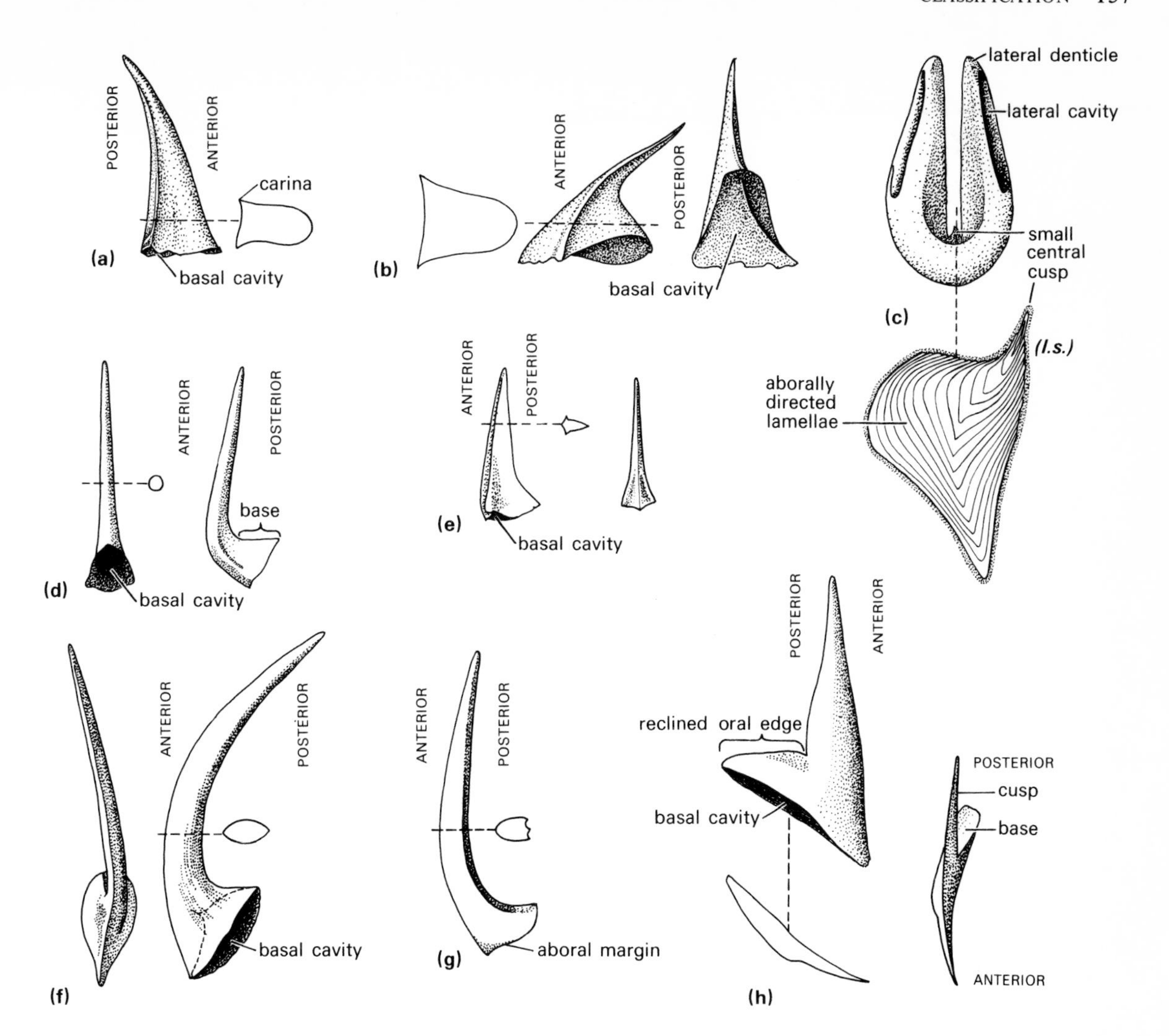

Figure 16.5 Protoconodonts, paraconodonts and simple cone euconodonts. (a) *Hertzina*, lateral view and cross section ×53; (b) *Furnishina*, cross section, lateral view and aboral view ×53; (c) *Westergaardodina*, above – lateral view ×37; below – section through central cusp and base ×80; (d) *Oneotodus*, aboral view, cross section and lateral views ×37; (e) *Distacodus*, lateral view, cross section and anterior views ×37; (f) *Drepanodus*, anterior view, lateral view and cross section ×37; (g) *Acontiodus*, lateral view and cross section ×37; (h) *Oistodus*, lateral, oral and aboral views ×37. ((a) & (b) after Müller 1962; (c) partly after Müller 1962 and after Müller & Nogami 1972; (d)–(h) after Lindström 1964)

a simple example, such as *Falodus* (L.–M. Ord., Fig. 16.6a), a denticle has been added to the anterior oral edge of an *Oistodus*-like cone. Conversely, in *Neoprioniodus* (M. Dev.–U. Carb., Fig. 16.6c) a series of denticles were added successively to the posterior oral edge, beginning to form a posterior process. More usually, growth proceeded unequally along two processes, as in *Prioniodina* (M. Sil.–M. Trias., Fig. 16.6b) in which the anterior bar is longer, and in *Hindeodella* (M. Ord.–M. Trias., Fig. 16.6f) in which the posterior bar is longer, supporting denticles of alternating sizes. *Ozarkodina* (M. Ord.–M. Trias., Figs. 16.1b, 16.6e) has blades of similar size and shape. *Hibbardella* (M. Ord.–M. Trias., Fig. 16.6d) with its two arched processes of equal length has an even stronger symmetry. *Chirognathus* (M. Ord., Fig. 16.6g) is not cancellate but fibrous and hyaline with long, thin and well separated denticles (especially on the anterior bar) and a wide, shallow basal cavity.

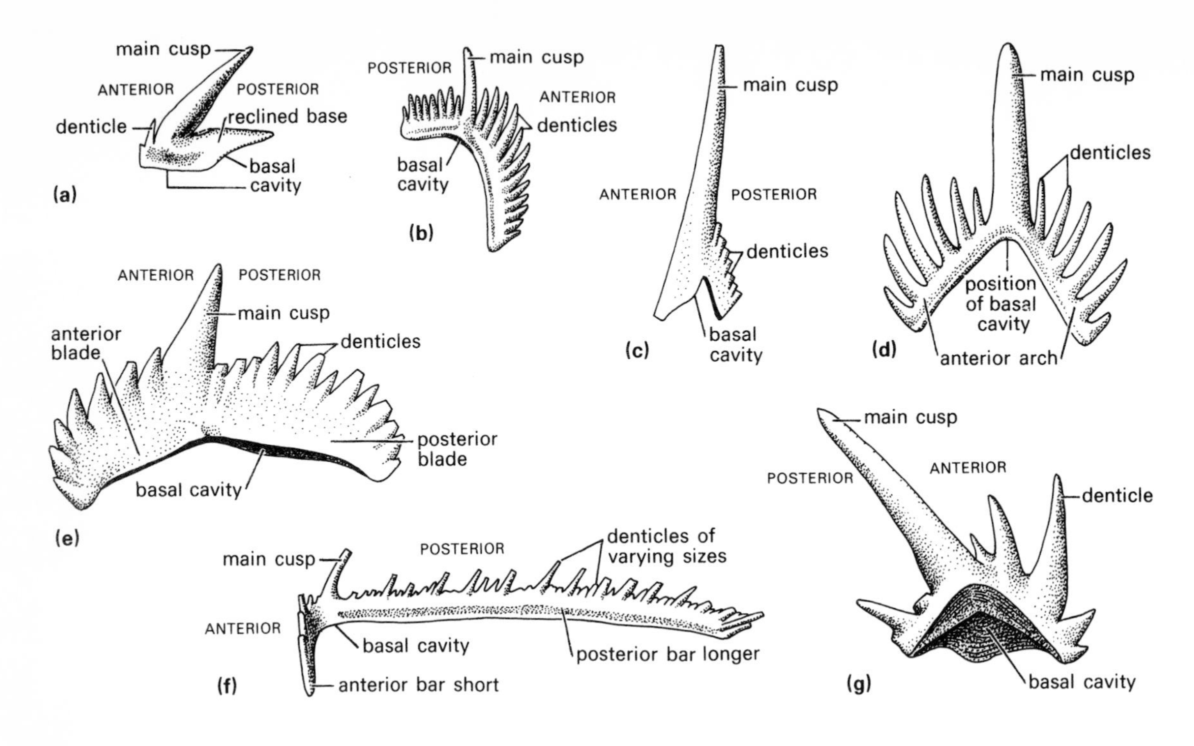

Figure 16.6 Compound euconodont element: bars and blades. (a) *Falodus*, lateral view ×37; (b) *Prioniodina*, lateral view ×17; (c) *Neoprioniodus*, lateral view ×37; (d) *Hibbardella*, anterior view ×17; (e) *Ozarkodina*, lateral view ×37; (f) *Hindeodella*, lateral view ×37; (g) *Chirognathus*, lateral view ×37. (After Lindström 1964 except (b) and (d) after Hass in Hass *et al.* 1962 from Ulrich and Bassler)

COMPOUND ELEMENTS – PLATFORMS. Platform conodonts were a further elaboration of the bar and blade model. They typify many Silurian to Triassic horizons, their variety and complexity making them useful as zonal fossils. A simple subdivision of the platforms by Hass *et al.* (1962) recognises two distinct kinds: 'Polygnathidae', in which the basal cavity is a tiny pit and the platforms develop from swollen lateral edges of the posterior process, and from lateral processes if present; and 'Idiognathodontidae', in which the basal cavity is large, with lips expanded into platforms that occupy much of the posterior end. The morphological nomenclature used here follows Lindström (1964), not Hass *et al.* (1962).

The denticles of polygnathids define a carina across the posterior platform which often continues into the anterior process as a denticulated blade. Lateral processes are also equipped with a carina and platforms. The basal cavity may extend as thin grooves along the aboral surface of the processes. For example, *Polygnathus* (L. Dev.–L. Carb., Fig. 16.7a) is leaf-shaped with platforms flanking most of the denticles, sculptured orally with tubercles and ridges. In *Amorphognathus* (M.–U. Ord., Fig. 16.7d) the virtually straight anterior and posterior processes bear lateral processes which may also branch. *Palmatolepis* (U. Dev., Fig. 16.7b, c) has a sinuous carina with an outer and an inner platform, the latter often lobate with a lateral process.

Idiognathodontids are generally leaf-shaped with an expanded basal cavity at the posterior end and a blade at the anterior end. The carina and blade are relatively straight and denticles are often weakly developed (e.g. *Gnathodus*, L.–U. Carb., Fig. 16.1c). In *Idiognathodus* (L.–U. Carb., Fig. 16.7e) denticles are replaced posteriorly by transverse ridges. *Cavusgnathus* (L. Carb.–L. Perm., Fig. 16.7f) also has a strong transverse sculpture in which the prominent denticles of the anterior blade pass into a lateral carina, flanked by a medially positioned second carina originating from the outer platform, with a

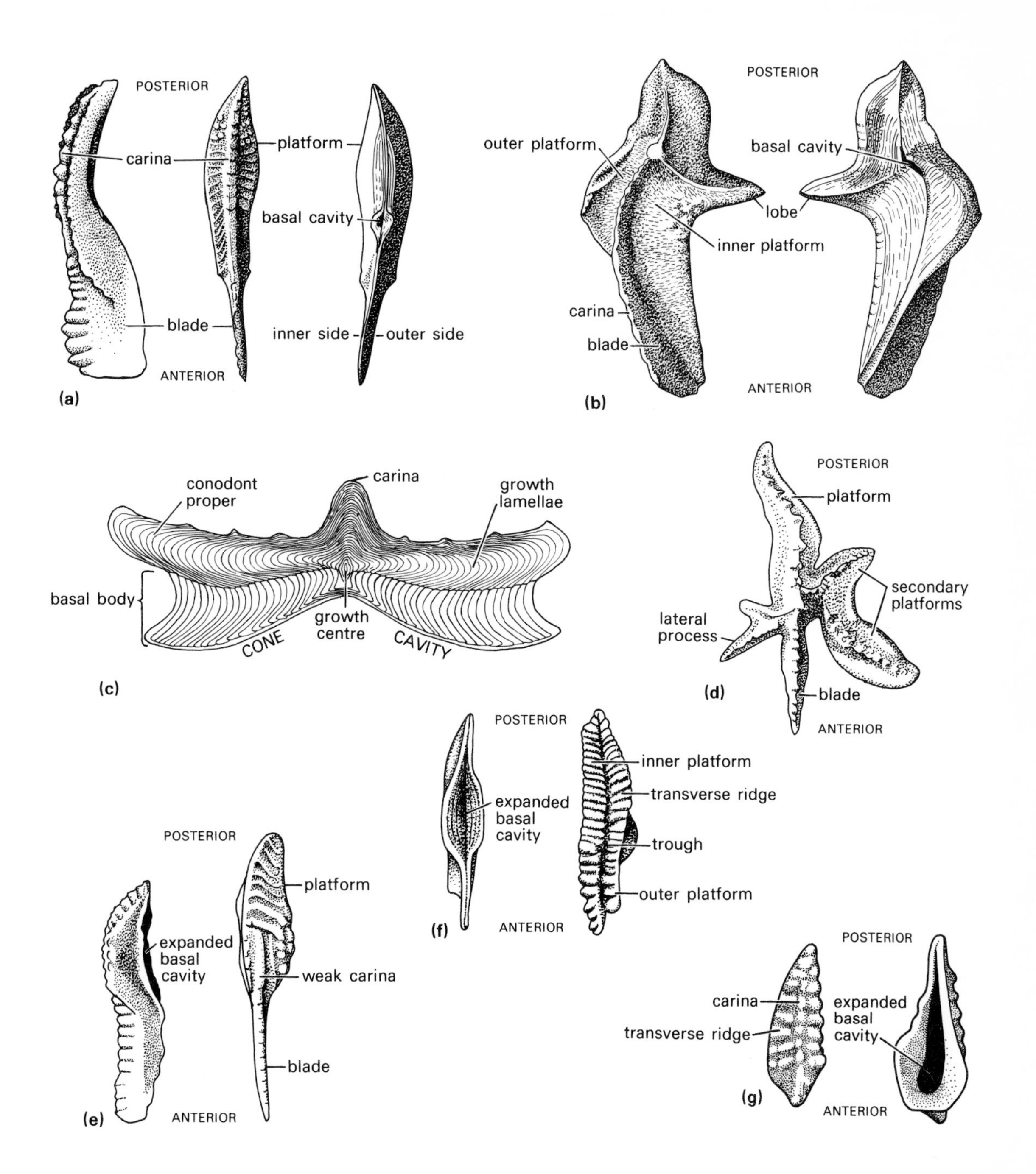

Figure 16.7 Compound euconodont elements: platforms. (a) *Polygnathus*, inner lateral, oral and aboral views ×37; (b) *Palmatolepis*, oral and aboral views ×37; (c) *Palmatolepis*, axial section through platform and basal body ×93; (d) *Amorphognathus*, oral view ×37; (e) *Idiognathodus*, inner lateral and oral views ×38; (f) *Cavusgnathus*, aboral and oral views ×30; (g) *Icriodus*, oral and aboral views (different specimens) ×25. ((a), (b), (d) & (e) after Lindström 1964; (c) after Lindström 1964 from Gross; (f) & (g) after Hass in Hass *et al.* 1962 from Ulrich & Bassler)

trough between. The large basal cavity is central in position. In *Icriodus* (L.–U. Dev., Fig. 16.7g) several rows of denticles also occur on the posterior process, whilst the anterior process is poorly developed. Curiously, the basal cavity has two apices.

General history of conodonts

Hertzina-like protoconodonts are known from rocks of probable late Precambrian age in Siberia (Matthews & Missarzhevsky 1975) and have been found in a variety of lower to upper Cambrian deposits. The paraconodonts probably developed from this early stock and both may be homologous with the basal body of the euconodonts and borne externally rather than internally (Bengtson 1976). Paraconodonts ranged from mid-Cambrian to early Ordovician times but were most typical of the late Cambrian, especially in pelagic trilobite limestones. Euconodonts also appeared at this time (Fig. 16.8) with an overall acme of element form genera in the middle of the Ordovician period. From this time onwards, element diversity tended to fall, except for a slight upswing in the late Devonian to early Carboniferous. Simple cones, bars and blades were at their most diverse in the Ordovician, especially in the mid-Ordovician when the fibrous hyaline kind also made a fleeting appearance. Platforms of polygnathid type are known from the Ordovician but may not be closely related to those which flourished in late Devonian to early Carboniferous times and provided such useful zone fossils. The idiognathodid type ranged from late Devonian to early Permian times, with an acme in the late Carboniferous (i.e. Pennsylvanian).

Curiously, the conodontophores survived the Permo–Triassic boundary extinctions without obvious injury, but became relatively small and scarce thereafter, eventually dying out during the Rhaetian age of the late Triassic. Other members of the nekto-benthos, such as the ammonoids also suffered extinctions, whilst novel phytoplankton elements (e.g. dinoflagellates and coccolithophores) began to diversify at this time. One suspects that an important and perhaps dramatic change in the pelagic environment connected with mid-oceanic rifting was the cause of this extinction. The reported Jurassic and Cretaceous conodonts are now thought to have been derived from older horizons.

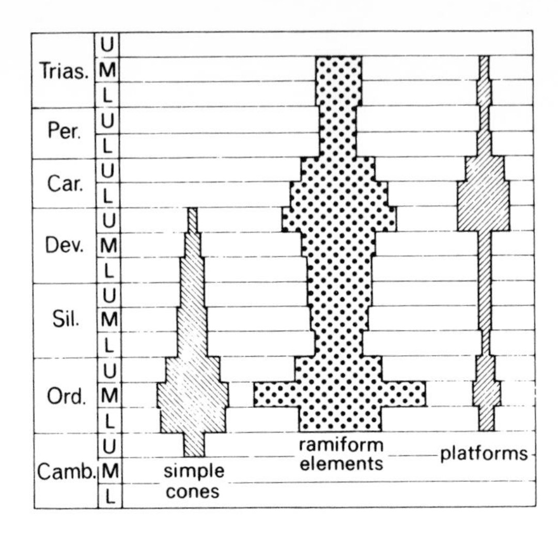

Figure 16.8 Diversity of euconodont form genera through time. Width of bars corresponds to the number of genera. L = Lower; M = Middle; U = Upper. (Largely based on data in Hass *et al.* 1962 and Lindström 1964)

Applications of conodonts

Despite the difficulties of interpreting the meaning of morphological changes and 'evolutionary' series in conodonts, there have been some detailed studies of these phenomena (Rhodes *et al.*, pp. 117–41; Austin, pp. 105–16 *in* Rhodes 1973; and Klapper & Johnson 1975). There are many examples of the value of conodont zones to stratigraphy, especially in the Ordovician to Permo-Trassic periods. Useful reviews are brought together by Sweet & Bergström (1971), Bergström (pp. 85–110) and Clark (pp. 111–33) *in* Swain (1977). The distinct 'European' and 'Midcontinent' conodont provinces of Ordovician times have often been discussed (see Barnes *et al.*, pp. 156–90 *in* Rhodes 1973, and Lindström 1974a) and lend some support to the hypothesis of seafloor spreading in the Palaeozoic. Studies of conodont palaeoecology (Barnes 1976) are of increasing value for estimation of depth of deposition and other environmental factors. An important paper by Epstein *et al.* (1977) shows that irreversible changes in the colour of conodonts, from pale amber to black, take place under the influence of temperature and time (i.e. in proportion to the depth and duration of burial). Such colour changes can now be used to assess the degree of chemical change that may have taken place in nearby hydrocarbon resources.

It has been suggested that the lamellae of conodonts may relate to astronomical events such as days, nights, tides or seasons (Tasch 1973, p. 801). If that were so they could also supplement those data obtained from corals and other invertebrates on the changing length of the Palaeozoic day, month and the year, but no such correlation has yet been demonstrated.

Further reading

The introductory sections to Hass *et al.* (1962) and Lindström (1964) are now rather outdated. Identification of specimens will, however, be aided by these works and by Ziegler (1973 to 1977). Some of the problems of conodont taxonomy are brought together by Lindström and Ziegler (1972). Palaeobiology and palaeoecology are discussed in the volumes edited by Rhodes (1973) and Barnes (1976), and biostratigraphic uses are reviewed in Sweet & Bergström (1971).

Hints for collection and study

Although conodonts may be retrieved from Palaeozoic mudstones and shales using methods C to E (see Appendix) they are more readily obtained from Palaeozoic limestones with method F. The residues should be washed over a 200-mesh sieve (method G) and concentrated with $C_2H_2Br_4$ (method K) or electromagnetically (method M). Pick and mount the conodonts as in method O. Conodonts are relatively less abundant than other microfossils so it may be necessary to prepare about 2 kg of limestone to yield twenty conodonts. Conodont extraction and concentration techniques are discussed by Lindström (1964) and by Collinson (pp. 94–102) *in* Kummel and Raup (1965).

Appendix – Reconnaissance methods

Before microfossils can be properly examined they must, of course, be collected from the rocks, prepared and then mounted. Each palaeontologist tends to have favourite methods for these procedures, some of which may be rather elaborate and tailored to particular needs or geared to laboratories with skilled technical assistance. There are, none the less, many simple, safe and inexpensive methods that can be used for more 'reconnaissance' investigations. To prepare your own material at all stages also has some advantages: it allows for greater flexibility and it increases the pleasure of discovery. Only these reconnaissance techniques will be dealt with here. Readers are referred to the chapters in Kummel and Raup (1965) for further information on techniques.

Sample collection

At surface outcrops either spot or channel sampling may be employed. **Spot** sampling consists of taking samples at predetermined stratigraphic levels. **Channel** samples are more continuous collections through longer stratigraphic intervals (say up to 3 m), which tend to blur the detailed story but avoid the risk of being totally barren. Subsurface outcrops may be sampled with various coring devices; simple manual ones such as the Dutch auger and the Hiller corer are suitable only for softer sediments. Retrieved cores can then be spot or channel sampled. More usually, though, the less expensive method of studying chippings brought up by subsurface drilling is followed. In studying drill chips care must be taken to avoid contamination from microfossils in the drilling mud and to take account of the phenomenon of 'caving' in which younger fragments have fallen downwards and become mixed with those of greater age. In this case the youngest rather than the oldest stratigraphic occurrences are more reliable for zoning.

When collecting from outcrops take care that the rock is unweathered and uncontaminated by recent vegetation or by hammers, chisels, trowels and the like. The sample bag (preferably polyethylene with a suitable closure device) should be absolutely clean inside. The amount of sample placed in the bag will depend, in part, on how much you can carry and store; it is always best to collect enough for resampling without having to return to the outcrop. For general reconnaissance studies about 500 g should be enough.

Labelling of samples is a very personal business. It is advisable, however, to note the sample number, rock unit, horizon above or below a known geological datum, locality (as accurately as possible), date and your name or initials. Some or all of this information can be placed on the outside of the sample bag and should also be placed on a card within the bag before sealing. At the same time the relevant data should be entered into the field note book.

Sample preparation

The sample must next be disaggregated to release the microfossils. Ideally this requires a small laboratory with an adequate sink (preferably with a sediment trap), Bunsen burners and tripods (or hotplates), an oven and, if possible, a fume cupboard. Other than the chemicals noted below, the laboratory will also need some heat-resistant bowls (e.g. stainless steel), evaporating dishes, petri dishes, pestle and mortar, glass measuring cylinders, glass beakers, filter funnels and filter papers, retort stands and clamps, flat-bottomed glass tubes, plastic buckets with lids and sets of sedimentological sieves, especially 325 (44 μm), 230 (63 μm), 200 (74 μm),

100 (149 μm), 80 (177 μm), 60 (250 μm), 40 (420 μm), 20 (840 μm) and 10 (2·0 mm) British or US Standard mesh sizes. If sieves are not available it is possible to make your own using good-quality nylon mesh fixed to a rigid frame (e.g. part of a plastic bowl, collander or bucket) with a strong glue or a clamp. The mesh size can then be measured with a graticule down a microscope or compared with the mesh of some standard sieves. The water taps should be fitted with moderate lengths of rubber tubing to facilitate wet sieving.

Many of the following sample processes require the rock to be broken into fragments. For small samples a pestle and mortar are sufficient, but larger and harder samples will require a rock splitter or a rock crusher. If these are not available, place the sample inside several clean polyethylene bags on a firm surface and strike with a hammer.

WARNING! Treat all chemicals and equipment with sober respect! Avoid inhaling and contact with the following chemicals, which should be kept stoppered in a safe place well away from naked flames, preferably in a fume cupboard: **acetone, alcohol, carbon tetrachloride, tetrabromoethane, paint cleaners/thinners,** and **all acids.**

Keep fingers out of an operating ultrasonic cleaner tank as they could suffer painful damage.

A. Pulverisation method. This simple and speedy technique may be used to extract coccoliths and organic-walled microfossils (e.g. spores, pollen, acritarchs) from well indurated rocks such as chalks and mudstones. It can also be used to liberate radiolarians and foraminifera with varying success.

(1) Place 5 to 20 g of fresh sample in a mortar. Add a few drops of distilled water and crush by pounding (not grinding) with a pestle until the largest fragments do not exceed 2 mm in diameter.
(2) Flush the sample into a jar or bottle with a jet of distilled water.
(3) If the sample is very argillaceous, the clay can be dispersed by placing the container in an ultrasonic cleaner device and letting it shake for two minutes to 2 hours, according to results. Note, however, that this can destroy some of the more delicate microfossil structures.
(4) Wash and concentrate as in methods G to L.

B. Scrubbing-brush method. This is another easy method useful for extracting calcareous and siliceous microfossils from partially indurated limestones (e.g. chalks and marly limestones), sandstones and shales.

(1) Fill a clean bucket with water.
(2) Take a fresh piece of the rock and scrub it under the water with a hard bristle toothbrush or a scrubbing brush. The action should be as gentle as possible to obtain a residue without damaging the microfossils.
(3) For larger calcareous and siliceous microfossils, strain the cloudy water through a 200-mesh sieve, flush out the residue into an evaporating dish, decant off the surplus water in the dish and dry the residue at a low temperature.
For organic walled microfossils, strain the water through a 325-mesh sieve and flush out the remaining residue into a glass bottle. Coccoliths can be obtained from the water flushed through these sieves if they are allowed to settle out for an hour or so.
(4) Concentrate the microfossils using methods H, J, K, L or M.

C. Solvent method. Partially indurated argillaceous rocks (excepting black and dark grey shales), marls and soft limestones can be disaggregated by this method.

(1) Break the fresh rock into fragments about 1 to 10 mm in diameter. The harder rocks will need smaller fragments with a greater surface area.
(2) Dry at a low temperature in an oven, remove and allow to cool.
(3) Pour on paint cleaning/thinning solvent (turpentine substitute) and allow to stand in a fume cupboard until the rock is saturated (usually from 30 minutes to 8 hours). **Handle with care.**
(4) Pour off the solvent. This can be collected for further use by straining through a filter paper.
(5) Pour on some hot water to cover the sample and allow to stand until the rock shows no

signs of breaking down further (usually from 5 to 30 minutes).

(6) If the disaggregation is only partial and further residue is required, repeat stages 2 to 5, or follow by method D.

(7) Wash and concentrate as in methods G to M.

D. Na_2CO_3 method. This washing-soda method is cheap, safe and effective with partially indurated argillaceous rocks, marls and soft limestones. However, it is not effective with black or dark grey shales, mudstones, chalks and porcellaneous limestones.

(1) Break the fresh rock into fragments about 1 to 10 mm in diameter. The harder rocks will require a smaller size of fragment.

(2) Place the rock fragments in a stable heat-resistant bowl or beaker, cover with water and add one or two large spoonfuls of Na_2CO_3.

(3) Set the liquid to boil and allow to simmer until the rock shows no further signs of breaking down. It is best to keep the water level topped up while boiling proceeds.

(4) Wash and concentrate as in methods G to M.

E. NaClO method. Ordinary domestic bleach (sodium hypochlorite) is a useful agent for disaggregating indurated carbonaceous black shales, mudstones, clays and coals for study of their organic-walled microfossils and conodonts. As it also bleaches dark organic tissues, making them clearer for microscopy, it can be used in conjunction with other methods. Disaggregation is relatively slow compared with the foregoing techniques, and a fume cupboard is advisable to reduce the smell of chlorine.

(1) Break the fresh rock into fragments from 1 to 10 mm in diameter and place in an evaporating dish, bowl or beaker.

(2) Cover with a 15–20% solution of NaClO in water and place a cover over it to prevent aerial contamination.

(3) Leave until a sufficient quantity of the rock has broken down (usually one day to several weeks). Top up the solution if evaporation occurs.

(4) Decant the supernatent liquid over a filter-lined funnel and cover the remaining residue with distilled water. Flush the filtrate with distilled water until no salt crystals are left. Remove the filter paper and flush this residue back into the evaporating dish.

(5) Wash and concentrate the residue as in methods G to M.

F. Formic acid method. Non-calcareous microfossils can be released from calcareous rocks by treatment with 10–15% commercial formic acid or non-glacial acetic acid. Although the former is more expensive, it has the advantage of being quick and suitable for both pure and argillaceous limestones and dolomites. The residues obtained in this way may contain conodonts, radiolarians, diatoms, organic-walled microfossils and archaeocopid ostracods. Silicified or phosphatised microfossils may also be liberated from limestones in this way.

(1) Break the rock into fragments about 10 to 30 mm in diameter.

(2) Place about 500 g in a 2 gallon (10 litre) plastic bucket and cover with about 1 litre of commercial grade formic acid (**Handle with care!**). Top this up to the 2 gallon mark with hot (*c.* 80°C) water. Smaller samples can be treated in smaller vessels with lesser quantities but the acid should again be diluted to about 10–15% and equal in volume more than two times the volume of the rock to be dissolved. Cover with a lid and place in a safe, well ventilated spot or in a fume cupboard.

(3) After 6 to 24 hours most of the limestones should have dissolved and the effervescence ceased. Any remaining rock fragments can be re-treated if necessary.

(4) Wash the sample as in method G. If microfossils smaller than 44 μm are to be studied, allow the sample to settle out through the spent acid and decant off the clear supernatent liquid. Filter the fine residue through a filter-lined funnel and then flush the filtrate into a stoppered bottle for further examination.

(5) Wash and concentrate as in methods G to M.

G. Washing and wet sieving. Once disaggregated,

many of the samples will now consist of muds (or silty and sandy muds). These clay minerals can obscure the microfossils. If they are removed by washing over a 325-, 230- or 200-mesh sieve, this procedure will also serve to concentrate some of the larger microfossils. Washing out the clay minerals, however, will cause the loss of coccoliths and the smallest spores, pollen, acritarchs and diatoms. The finest sieve size should therefore be chosen with care. For general purposes the following procedure may be followed.

(1) Put aside about 20 cc of wet sample in water if it is intended to study the smaller spores, pollen, microplankton and coccoliths (see method H).
(2) Wash the bulk of the sample gradually through a clean, fine mesh sieve with a gentle jet of water from the tap; (for diatoms, spores and pollen, use distilled water). A 300-mesh sieve will retain many small diatoms and organic-walled microfossils whilst a 230- or 200-mesh sieve retains most of the smaller radiolarians, foraminifera and conodonts and a 60-mesh sieve the ostracods. If there are many shells or clasts greater than 2 mm in diameter, place a 20-mesh sieve over the top to retrieve the coarser material. If the clays are difficult to disperse from the residue, it may be necessary to boil the sample with Na_2CO_3 (method D), but it is quicker to place the sample in a beaker of water within an ultrasonic cleaner device for several minutes.
(3) Flush the residue into an evaporating dish with a jet of distilled water. For coccoliths, diatoms, radiolarians, silicoflagellates and organic-walled microfossils, place this residue in a clean, stoppered bottle with distilled water. For foraminifera, radiolarians, silicoflagellates, tintinnids, conodonts and ostracods, decant off the supernatent liquid and set residue to dry at a low temperature in an oven.
(4) Sort and concentrate as in methods H to L.

Sorting and concentration

The discovery and analysis of microfossils is greatly speeded by sorting them into size classes and separating them according to their specific gravities. A variety of methods can be employed here.

H. Decanting. Decanting is almost as quick as centrifugation and can be done with a minimum of facilities. It is especially suitable for sorting and concentrating the coccoliths and organic-walled microfossils.

(1) Place six clean, flat-bottomed glass tubes in a row on a stable surface. Before each place six clean glass slides and have ready glass cover slips, distilled water, a packet of drinking straws and a watch or a clock with a second hand.
(2) Take the bottle of sample 'fines' in water (which have been pulverised, scrubbed or otherwise disaggregated and are either washed or unwashed), swirl them around gently to place the fine material in suspension and then decant into tube 1.
(3) Allow to settle for 30 seconds and then carefully decant the supernatent liquid into tube 2.
(4) Allow to settle for 60 seconds and then decant the supernatent liquid into tube 3.
(5) Allow this to settle for 2 minutes and then decant the supernatent liquid into tube 4.
(6) Allow to settle for 5 minutes and then decant the supernatent liquid into tube 5.
(7) Allow to settle for 10 minutes and then decant the supernatent liquid into tube 6.
(8) Allow to settle for 20 minutes and then decant the supernatent liquid back into the original bottle.
(9) As the decanting proceeds, spare moments can be used to make temporary mounts on the glass slides. With a clean drinking straw, draw up a little of the residue from each tube, drop some on to the glass slide and cover with a cover slip.
(10) To make permanent mounts, remove the cover slip (as if opening the lid of a hinged box) and allow the slide and residue to dry at a low temperature away from possible sources of contamination. Place a drop of mounting medium (e.g. Canada Balsam) on a clean cover slip and drop this over the residue. Allow to dry before examining with transmitted light.

I. Dry sieving. Calcareous, siliceous and phosphatic microfossils can vary greatly in size so it is useful to sieve the dried residue into various fractions using, for example, 200, 100, 80, 40 and 20 mesh sizes. Each fraction should then be placed in a bag or bottle with a label bearing relevant data and examined separately (see method O).

Separation by heavy liquids. Microfossils in washed residues can be concentrated by treatment with a variety of heavy liquids. **Care** should be taken to avoid both breathing and touching these liquids as they are toxic. A fume cupboard and careful preparation of the equipment are therefore necessary.

J. Carbon tetrachloride (CCl_4; SG 1·58). This can be used to concentrate buoyant foraminifera, radiolarians and diatoms. Unfortunately it does not concentrate thick-shelled, infilled or fragmentary microfossils or ostracods.

(1) Place the washed and dried sample residue in a beaker.
(2) Add two or three times this volume of CCl_4 and stir vigorously with a clean glass rod or a disposable tooth pick. The lighter microfossils mentioned above will float to the surface if present. (**Handle with care!**)
(3) Pour this 'float' into a filter-lined funnel arranged over a collecting vessel.
(4) Stir the sample again (adding more CCl_4 if necessary) and pour off the float, repeating the process until none is left in the beaker.
(5) Allow the filter paper and residue to dry until odourless. Filter off the remaining sediment and allow this to dry until this is also odourless. Return the filtered CCl_4 to the bottle.
(6) Put the float and remaining sediment in separate containers for subsequent microscopic study (method O).

K. Carbon tetrachloride (second method). To concentrate organic-walled microfossils, the following variant of the above method may be employed.

(1) Filter the water from the sample through a filter-lined funnel or through a fine meshed sieve.
(2) Remove the $CaCO_3$ from the residue with formic or acetic acid (see method F).
(3) Rinse the sample with distilled water through a filter-lined funnel or a fine-meshed sieve.
(4) Flush the filtrate off the filter paper or sieve into a beaker using a fine jet of acetone (**Care!**). Decant the acetone plus water into a filter-lined funnel and repeat until all the water has been removed. Flush the filtrate back into the beaker again with as little acetone as possible and let this evaporate off in a fume cupboard.
(5) Pour some CCl_4 into the beaker and stir gently until the residue is well dispersed. Cover the beaker and allow it to stand for at least 2 hours. **Handle with care!**
(6) Decant off the float (bearing the organic residue) into a filter-lined funnel over a beaker. When the CCl_4 has filtered through, flush the light residue with a little jet of acetone into a bottle of distilled water. Allow the acetone to evaporate off.
(7) Prepare temporary and permanent mounts as in method H.

L. Tetrabromoethane ($C_2H_2Br_4$; SG 2·96). This can be used to concentrate conodonts as well as calcareous and siliceous microfossils, the former separating in the heavy fraction and the latter two in the lighter fraction.

(1) Prepare a retort stand with brackets to hold two filter funnels, one a short distance above the other. Place a clean beaker below the lower funnel, which should be lined with a fast, strong quality filter paper. The top funnel must be fitted with a short length of tubing (preferably clear plastic) attached at the bottom, with a hose clamp over it to control the flow.
(2) To reduce the specific gravity of tetrabromoethane to about 2·75, pour some into the top funnel and then place a large piece of calcite (e.g. a calcite rhomb) in it. Dilute the liquid with a little CCl_4 or absolute alcohol until the calcite begins to sink slightly. Now top up the solution with a little more $C_2H_2Br_4$ until the calcite begins to float again. **Handle these liquids with care.**

(3) Remove the calcite and pour in the washed and dried sample residue. Stir gently until it is evenly dispersed within the funnel. Allow the sample to separate out (which may take only 5 minutes if diluted, as above, but up to 15 hours if undiluted).
(4) Release the hose clamp and allow the heavy residue (with the conodonts) to drain into the filter-lined funnel beneath. The heavy liquid that filters into the beaker below can now be put aside in a bottle for further use.
(5) Release the hose clamp again to drain off and filter the light fraction (with foraminifera, ostracods, radiolarians and organic-walled microfossils). Retrieve the heavy liquid as before.
(6) Partially fill three beakers with commercial acetone. Gather up each filter paper (with residue) and place it successively in the beakers for about 2 minutes each. This will remove more of the heavy liquid.
(7) Spread the filter papers out to dry in the fume cupboard and let them dry until odourless. To retrieve the $C_2H_2Br_4$ from the acetone, place the beakers at the back of the fume cupboard and allow them to evaporate off the acetone until a calcite rhomb will again float freely on the liquid.

M. Electromagnetic separation. If available this apparatus can be used to concentrate various kinds of microfossils, especially conodonts. Some hints on its use are given by Dow (pp. 263–7 *in* Kummel & Raup 1965).

N. Stained acetate peels. Well indurated limestones are generally difficult to disaggregate so that their microfossils (e.g. larger foraminifera, calcareous algae, radiolarians) are often studied in petrographic thin sections. A quicker method that destroys less material is the acetate peel technique. This takes a detailed impression of an acid-etched surface, but although it gives a clear indication of the gross morphology of a microfossil, the ultrastructure is usually better seen in proper thin sections.

(1) Cut the limestone vertically into slabs with a rock saw, each slab about 10 mm thick.
(2) Polish the faces with successively finer grades of corundum powder and water on a glass plate until the surface is quite smooth.
(3) Rinse the limestone slabs with distilled water and allow to dry. **Do not touch these polished faces.**
(4) Prepare the following solutions; a) 0·2 g of Alizarin Red dissolved in 100 ml of 1·5% HCl; b) 2·0 g of potassium ferricyanide dissolved in 100 ml of 1·5% HCl. Store these two separately, and **handle with care.**
(5) Mix the above together in the ratio a:b = 3:2 just before use. Pour into a stable receptacle wide enough to allow the submersion of one polished face of limestone.
(6) Immerse one face of the limestone in the solution, agitating it slightly whilst holding between finger and thumb. Etching may take from 15 to 60 seconds according to the age and induration of the limestone. At this stage calcite and aragonite should stain pink to red, ferroan calcite stain royal blue or mauve and dolomite remain unstained. Ferroan dolomite will stain a pale to deep turquoise blue.
(7) Wash the stained and etched slab gently in a dish of distilled water and allow to dry. Drying can be hastened by flushing with acetone. **Take care!**
(8) The following stage will require a well ventilated place, preferably a fume cupboard. Place the etched and stained limestone slab face uppermost on a sheet of rubber foam underlain by a tray to collect residual liquid. Make sure this face is horizontal. Also have ready some pre-cut strips of clean and transparent cellulose acetate sheeting. This can be bought in rolls of varying thicknesses but it should be both flexible and strong enough not to tear during handling.
(9) Squirt a thin layer of acetone over the whole of the etched face; the first coating usually evaporates quickly and it is better to wait until the surface is nearly dry again before squirting on another layer of acetone. **Handle this with care.**
(10) Take a pre-cut acetate strip in both hands and quickly align it with the polished slab. Bring the longest edge of the strip and the slab into contact and, with moderate speed,

bring the strip down to rest over the polished and etched face, avoiding the development of bubbles by pushing a little wall of acetone in front.

(11) Allow to dry for at least 3 minutes, during which time the peel must not be touched. To remove the peel, take one free corner of the strip and peel back with a firm, even pressure. Trim off any surplus acetate sheet from around the peel immediately after pulling (to prevent wrinkling). Place the peel between blotting paper and press for about 30 minutes under a pile of books (to prevent curling).

(12) Several peels can be taken from the same prepared surface without re-etching. Serial sections can be prepared by grinding the limestone face down a little further between each peel. In this case, re-etching and staining will be necessary every time.

(13) The peels can be labelled by writing directly on the 'etched' side with Indian ink or a biro. Store them in labelled envelopes and examine with transmitted light between glass slides taped together.

O. Picking and mounting dried samples. Dried residues are best scanned on a picking tray. This should be flat with a black surface (for calcareous and siliceous microfossils) or white surface (for dark phosphatic microfossils) divided into squares about 5 × 5 mm, each one preferably numbered. It should also be easy to clean to reduce the risk of contamination. Painted tobacco-tin lids or card-backed petri dishes are often used.

A small portion of the sample should be gently tapped from its container and scattered lightly and evenly over the tray. The grains and microfossils can be manoeuvred with the aid of a good quality 000 sable hair paint brush and removed to a Franke slide for storage and examination. To do this, either entangle the microfossil in the hairs at the tip of the brush and dislodge it by stroking the brush against the margin of the slide cavity, or pick it up with the fine point of a water-moistened brush and dislodge it by stroking the brush gently on the tacky mounting surface of the slide. Adhesion of microfossils is improved by brushing the slide's surface beforehand with a weak solution of Gum Tragacanth to which a drop of Clove Oil has been added (to reduce fungal growth).

Franke slides, which are the most popular means of storing dried microfossils, can be purchased commercially. If only a few are needed, however, it is possible to prepare and tailor them to your particular needs. Simple ones may be quickly made as follows: Trace the outline of a glass slide on to a piece of card about 2 mm thick and cut it out. Also cut out a rectangle or square (or several of these) from the middle of this cardboard mount. Small circular holes can be cut out by means of a paper punch. Apply a strong adhesive to one side of the card and affix it either to a glass slide or to a piece of black faced card of similar size and shape. The cavity (or cavities) so formed will therefore be clear for transmitted light viewing or black for viewing with reflected light. To make a cover, place a glass slide on top, held in place with thin strips of adhesive brown paper tape along the edges or by clips or hoops at either end. A neater glass cover that will slide on and off can be made by glueing to it a strong paper sleeve that will cover both the undersurface and sides.

Bibliography

Abbreviations:
IRDSDP = *Initial Reports of the Deep-Sea Drilling Project,* Washington (U.S. Govt Printing Office).
RPP = *Review of Palaeobotany and Palynology*, Amsterdam.

Abbot, M. L. 1954. Revision of the Palaeozoic genus *Oligocarpia*. *Palaeontographica B* **96**, 39–65.

Abushik, A. F. 1971. Orientation in Leperditiida. In *Fossil Ostracoda*, O. S. Vyalov (ed.), 102–105. Jerusalem: Israel Program for Scientific Translations.

Anderson, F. W. 1963. Correlation of the Upper Purbeck Beds of England with the German Wealden. *L'pool Manchr Geol J.* 3, 21–32.

Andreev, Yu. N. 1971. Sexual dimorphism of the Cretaceous ostracods of the Gissaro–Tadzhik region. In *Fossil Ostracoda*, O. S. Vyalov (ed.), 56–70. Jerusalem: Israel Program for Scientific Translations.

Andrews, H. N. 1947. *Ancient plants and the world they lived in.* New York: Comstock.

Angel, M. V. 1969. Planktonic ostracods from the Canary Island region; their depth distributions, diurnal migrations, and community organization. *J. Mar. Biol Ass. UK.* **49**, 515–33.

Bandy, O. L. 1964. General correlation of foraminiferal structure with environment. In *Approaches to paleoecology*, J. Imbrie and N. D. Newell (eds), 75–90. New York: John Wiley.

Banner, F. T. and E. Williams 1973. Test structure, organic skeleton and extrathalmous cytoplasm of *Ammonia* Brünnich. *J. Foramin. Res.* **3**, 49–69.

Barghoorn, E. S. and S. A. Tyler 1965. Micro-organisms from the Gunflint chert. *Science* **147**, 563–77.

Barghoorn, E. S. and J. W. Schopf 1966. Micro-organisms three billion years old from the Precambrian of South Africa. *Science* **152**, 758–63.

Barnes, C. R. (ed.) 1976. *Conodont paleoecology*. Spec. Pap. Geol. Ass. Can., no. 15.

Barnes, C. R. and L. E. Fåhraeus 1975. Provinces, communities and the proposed nekto-benthonic habit of Ordovician conodontophorids. *Lethaia* **8**, 133–50.

Barnes, R. D. 1968. *Invertebrate zoology*. Philadelphia: W. B. Saunders.

Barr, F. T. 1970. The foraminiferal genus *Bolivinoides* from the upper Cretaceous of Libya. *J. Paleont.* **44**, 642–54.

Barron, J. A. 1976. Revised Miocene and Pliocene diatom biostratigraphy of upper Newport Bay, Newport Beach, California. *Mar. Micropaleont.* **1**, 27–63.

Baschnagel, R. A. 1966. New fossil algae from the Middle Devonian of New York. *Trans Am. Microsc. Soc.* **85**, 297–302.

Bate, R. H. 1965. Freshwater ostracods from the Bathonian of Oxfordshire. *Palaeontology* **8**, 749–59.

Bate, R. H. 1972. Phosphatized ostracods with appendages from the lower Cretaceous. *Palaeontology* **15**, 379A–93.

Bate, R. H. and B. A. East 1972. The structure of the ostracode carapace. *Lethaia* **6**, 177–94.

Bate, R. H. and J. E. Robinson 1978. *A stratigraphical index of British Ostracoda.* Spec. Iss. Geol. J., no. 8.

Belyaeva, N. V. 1963. The distribution of planktonic foraminifers over the Indian Ocean bottom. *Voprosi Mikropaleontologii* **7**, 209–22. (In Russian)

Bengtson, S. 1976. The structure of some Middle Cambrian conodonts and the early evolution of conodont structure and function. *Lethaia* **9**, 185–206.

Benson, R. H. 1959. Ecology of Recent ostracodes of the Todos Santos Bay region, Baja California, Mexico. *Paleont. Contr. Univ. Kansas* **23**, 1–80.

Benson, R. H. 1972. The *Bradleya* problem, with descriptions of two new psychrospheric ostracode genera, *Agrenocythere* and *Poseidonamicus* (Ostracoda: Crustacea). *Smithson. Contr. Paleobiol.*, no. 12.

Benson, R. H. 1975. The origin of the psychrosphere as recorded in changes of deep-sea ostracode assemblages. *Lethaia* **8**, 69–84.

Benson, R. H. *et al.* (1961). Ostracoda. In *Treatise on invertebrate paleontology. Part Q Arthropoda 3: Crustacea*, R. C. Moore (ed.) Geol Soc. Am. and Univ. Kansas Press.

Berger, W. H. 1971. Planktonic foraminifera: sediment production in an oceanic front. *J. Foramin. Res.* **1**, 95–118.

Berger, W. H. 1973. Deep-sea carbonates: Pleistocene dissolution cycles. *J. Foramin. Res.* **3**, 187–95.

Bettenstaedt, F. 1958. Phylogenetische Beobachtungen in der Mikropaläontologie. *Paläontol. Z.* **32**, 115–40.

Black, M. 1972. *British Lower Cretaceous coccoliths. 1. Gault Clay, Part 1.* Palaeontogr. Soc. Monogr.

Bloeser, B., J. W. Schopf, R. J. Horodyski and J. W. Breed. 1977. Chitinozoans from the late Precambrian Chuar Group of the Grand Canyon, Arizona. *Science* **195**, 676–9.

Boltovskoy, E. 1973. Reconstruction of post-Pliocene climatic changes by means of planktonic foraminifera. *Boreas* **2**, 55–68.

Boltovskoy, E. and R. Wright. 1976. *Recent foraminifera.* The Hague: W. Junk.

Borza, K. 1969. *Die Mikrofazies und Mikrofossilien des Oberjuras und der Unterkreide der Klippenzone der Westkarpaten.* Bratislava: Slovak Academy of Sciences.

Bosence, D. W. J. 1976. Ecological studies on two unattached coralline algae from western Ireland. *Palaeontology* **19**, 365–95.

Boulter, M. C. and G. C. Wilkinson. 1977. A system of group names for some Tertiary pollen. *Palaeontology* **20**, 559–80.

Bradbury, J. P. 1975. *Diatom stratigraphy and human settlement in Minnesota.* Spec. Pap. Geol Soc. Am., no. 171.

Bramlette, M. N. 1958. Significance of coccolithophorids in calcium carbonate deposition. *Bull. Geol Soc. Am.* **69**, 121–6.

Bramlette, M. N. and E. Martini. 1964. The great change in calcareous nannoplankton fossils between the Maastrichtian and the Danian. *Micropaleontology* **10**, 291–322.

Bramlette, M. N. and W. R. Riedel. 1954. Stratigraphic value of discoasters and some other microfossils related to Recent coccolithophores. *J. Paleont.* **28**, 385–403.

Brasier, M. D. 1975. Morphology and habitat of living benthonic foraminiferids from Caribbean carbonate environments. *Revta Esp. Micropaleont.* **7**, 567–78.

Brasier, M. D. 1975a. An outline history of seagrass communities. *Palaeontology* **18**, 681–702.

Brasier, M. D. 1977. An early Cambrian chert biota and its implications. *Nature* **268**, 719–20.

Brasier, M. D. 1979. The Cambrian radiation event. In *The origin of major invertebrate groups*, M. R. House (ed.). London: Academic Press.

Brönnimann, P. and H. H. Renz (eds) 1969. *Proceedings of the first international conference on planktonic microfossils, Geneva 1967.* Vol. 1, 422pp; vol. 2, 745pp. Leiden: E. J. Brill.

Bukry, D. 1971. *Discoaster* evolutionary trends. *Micropaleontology* **17**, 43–52.

Bukry, D. 1973. Coccolith and silicoflagellate stratigraphy, Tasman Sea and southwestern Pacific Ocean. *IRDSDP* **21**, 885–91.

Bukry, D. 1975. Coccolith and silicoflagellate stratigraphy, northwestern Pacific Ocean, DSDP Leg 32. *IRDSDP* **32**, 677–701.

Bukry, D. 1976. Silicoflagellate and coccolith stratigraphy, southeastern Pacific Ocean, DSDP Leg 34, *IRDSDP* **34**, 715–35.

Bukry, D. and J. H. Foster 1973. Silicoflagellate and diatom stratigraphy, DSDP Leg 16, *IRDSDP* **16**, 815–71.

Campbell, A. S. 1954. Tintinnina. In *Treatise on invertebrate paleontology. Part D. Protista 3: Protozoa (chiefly Radiolaria and Tintinnina)*, R. C. Moore (ed.) 166–80. Geol Soc. Am. and Univ. Kansas Press.

Campbell, A. S. 1954a. Radiolaria. In *Treatise on invertebrate paleontology. Part D, Protista 3: Protozoa (chiefly Radiolaria and Tintinnina)*, R. C. Moore (ed.), 11–163. Geol Soc. Am. and Univ. Kansas Press.

Carr, N. G. and B. A. Whitton (eds) 1973. *The biology of blue-green algae.* Oxford: Blackwell Scientific Publications.

Chaloner, W. G. 1967. Spores and land-plant evolution. *RPP* **1**, 83–93.

Chapman, V. J. and D. J. Chapman 1973. *The Algae.* London: Macmillan.

Ciesielski, P. F. 1975. Biostratigraphy and paleoecology of Neogene and Oligocene silicoflagellates from cores recovered during Antartic Leg 28, DSDP. *IRDSDP* **28**, 625–91.

Clarkson, E. N. K. 1979. *Invertebrate palaeontology and evolution*, London: George Allen & Unwin.

Cloud, P. 1976. Beginnings of biospheric evolution and their biochemical consequences. *Paleobiology* **2**, 351–87.

Colom, G. 1948. Fossil tintinnids; loricated infusoria of the Order Oligotricha. *J. Paleont.* **22**, 233–63.

Combaz, A. 1967. Leiosphaeridaceae Eisenack 1954, et Protoleiosphaeridae Timofeev 1959 – leurs affinitiés, leur rôle sédimentologique et géologique. *RPP* **1**, 309–21.

Conway-Morris, S. 1976. A new Cambrian lophophorate from the Burgess Shale of British Columbia. *Palaeontology* **19**, 199–222.

Cornell, W. C. 1970. The chrysomonad cyst-families Chrysostomataceae and Archaeomonadaceae: their status in paleontology. *Proc. N. Am. Paleont. Conv. 1969*, Part G, 958–94.

Cornell, W. C. 1974. Silicoflagellates as paleonvironmental indicators in the Modelo Formation. *J. Paleont.* **48**, 1018–29.

Costa, L. I. and C. Downie 1976. The distribution of the dinoflagellate *Wetzeliella* in the Palaeogene of northwestern Europe. *Palaeontology* **19**, 591–614.

Cramer, F. H. 1973. Middle and upper Silurian chitinozoan succession in Florida subsurface. *J. Paleont.* **47**, 279–88.

Cramer, F. H. and M. del C. R. Diez 1970. Rejuvenation of Silurian chitinozoans from Florida. *Revta Esp. Micropaleont.* **2**, 45–54.

Cramer, F. H. and M. del C. R. Diez 1974. Silurian acritarchs, distribution and trends. *RPP* **18**, 137–54.

Croft, W. N. and E. A. George 1959. Blue-green algae from the Middle Devonian of Rhynie, Aberdeenshire. *Bull. Brit. Mus. Nat. Hist. (Geol.)* **3**, 341–53.

Cummings, R. H. 1955. *Nodosinella* Brady, 1876, and associated Upper Palaeozoic genera. *Micropaleontology* **1**, 221–38.

Currey, D. 1966. Problems of correlation in the Anglo–Paris–Belgian basin. *Proc. Geol. Ass.* **77**, 437–65.

Cushman, J. A. 1948. *Foraminifera, their classification and economic use*, 4th edn Cambridge, Mass.: Harvard Univ. Press.

Dale, B. 1976. Cyst formation, sedimentation and preservation: factors affecting dinoflagellate assemblages in Recent sediments from Trondheimsfjord, Norway. *RPP* **22**, 39–60.

Davis, M. B. 1967. Pollen accumulation rates at Rogers Lake, Connecticut, during late- and postglacial time. *RPP* **2**, 219–30.

Diez, M. del C. R. and F. H. Cramer 1974. Range chart of selected Lower Paleozoic acritarch taxa. *RPP* **18**, 155–70.

Dimbleby, G. W. 1969. Pollen analysis. In *Science in archaeology, a survey of progress and research*, D. Brothwell and E. Higgs (eds), 167–77. London: Thames & Hudson.

Dodge, J. D. and R. M. Crawford 1970. A survey of thecal fine structure in the Dinophyceae. *J. Linn. Soc. Bot.* **63**, 53–67.

Downie, C. 1973. Observations on the nature of the acritarchs. *Palaeontology* **16**, 239–59.

Downie, C. 1974. Acritarchs from near the Pre-cambrian/Cambrian boundary – a preliminary account. *RPP* **18**, 57–60.

Downie, C., W. R. Evitt and W. A. S. Sarjeant 1963. Dinoflagellates, hystrichospheres and the classification of the acritarchs. *Stanford Univ. Publs* (*Geol Sci.*) **7**, 1–16.

Downie, C., M. A. Hussain and G. L. Williams 1971. Dinoflagellate cyst and acritarch associations in the Paleogene of southeast England. *Geoscience and Man* **3**, 29–35.

Downie, C., T. R. Lister, A. L. Harris and D. J. Fettes 1971. *A palynological investigation of the Dalradian rocks of Scotland.* Rep. Inst. Geol Sci., no. 71/9.

Druce, E. C., F. H. T. Rhodes and R. L. Austin 1974. Recognition, evolution and taxonomy of lower Carboniferous conodont assemblages. *J. Paleont.* **48**, 387–402.

Echols, R. J. and G. A. Fowler 1973. Agglutinated tinitinnid loricae from some Recent and late Pleistocene shelf sediments. *Micropaleontology* **19**, 431–43.

Edwards, A. R. 1973. Calcareous nannofossils from the southwest Pacific deep sea drilling project, Leg 21. *IRDSDP* **21**, 641–61.

Eisenack, A. and G. Kjellström 1964 to date. *Katalog der fossilien Dinoflagellaten, Hystrichophären und verwandten Microfossilien.* Stuttgart, W. Germany: Schweizerbart'sche. (Over 4 volumes in loose-leaf)

Elliott, G. F. 1975. Transported algae as indicators of different marine habitats in the English middle Jurassic. *Palaeontology* **18**, 351–66.

Ellis, B. F. and A. R. Messina 1940 to date. *Catalogue of Foraminifera*. Spec. Pubs Am. Mus. Nat. Hist. (Over 50 volumes in loose-leaf)

Ellis, B. F. and A. R. Messina 1952 to date. *Catalogue of Ostracoda*. Spec. Pubs Am. Mus. Nat. Hist. (Over 23 volumes in loose-leaf)

Emiliani, C. 1966. Isotopic paleotemperatures. *Science* **154**, 851–7.

Epstein, A. G., J. B. Epstein and L. D. Harris 1977. *Conodont colour alteration – an index to organic metamorphism.* Prof. Pap. US Geol Surv., no. 995.

Erdtman, G. 1943. *An introduction to pollen analysis.* Waltham, Mass.: Chronica Botanica.

Evitt, W. R. 1961. Observations on the morphology of fossil dinoflagellates. *Micropaleontology* **7**, 385–420.

Evitt, W. R. 1963. A discussion and proposals concerning fossil dinoflagellates, hystrichospheres and acritarchs. *Proc. Natn. Acad. Sci.* **49**, 158–64, 298–302.

Evitt, W. R. 1963a. Occurrence of freshwater alga *Pediastrum* in Cretaceous marine sediments. *Am. J. Sci.* **261**, 890–3.

Faegri, K. and J. Iversen 1975. *Textbook of pollen analysis.* New York: Hafner Press.

Fairchild, T., J. W. Schopf and R. L. Folk 1973. Filamentous algal microfossils from the Caballos Novaculite, Devonian of Texas. *J. Paleont.* **47**, 946–52.

Farinacci, A. 1969 to date. *Catalogue of calcareous nannofossils.* Rome: Edizioni Tecnoscienza. (Over 7 volumes)

Ferguson, I. K. and J. Muller 1976. *The evolutionary significance of the exine.* Symp. Ser. Linn Soc., no. 1. London: Academic Press.

Feyling-Hanssen, R. W., J. A. Jorgensen, K. L. Knudsen and A.-L. L. Andersen 1971. Late Quaternary Foraminifera from Vendyssel, Denmark and Sandnes, Norway. *Bull. Geol Soc. Denmark* **21**, 67–317.

Fiest-Castel, M. 1977. Evolution of the charophyte floras in the upper Eocene and lower Oligocene of the Isle of Wight. *Palaeontology* **20**, 143–57.

Flügel, E. 1977. *Fossil Algae. Recent results and developments.* Berlin: Springer-Verlag.

Fogg, G. E., W. D. P. Stewart, P. Fay and A. E. Walsby 1973. *The blue-green algae.* London: Academic Press.

Foreman, H. P. 1963. Upper Devonian Radiolaria from the Huron member of the Ohio shale. *Micropaleontology* **9**, 267–304.

Foreman, H. P. 1975. Radiolaria from the North Pacific, Deep Sea Drilling Project, Leg 32. *IRDSDP* **32**, 579–673.

Foreman, H. P. and W. R. Riedel 1972 to date. *Catalogue of polycystine Radiolaria.* New York: Micropaleontology Press, Am. Mus. Nat. Hist.

Frerichs, W. E., M. E. Heiman, L. E. Borgman and A. W. H. Bé 1972. Latitudinal variations in planktonic test porosity. *J. Foramin. Res.* **2**, 6–13.

Fuller, H. J. and O. Tippo 1949. *College Botany.* New York: Henry Holt.

Funnell, B. M. 1967. Foraminifera and Radiolaria as depth indicators in the marine environment. *Mar. Geol.* **5**, 333–47.

Funnell, B. M. and W. R. Riedel (eds) 1971. *The micropalaeontology of oceans.* Cambridge: Cambridge Univ. Press.

Galloway, J. J. 1933. *A manual of foraminifera.* Indiana: Bloomington, Prinicipia Press.

Gardner, J. V. and L. H. Burckle 1975. Upper Pleistocene *Ethmodiscus rex* oozes from the eastern equatorial Atlantic. *Micropaleontology* **21**, 236–42.

Gartner Jr, S. 1970. Phylogenetic lineages in the lower Tertiary coccolith genus *Chiasmolithus. Proc. N. Am. Paleont. Conv.* 1969, Part G. 930–57.

Ginsburg, R. N. (ed.) 1975. *Tidal deposits.* Berlin: Springer-Verlag.

Ginsburg, R. N., R. Rezak and J. L. Wray 1971. Geology of calcareous algae (notes for a short course). *Sedimenta* **1** (loose-leaf), 61pp.

Glaessner, M. F. 1945. *Principles of micropalaeontology.* New York: Hafner Press.

Glintzboeckel, C. and J. Magné 1959. Répartition des microfaunes à plancton et à ostracodes dans le Crétacé Supérieur de la Tunisie et de l'est Algérien. *Revue Micropaléont.* **2**, 57–67.

Godwin, H. 1967. Pollen analytic evidence for the cultivation of Cannibis in England. *RPP* **4**, 71–80.

Goll, R. M. 1972. Leg 9 synthesis, Radiolaria. *IRDSDP* **9**, 947–1058.

Goll, R. M. and K. R. Bjørklund 1974. Radiolaria in the surface sediments of the South Atlantic. *Micropaleontology* **20**, 38–75.

Gombos Jr, A. M. 1975. Fossil diatoms from Leg 7, Deep Sea Drilling Project. *Micropaleontology* **21**, 306–33.

Gombos Jr, A. M. 1977. Archaeomonads as Eocene and Oligocene guide fossils in marine sediments. *IRDSDP* **36**, 689–95.

Gombos Jr, A. M. 1977a. Paleogene and Neogene diatoms from the Falkland Plateau and Malvinas outer basin: Leg 36, Deep Sea Drilling Project. *IRDSDP* **36**, 575–687.

Grignani, D. 1967. Correlation with Chitinozoa in the Devonian and Silurian in some Tunisian well samples. *RPP* **5**, 315–25.

Habib, D. 1968. Spore and pollen paleoecology of the Redstone seam (Upper Pennsylvanian) of West Virginia. *Micropaleontology* **14**, 199–220.

Habib, D. 1975. Neocomian dinoflagellate zonation in the western Atlantic. *Micropaleontology* **21**, 373–92.

Haq, B. U. and A. Boersma (eds) 1978. *Introduction to marine micropaleontology.* Amsterdam: Elsevier.

Haq, B. U. and G. P. Lohmann 1976. Early Cenozoic calcareous nannoplankton biogeography of the Atlantic Ocean. *Mar. Micropaleont.* **1**, 20–97.

Haq, B. U. and G. P. Lohmann 1977. Calcareous nannoplankton biogeography and its paleoclimatic implications. Cenozoic of the Falkland Plateau (DSDP Leg 36) and Miocene of the Atlantic Ocean. *IRDSDP* **36**, 745–59.

Harker, S. D. and W. A. Sarjeant 1975. The stratigraphic distribution of the organic-walled dinoflagellate cysts in the Cretaceous and Tertiary. *RPP* **20**, 217–316.

Harland, R. 1971. A summary review of the morphology and classification of the fossil Peridiniales (dinoflagellates) with respect to their modern representatives. *Geophytology* **1**, 135–50.

Harland, W. B. *et al.* 1967. *The fossil record. A symposium with documentation.* London: Geol Soc. Lond.

Harris, T. M. 1938. *The British Rhaetic flora.* London: Brit. Mus. (Nat. Hist.).

Harris, T. M. 1964. *The Yorkshire Jurassic flora 2: Caytoniales, Cycadales and Pteridosperms.* London: Brit. Mus. (Nat. Hist.).

Hart, M. B. and D. J. Carter 1975. Some observations on the Cretaceous Foraminiferida of southeast England. *J. Foramin. Res.* **5**, 114–26.

Hartmann, G. (ed.) 1976. *International symposium on evolution of post-Paleozoic Ostracoda.* Naturwissenschaflicher Verein Hamburg, Abh. u. Ver. (N.F.) 18/19.

Hass, W. H., W. Hantzschel, D. W. Fisher, B. F. Howell, F. H. T. Rhodes, K. J. Muller and R. C. Moore 1962. Miscellanea. In *Treatise on invertebrate paleontology, Part W*, R. C. Moore (ed.), 3–98. Geol Soc. Am. and Univ. Kansas Press.

Haynes, J. 1965. Symbiosis, wall structure and habitat in Foraminifera. *Cushman Fdn Foramin. Res.* **16**, 40–3.

Hays, J. D. 1970. Stratigraphy and evolutionary trends of Radiolaria in North Pacific deep sea sediments. In *Geological investigations of the North Pacific*, J. D. Hays (ed.), 185–218. Mem. Geol Soc. Am., no. 126.

Hedberg, H. D. (ed.) 1976. *International stratigraphic guide.* New York: Wiley–Interscience.

Hedley, R. H. and C. G. Adams 1974. *Foraminifera*, vol. 1. London: Academic Press.

Hedley, R. H. and C. G. Adams 1976. *Foraminifera*, vol. 2. London: Academic Press.

Hekel, H. 1973. Nannofossil biostratigraphy, Leg 20, Deep Sea Drilling Project. *IRDSDP* **20**, 221–47.

Hendey, N. I. 1964. *An introductory account of the smaller algae of British coastal waters. Part V: Bacillariophyceae (Diatoms).* Fisheries Investigations Ser. IV. London: HMSO.

Hitchings, V. H. and A. T. S. Ramsay 1978. Conodont assemblages: a new functional model. *Palaeogeog., Palaeoclimatol., Palaeoecol.* **24**, 137–50.

Holdsworth, B. K. 1969. The relationship between the genus *Albaillella* Deflandre and the ceratoikiscid Radiolaria. *Micropaleontology* **15**, 230–6.

Holdsworth, B. K. and B. M. Harker 1975. Possible indicators of degree of radiolarian dissolution in calcareous sediments of the Ontong–Java Plateau. *IRDSDP* **30**, 489–95.

Honjo, S. 1976. Coccoliths: production, transportation and sedimentation. *Mar. Micropaleont.* **1**, 65–79.

Honjo, S. and H. Okada 1974. Community structure of coccolithophores in the photic layer of the mid-Pacific. *Micropaleontology* **20**, 209–30.

Hopping, C. A. 1967. Palynology and the Oil Industry. *RPP* **2**, 23–48.

Hornibrook, N. de B. 1949. A new family of living Ostracoda with striking resemblances to some Palaeozoic Beyrichiidae. *Trans Roy. Soc. NZ* **77**, 469–71.

Hovasse, R. 1934. Ebriacées, Dinoflagellés et Radiolaires. *C.r. Hebd. Séanc. Acad. Sci.* **198**, 402–4.

Hughes, N. F. and C. A. Croxton 1973. Palynologic correlation of the Dorset 'Wealden'. *Palaeontology* **16**, 567–601.

Hughes, N. F. and J. C. Moody-Stuart 1967. Palynological facies and correlation in the English Wealden. *RPP* **1**, 259–68.

Hughes, N. F. and B. Pacltová 1972. Freshwater to marine time-correlation potential of Cretaceous and

Tertiary palynomorphs. *24th Int. Geol Congr., Montreal* **7**, 397–401.

Hustedt, F. 1930. Bacillariophyta (Diatoms). In *Die Susswasserflora Mitteleuropas*, A. Pascher (ed.). Jena: G. Fischer.

Ingle Jr, J. C. 1973. Summary comments on Neogene biostratigraphy, physical stratigraphy and paleo-oceanography in the marginal northeastern Pacific Ocean. *IRDSDP* **18**, 949–60.

Jansonius, J. 1967. Systematics of the Chitinozoa. *RPP* **1**, 345–60.

Jansonius, J. 1970. Classification and stratigraphic application of Chitinozoa. *Proc. N. Am. Paleont. Conv.* 1969, Part G, 789–808.

Jenkins, D. G. 1965. Planktonic foraminifera and Tertiary intercontinental correlations. *Micropaleontology* **11**, 265–77.

Jenkins, D. G. 1970. Foraminiferida and New Zealand Tertiary biostratigraphy. *Revta Esp. Micropaleont.* **2**, 13–26.

Jenkins, W. A. M. 1969. *Chitinozoa from the Ordovician Viola and Fernvale Limestones of the Arbuckle Mountains, Oklahoma.* Spec. Paps Palaeont., no. 5.

Jenkins, W. A. M. 1970. Chitinozoa. *Geoscience and Man* **1**, 1–20.

Jeppson, L. 1971. Element arrangement of conodont apparatuses of *Hindeodella* type and in similar forms. *Lethaia* **4**, 101–23.

Jodry, R. L. and D. E. Campau 1961. Small pseudochitinous and resinous microfossils, new tools for the subsurface geologist. *Bull. Am. Assoc. Petrol. Geol.* **45**, 1378–91.

Johnson, J. H. 1971. An introduction to the study of organic limestones. *Colo. Sch. Mines Q.* **66**.

Jones, D. J. 1956. *Introduction to microfossils.* New York: Harper Brothers.

Jost, M. 1968. Microfossils of problematic systematic position from Precambrian rocks at White Pine, Michigan. *Micropaleontology* **14**, 365–8.

Kafescioglu, I. A. 1971. Specific diversity of planktonic foraminifera on the continental shelves as a paleobathymetric tool. *Micropaleontology* **17**, 455–70.

Kapp, R. O. 1969. *How to know pollen and spores.* Dubuque, Iowa: Brown & Co.

Karkhanis, S. N. 1976. Fossil iron bacteria may be preserved in Precambrian ferroan carbonate. *Nature* **261**, 406–7.

Kazmierczak, J. 1976. Devonian and modern relatives of the Precambrian *Eosphaera*: possible significance for the early eukaryotes. *Lethaia* **9**, 39–50.

Kazmierczak, J. 1976a. Volvocacean nature of some Palaeozoic non-radiosphaerid calcispheres and parathuramminid 'Foraminifera'. *Acta Palaeont. Pol.* **21**, 245–58.

Kazmierczak, J. and S. Golubić 1976. Oldest organic remains of boring algae from Polish Upper Silurian. *Nature* **261**, 404–6.

Kedves, M. 1960. Études palynologiques dans le Bassin de Dorog-I-. *Pollen et Spores* **2**, 89–118.

Keen, M. C. 1972. Evolutionary patterns of Tertiary ostracods and their use in defining stage and epoch boundaries in Western Europe. *24th Int. Geol Congr., Montreal* **7**, 190–7.

Keen, M. C. 1977. Ostracod assemblages and the depositional environments of the Headon, Osborne and Bembridge Beds (upper Eocene) of the Hampshire Basin. *Palaeontology* **20**, 405–46.

Kemp, E. M. 1968. Probable angiosperm pollen from the British Barremian to Albian strata. *Palaeontology* **11**, 421–34.

Kennet, J. P. 1973. Middle and late Cenozoic planktonic foraminiferal biostratigraphy of the southwestern Pacific. DSDP Leg 21. *IRDSDP* **21**, 575–639.

Kesling, R. V. 1951. Terminology of ostracode carapaces. *Contrib. Mus. Paleont. Univ. Mich.* **9**, 93–171.

Kilenyi, T. I. and N. W. Allen 1968. Marine–brackish bands and their microfauna from the lower part of the Weald Clay of Sussex and Surrey. *Palaeontology* **11**, 141–62.

Klapper, G. and D. B. Johnson 1975. Sequence in conodont genus *Polygnathus* in Lower Devonian at Lone Mountain, Nevada. *Geologica et Palaeontologica* **9**, 65–83.

Klapper, G. and G. M. Philip 1971. Devonian conodont apparatus and their vicarious skeletal elements. *Lethaia* **4**, 429–52.

Knoll, A. H. and D. A. Johnson 1975. Late Pleistocene evolution of the collosphaerid radiolarian *Buccinosphaera invaginata. Micropaleontology* **21**, 60–8.

Kobluk, D. R. and M. J. Risk 1977. Micritization and carbonate-grain binding by endolithic algae. *Bull. Am. Ass. Petrol. Geol.* **61**, 1069–82.

Kofoid, C. A. and A. S. Campbell 1939. The Tintinnoinea. *Bull. Mus. Comp. Zool. Harv.* **84**, 1–473.

Kofoid, C. A. and O. Swezy 1921. *The free-living unarmoured Dinoflagellata*. Berkeley, California: Univ. California Press.

Koizumi, I. 1975. Neogene diatoms from the western margin of the Pacific Ocean, Leg 31, Deep Sea Drilling Project. *IRDSDP* **31**, 779–819.

Koizumi, I. 1975a. Neogene diatoms from the northwestern Pacific Ocean, Deep Sea Drilling Project. *IRDSDP* **32**, 865–89.

Kremp, G. O. W. 1965. *Morphologic encyclopedia of palynology*. Tucson, Arizona: Univ. Arizona Press.

Kremp, G. O. W. 1967. Tetrad markings of pteridophytic spores and their evolutionary significance. *RPP* **3**, 311–23.

Krutak, P. R. 1972. Some relationships between grain size of substrate and carapace size in modern brackish-water Ostracoda. *Micropaleontology* **18**, 153–9.

Kummel, B. and D. Raup (eds) 1965. *Handbook of paleontological techniques.* San Francisco: W. H. Freeman.

Kuprianova, L. A. 1969. On the evolutionary levels in the morphology of pollen grains and spores. *Pollen et Spores* **11**, 333–51.

Kutznetsov, S. I., M. V. Ivanov and N. N. Lyalikova 1963. *Introduction to geological microbiology*. New York: McGraw Hill.

Landing, E. 1977. *'Prooneotodus' tenuis* (Muller 1959) apparatuses from the Taconic allocthon, eastern New York: construction, taphonomy and the protoconodont 'supertooth' model. *J. Paleont.* **51**, 1072–84.

Laufeld, S. 1974. Silurian Chitinozoa from Gotland. *Fossils and Strata* **5**.

Lewis, K. B. 1970. A key to the Recent genera of the Foraminiferida. *Mem. NZ Oceanogr. Inst.* **196**.

Lindström, M. 1964. *Conodonts*. Amsterdam: Elsevier.

Lindström, M. 1970. A suprageneric taxonomy of the conodonts. *Lethaia* **3**, 427–45.

Lindström, M. 1974. The conodont apparatus as a food-gathering mechanism. *Palaeontology* **17**, 729–44.

Lindström, M. 1974a. Conodont palaeogeography of the Ordovician. In *The Ordovician System*, M. G. Bassett (ed.), 501–22. Cardiff: Univ. Wales Press and Nat. Mus. Wales.

Lindström, M. and W. Ziegler (eds) 1972. Symposium on conodont taxonomy. *Geologica et Palaeontologica*, Sonderband 1.

Ling, H. Y. 1972. Upper Cretaceous and Cenozoic silicoflagellates and ebridians. *Bull. Am. Paleont.* **62**, 135–229.

Ling, H. Y. 1975. Silicoflagellates and ebridians from Leg 31. *IRDSDP* **31**, 763–73.

Lipps, J. H. 1970. Ecology and evolution of silicoflagellates. *Proc. N. Am. Paleont. Conv.* 1969, Part G, 965–93.

Lipps, J. H. and J. W. Valentine 1970. The role of foraminifera in the trophic structure of marine communities. *Lethaia* **3**, 279–86.

Lister, T. R. 1970. *The acritarchs and Chitinozoa from the Wenlock and Ludlow Series of the Ludlow and Millichope areas; Shropshire, Part 1.* Palaeontogr. Soc. Monogr., no. 124, 1–100.

Loeblich Jr, A. R. 1970. Morphology, ultrastructure and distribution of Paleozoic acritarchs. *Proc. N. Am. Paleont. Conv.* 1969, Part G, 705–88.

Loeblich Jr, A. R. and H. Tappan 1964. Protista 2; Sarcodina, chiefly 'Thecamoebians' and Foraminiferida. In *Treatise on invertebrate paleontology, Part C*, R. C. Moore (ed.), (2 vols). Geol Soc. Am. and Univ. Kansas Press.

Loeblich III, A. R., L. A. Loeblich, H. Tappan and A. R. Loeblich Jr 1968. *Annotated index of fossil and Recent silicoflagellates and ebridians with descriptions and illustrations of validly proposed taxa.* Mem. Geol Soc. Am., no. 106.

Löffler, H. and D. Danielpol (eds) 1977. *Aspects of ecology and zoogeography of Recent and fossil Ostracoda.* The Hague: W. Junk.

Männil, R. 1972. The zonal distribution of Ordovician chitinozoans in the eastern Baltic area. *24th Int. Geol Congr., Montreal* **7**, 569–71.

Manton, S. M. 1977. *The Arthropoda. Habits, functional morphology and evolution.* Oxford: Oxford Univ. Press.

Manum, S. B, 1976. Dinocysts in Tertiary Norwegian–Greenland Sea sediments (Deep Sea Drilling Project, Leg 38) with observations on palynomorphs and palynodebris in relation to environment. *IRDSPD* **38**, 897–919.

Margulis, L. 1970. *Origin of eucaryotic cells.* New Haven, Conn.: Yale Univ. Press.

Marshall, A. E. and A. H. V. Smith 1964. Assemblages of miospores from some upper Carboniferous coals and their associated sediments in the Yorkshire coalfield. *Palaeontology* **7**, 656–73.

Marshall, S. M. 1934. The Silicoflagellata and Tintinnoinea. *Brit. Mus. (Nat. Hist.) Great Barrier Reef Expedition 1928–29, Scientific Reports*, **4**, 623–64.

Marszalek, D. S., R. C. Wright and W. W. Hay 1969. Function of the test in the Foraminifera. *Trans Gulf Coast Ass. Geol Socs* **19**, 341–52.

Martin, A. R. H. 1976. Upper Palaeocene Salvinaceae from the Woolwich/Reading Beds near Cobham, Kent. *Palaeontology* **19**, 173–84.

Martini, E. and C. Müller 1976. Eocene to Pleistocene silicoflagellates from the Norwegian–Greenland Sea (DSDP, Leg 38). *IRDSDP* **38**, 857–95.

Matthews, S. C. and V. V. Missarzhevsky 1975. Small shelly fossils of late Precambrian age: a review of recent work. *J. Geol Soc.* **131**, 289–304.

McAndrews, J. H. 1968. Pollen evidence for the protohistoric development of the 'Big Woods' in Minnesota (USA). *RPP* **7**, 201–11.

Mikkelsen, N., L. Labeyris Jr and W. H. Berger 1978. Silica oxygen isotopes in diatoms, a 20000 yr record in deep-sea sediments. *Nature* **271**, 536–8.

Millay, M. A. and T. N. Taylor 1976. Evolutionary trends in fossil gymnosperm pollen. *RPP* **21**, 65–91.

Moore, R. C. 1954. Heliozoa. In *Treatise on invertebrate paleontology. Part D, Protista 3: Protozoa (chiefly Radiolaria and Tintinnina)*, R. C. Moore (ed.), 7–10. Geol Soc. Am. and Univ. Kansas Press.

Moore Jr, T. C. 1972. Mid-Tertiary evolution of the radiolarian genus *Calocycletta*. *Micropaleontology* **18**, 144–52.

Morley Davies, A. 1971. *Tertiary faunas*, 2nd edn. (Revised by F. E. Eames and R. J. G. Savage). Vol I. *The composition of Tertiary faunas.* London: George Allen & Unwin.

Muir, M. D. 1967. Reworking in Jurassic and Cretaceous spore assemblages. *RPP* **5**, 145–54.

Müller, K. 1962. Supplement to systematics of conodonts. In *Treatise on invertebrate paleontology, Part W*, R. C. Moore (ed.) 246–9. Geol Soc. Am. and Univ. Kansas Press.

Müller, K. J. and Y. Nogami 1972. Growth and function of conodonts. *24th Int. Geol Congr., Montreal* **7**, 20–7.

Murray, J. W. 1973. *Distribution and ecology of living benthic foraminiferids.* London: Heinemann.

Murray, J. W. and C. A. Wright 1974. *Palaeogene Foraminiferida and palaeoecology, Hampshire and Paris Basins and the English Channel.* Spec. Paps Palaeont., no. 14.

Musich, L. F. 1973. Pollen occurrence in eastern north Pacific sediments. Deep Sea Drilling Project, Leg 18. *IRDSDP* **18**, 799–815.

Nair, P. K. K. 1968. A concept on pollen evolution

in 'primitive' angiosperms. *J. Palynol.* **4**, 15–20.
Navale, G. K. B. and R. S. Tiwari 1968. Palynological correlation of coal seams, their nature and formation in Rampur coalfield, Lower Gondwana (India). *RPP* **6**, 155–69.
Nazarov, B. B. 1975. Lower and middle Paleozoic radiolarians of Kazakhstan. *Trudy̆ Geol Inst. Akad. Nauk. SSSR* **275**, 202pp. (In Russian)
Neale, J. W. (ed.) 1969. *The taxonomy, morphology and ecology of Recent Ostracoda.* Edinburgh: Oliver & Boyd.
Nigrini, C. 1970. Radiolarian assemblages in the North Pacific and their applications to a study of Quaternary sediments in core V20–130. In *Geological investigations of the North Pacific*, J. D. Hays (ed.), 139–183. Mem. Geol Soc. Am., no. 126.
Norris, G. and W. A. S. Sarjeant 1965. A descriptive index of genera of fossil Dinophyceae and Acritarcha. *NZ Geol Surv. Paleont. Bull.* **40**.
Oertli, H. J. (ed.) 1971. *Colloque sur la paléoecologie des ostracodes.* Pau, France: Bulletin du Centre de Recherches Pau – SNPA.
Oldfield, F. 1970. Some aspects of scale and complexity in pollen-analytically based palaeoecology. *Pollen et Spores* **12**, 163–71.
Olsson, R. K. 1970. Paleocene planktonic foraminiferal biostratigraphy and paleozoogeography of New Jersey. *J. Paleont.* **44**, 589–98.
Pannella, G. 1972. Precambrian stromatolites as paleontological clocks. *24th Int. Geol Congr., Montreal* **1**, 50–7.
Patriquin, D. J. 1972. Carbonate mud production by epibionts on *Thalassia*. An estimate based on leaf growth rate data. *J. Sedim. Petrol.* **42**, 687–9.
Peck, R. E. 1957. North American Mesozoic Charophyta. *Prof. Paps US Geol Surv.* **294–A**, 1–44.
Perkins, R. D., M. D. McKenzie and P. L. Blackwelder 1972. Aragonite crystals within codiacean algae: distinctive morphology and sedimentary implications. *Science* **175**, 624–6.
Pessagno, E. A. and K. Miyano 1968. Notes on the wall structure of the Globigerinacea. *Micropaleontology* **14**, 38–50.
Petruchevskaya, M. G. 1971. On the natural system of polycystine Radiolaria (Class Sarcodina). *Proc. 2nd Int. Plankt. Conf., Rome* 1970, 981–92. Rome: Edizioni Tecnoscienza.
Petruchshevskaya, M. G. 1975. Cenozoic radiolarians of the Antarctic, Leg. 29, DSDP. *IRDSDP* **29**, 541–675.
Phillips, L. 1974. Reworked Mesozoic spores in Tertiary leaf-beds on Mull, Scotland. *RPP* **17**, 221–32.
Phleger, F. B. 1960. *Ecology and distribution of Recent Foraminifera.* Baltimore: John Hopkins Press.
Phleger, F. B. and A. Soutar 1973. Production of benthic Foraminifera in three east Pacific oxygen minima. *Micropaleontology* **19**, 110–15.
Pocock, S. A. J. 1972. Dating and correlation of the Jurassic strata of western Canada by means of plant microfossils. *24th Int. Geol Congr., Montreal* **7**, 402–11.
Poignant, A.–F. 1974. Les algues calcaires fossiles; leur intérêt stratigraphique. *Newsl. Stratigr.* **3**, 181–92.
Pokorný, V. 1958. *Grundzüge der Zoologischen Mikropaläontologie.* Berlin: VEB Deutscher Verlag der Wissenschaften.
Pokorný, V. 1963. *Principles of zoological micropalaeontology.* (English translation edited by J. W. Neale). Vol. 1. Oxford: Pergamon Press.
Pollard, J. E. 1966. A non-marine ostracod fauna from the coal measures of Durham and Northumberland. *Palaeontology* **9**, 667–97.
Postuma, J. A. 1971. *Manual of planktonic Foraminifera.* Amsterdam: Elsevier.
Potonié, H. 1962. Synopsis der sporae *in situ. Bei. Geol Jb.* **52**, 1–204.
Potonié, H. and G. O. W. Kremp 1955, 1956. Die sporae dispersae des Ruhrkarbons, ihre Morphographie und Stratigraphie mit Ausblicken auf Arten anderer Gebiete und Zeitabschnitte. *Palaeontographica B* **98**, 1–156; **99**, 85–191; **100**, 65–121.
Preiss, W. V. 1977. The biostratigraphic potential of Precambrian stromatolites. *Precambrian Research* **5**, 207–19.
Priddle, J. 1974. The function of conodonts. *Geol Mag* **111**, 255–7.
Puri, H. S. (ed.) 1964. *Ostracods as ecological and palaeoecological indicators.* Publ. Staz. Zool. Napoli, suppl., no. 33.
Raaben, M. E. 1969. Columnar stromatolites and late Precambrian stratigraphy. Am. J. Sci. **267**, 1–18.
Raistrick, A. 1934–5. The correlation of coals by microspore content. Part 1. The seams of Northumberland. *Trans Instn Min. Engrs* **88**, 142–53 and 259–64.
Ramsay, A. T. S. 1973. A history of organic siliceous sediments in oceans. In *Organisms and continents through time*, N. F. Hughes (ed.) 199–234. Spec. Paps Palaeont., no. 12.
Ramsay, A. T. S. (ed.) 1977. *Oceanic micropalaeontology.* 2 vols. London: Academic Press.
Resig, J. M. 1976. Benthic foraminiferal stratigraphy, eastern margin, Nazca plate. *IRDSDP* **34**, 743–59.
Rhodes, F. H. T. 1954. The zoological affinities of the conodonts. *Biol Revs* **29**, 419–52.
Rhodes, F. H. T. (ed.) 1973. *Conodont Paleozoology.* Spec. Paps Geol Soc. Am., no. 141.
Richardson, J. B. 1967. Some British lower Devonian spore assemblages and their stratigraphic significance. *RPP* **1**, 111–29.
Riedel, W. R. 1967. Radiolarian evidence consistent with spreading of the Pacific floor. *Science* **157**, 540–2.
Riedel, W. R. and A. Sanfilippo 1971. Cenozoic Radiolaria from the western tropical Pacific. *IRDSDP* **7**, 1529–672.
Riegel, W. 1974. Phytoplankton from the upper Emsian and Eifelian of the Rhineland, Germany. A preliminary report. *RPP* **18**, 29–39.
Rietschel, S. 1973. Zur Deutung der Conodonten. *Natur Mus., Frankf.* **103**, 409–18.
Rogl, F. and H. M. Bolli 1973. Holocene to Pleistocene

planktonic foraminifera of Leg 15, Site 147 (Cariaco Basin (Trench), Caribbean Sea) and their climatic interpretation. *IRDSDP* **15**, 553–615.

Round, F. E. 1961. The diatoms of a core from Esthwaite water. *New Phytol.* **60**, 43–59.

Ross, C. A. 1972. Paleobiological analysis of fusilinacean (Foraminiferida) shell morphology. *J. Paleont.* **46**, 719–28.

Rossignol-Strick, M. 1973. Pollen analysis of some sapropel layers from the deep sea floor of the eastern Mediterranean. *IRDSDP* **13**, (2), 971–85.

Roth, P. H. 1973. Calcareous nannofossils – Leg 17, Deep Sea Drilling Project. *IRDSDP* **17**, 695–795.

Saidova, Kh. M. 1967. Sediment stratigraphy and paleogeography of the Pacific Ocean by benthonic Foraminifera during the Quaternary. *Progress in Oceanography* **4**, 143–51.

Sarjeant, W. A. S. 1967. The stratigraphical distribution of fossil dinoflagellates. *RPP* **1**, 323–43.

Sarjeant, W. A. S. 1974. *Fossil and living dinoflagellates.* London: Academic Press.

Scagel, R. F., R. J. Bandoni, G. E. Rouse, W. B. Schofield, J. R. Stein and T. M. C. Taylor 1965. *An evolutionary survey of the plant kingdom.* London: Blackie.

Schafer, C. T. and B. R. Pelletier 1976. *First international symposium on benthonic foraminifera of continental margins.* Spec. Publs Marit. Sediments, no. 1, (2 vols).

Schopf, J. M., E. G. Ehlers, D. V. Stiles and J. D. Birle 1965. Fossil iron bacteria preserved in pyrite. *Proc. Am. Phil. Soc.* **109**, 288–308.

Schopf, J. W. 1970. Precambrian micro-organisms and evolutionary events prior to the origin of vascular plants. *Biol Revs* **45**, 319–52.

Schopf, J. W. 1971. Organically preserved Precambrian microfossils. *Proc. N. Am. Paleont. Conv.* 1969, Part H, 1013–57.

Schopf, J. W. 1972. Evolutionary significance of the Bitter Springs (Late Precambrian) microflora. *24th Int. Geol Congr., Montreal* **1**, 68–77.

Schopf, J. W. 1977. Biostratigraphic usefulness of stromatolitic Precambrian microbiotas: a preliminary analysis. *Precambrian Research* **5**, 143–73.

Schopf, J. W. and E. S. Barghoorn 1967. Alga-like fossils from the Early Precambrian of South Africa. *Science* **156**, 508–12.

Schopf, J. W., B. N. Haugh, R. E. Molnar and D. F. Satterthwait 1973. On the development of metaphytes and metazoans. *J. Paleont.* **47**, 1–9.

Schrader, H.-J. 1973. Stratigraphic distribution of marine species of the diatom *Denticula* in Neogene North Pacific sediments. *Micropaleontology* **19**, 417–30.

Schrader, H.-J. and J. Fenner 1976. Norwegian Sea Cenozoic diatom biostratigraphy and taxonomy. *IRDSDP* **38**, 921–1099.

Scott, D. S. and F. S. Medioli 1978. Vertical zonation of marsh foraminifera as accurate indicators of former sea levels. *Nature* **272**, 528–31.

Seddon, G. and W. C. Sweet 1971. An ecologic model for conodonts. *J. Paleont.* **45**, 869–80.

Seiglie, G. A. 1975. Foraminifers of Guayanilla Bay and their use as environmental indicators. *Revta Esp. Micropaleont.* **7**, 453–87.

Setty, M. G. A. P. 1966. Preparation and method of study of fossil diatoms. *Micropaleontology* **12**, 511–14.

Shackleton, N. J. and J. P. Kennet 1975. Paleotemperature history of the Cenozoic and the initiation of Antarctic glaciation: oxygen and carbon isotope analyses in Deep Sea Drilling Project sites 277, 279 and 281. *IRDSDP* **29**, 743–55.

Shafik, S. 1975. Nannofossil biostratigraphy of the south west Pacific Deep Sea Drilling Project, Leg 30. *IRDSDP* **30**, 549–98.

Sieburth, J. M. 1975. *Microbial seascapes. A pictorial essay on marine micro-organisms and their environments.* Baltimore: University Park Press.

Sliter, W. V. 1972. Upper Cretaceous planktonic foraminiferal zoogeography and ecology – eastern Pacific margin. *Palaeogeogr., Palaeoclimatol., Palaeoecol.* **12**, 15–31.

Sliter, W. V. and R. A. Baker 1972. Cretaceous bathymetric distribution of benthic foraminifers. *J. Foramin. Res.* **2**, 167–83.

Smith, A. H. V. 1962. The palaeoecology of Carboniferous peats based on the miospores and petrography of bituminous coals. *Proc. Yorks. Geol Soc.* **33**, 423–74.

Smith, A. H. V. and M. A. Butterworth 1967. *Miospores in the coal seams of the Carboniferous of Great Britain.* Spec. Paps Palaeontol., no. 1.

Smith, N. D. and R. S. Saunders 1970. Paleoenvironments and their control of acritarch distribution: Silurian of east-central Pennsylvania. *J. Sedim. Petrol.* **40**, 324–33.

Stanley, E. A. 1966. The problem of reworked pollen and spores in marine sediments. *Mar. Geol.* **4**, 397–408.

Stanley, E. A. 1967. Palynology of six ocean-bottom cores from the south-western Atlantic Ocean. *RPP* **2**, 195–203.

Staplin, F. L. 1961. Reef-controlled distribution of Devonian microplankton in Alberta. *Palaeontology* **4**, 392–424.

Stradner, H. 1973. Catalogue of calcareous nannoplankton from sediments of Neogene age in the eastern North Atlantic and Mediterranean Sea. *IRDSDP* **13** (2), 1137–99.

Stehli, F. G. and W. B. Creath 1964. Foraminiferal ratios and regional environments. *Bull. Am. Ass. Petrol. Geol* **48**, 1810–27.

Swain, F. M. (ed.) 1977. *Stratigraphic micropaleontology of Atlantic Basin and Borderlands.* Amsterdam: Elsevier.

Swain, F. M., L. A. Kornicker and R. F. Lunndin (eds) 1975. Biology and paleobiology of ostracods. *Bull. Am. Paleont.* **65**.

Sweet, W. C. and S. M. Bergström (eds) 1971. *Symposium on conodont biostratigraphy.* Mem. Geol Soc. Am., no. 127.

Sylvester-Bradley, P. C. 1956. The structure, evolution

and nomenclature of the ostracod hinge. *Bull. Brit. Mus. (Nat. Hist.) A*, **3**, 1–21.

Tappan, H. and A. R. Loeblich Jr 1968. Lorica composition of modern and fossil Tinitinnida (ciliate Protozoa), systematics, geologic distribution and some new Tertiary taxa. *J. Paleont.* **42**, 1378–94.

Tappan, H. and A. R. Loeblich Jr 1971. Surface sculpture of the wall in Lower Paleozoic acritarchs. *Micropaleontology* **17**, 385–410.

Tappan, H. and A. R. Loeblich Jr 1972. Fluctuating rates of protistan evolution, diversification and extinction. *24th Int. Geol Congr., Montreal* **7**, 205–13.

Tappan, H. and A. R. Loeblich Jr 1973. Evolution of the ocean plankton. *Earth Sci. Revs* **9**, 207–40.

Tasch, P. 1973. *Paleobiology of the invertebrates.* New York: Wiley.

Taugourdeau, P. 1961. Les chitinozoaires: techniques d'études, morphologie et classification. *Mem. Soc. Géol Fr., N. Ser.* **104**.

Taugourdeau, P. *et al.* 1967. *Microfossiles organiques du Paléozoic. Les Chitinozoaires (1). Analyse bibliographique illustrée.* Paris: Centre National de la Recherche Scientifique.

Taugourdeau, P. and P. de Jekhowsky 1960. Répartition et description des chitinozoaires Siluro-Dévoniens de quelques sondages de la CREPS, de la CFPA et de la SN REPAL au Sahara. *Révue Inst. Fr. Pétrole* **15**, 1199–260.

Traverse, A. 1955. Occurrence of the oil-forming *Botryococcus* in lignites and other Tertiary sediments. *Micropaleontology* **1**, 343–50.

Traverse, A., H. T. Ames and W. Spackman 1957 to date. *Catalog of fossil spores and pollen.* Palynological labs, Pennsylvania State Univ. (Over 40 vols)

Tschudy, R. H. and R. A. Scott (eds) 1969. *Aspects of palynology.* New York: Wiley–Interscience.

Tynan, E. J. 1971. Geologic occurrence of the archaeomonads. *Proc. 2nd Int. Plankt. Conf., Rome 1970*, 1225–30. Rome: Edizioni Tecnoscienza.

Uchio, T. 1960. *Ecology of living benthonic foraminifera from the San Diego area.* Spec. Paps Cushman Fdn Foramin. Res., (Sharon, Mass.) no. 5.

van der Hammen, T., J. H. Werner and H. van Dommelen 1973. Palynological record of the upheaval of the northern Andes. A study of the Pliocene and lower Quaternary of the Columbian eastern Cordillera and the early evolution of its high-Andean biota. *RPP* **16**, 1–122.

van der Werff, A. and H. Huls 1957–63. *Diatomeeenflora van Nederland.* (In 7 parts)

van Geel, B. 1976. Fossil spores of Zygemataceae in ditches of a prehistoric settlement in Hoogkarspel. *RPP* **22**, 337–44.

van Konijnenburg, J. H. A. 1971. *In situ* gymnosperm pollen from the Middle Jurassic of Yorkshire. *Acta Bot. Neerl.* **20**, 1–96.

van Landingham, S. L. 1967 to date. *Catalogue of the fossil and recent genera and species of diatoms and their synonyms.* (A revision of F. W. Mills, *An index to the genera and species of the Diatomaceae and their synonyms*) Lehre, W. Germany: J. Cramer Verlag.

van Morkhoven, F. 1962–3. *Post-Palaeozoic Ostracoda. Their morphology, taxonomy and economic use.* Vol. 1 (1962, general); vol. 2 (1963, generic descriptions). Amsterdam: Elsevier.

van Oyen, F. G. and F. Calandra 1963. Note sur les chinitozoaires. *Revue Micropaléont.* **6**, 13–18.

van Zeist, W. 1967. Archaeology and palynology in the Netherlands. *RPP* **4**, 45–65.

Vavrdová, M. 1974. Geographical differentiation of Ordovician acritarch assemblages in Europe. *RPP* **18**, 171–5.

Vidal, G. 1976. Late Precambrian microfossils from the Visingsö Beds in southern Sweden. *Fossils and Strata* **9**.

von Koenigswald, G. H. R., J. D. Emeis, W. L. Buning and C. W. Wagner 1963. *Evolutionary trends in Foraminifera.* Amsterdam: Elsevier.

Wagner, C. W. 1964. *Manual of larger Foraminifera.* The Hague: Bataafse Internationale Petroleum, Maatschappij N.V.

Wall, D. 1962. Evidence from Recent plankton regarding the biological affinities of *Tasmanites* Newton 1875 and *Leiosphaeridia* Eisenack 1958. *Geol Mag.* **99**, 353–63.

Wall, D. 1965. Microplankton, pollen and spores from the Lower Jurassic in Britain. *Micropaleontology* **11**, 151–90.

Wall, D. 1971. Quaternary dinoflagellate micropaleontology: 1959 to 1969. *Proc. N. Am. Paleont. Conv.* 1969, Part G, 844–66.

Wall, D. and B. Dale 1968. Early Pleistocene dinoflagellates from the Royal Society borehole at Ludham, Norfolk. *New Phytol.* **67**, 315–26.

Wall, D. and B. Dale 1968a. Modern dinoflagellate cysts and evolution of the Peridiniales. *Micropaleontology* **14**, 265–304.

Wall, D., B. Dale, G. P. Lohmann and W. K. Smith 1977. The environmental and climatic distribution of dinoflagellate cysts in modern marine sediments from regions in the North and South Atlantic Oceans and adjoining seas. *Mar. Micropaleont.* **2**, 121–200.

Walter, M. W. 1972. *Stromatolites and the biostratigraphy of the Australian Precambrian and Cambrian.* Spec. Paps Palaeont., no. 11.

Walter, M. R. (ed.) 1976. *Stromatolites.* Amsterdam: Elsevier.

Walter, M. R., J. Bauld and T. D. Brock 1972. Siliceous algal and bacterial stromatolites in hot springs and geyser effluents of Yellowstone National Park. *Science* **178**, 402–5.

Walter, M. R., A. D. T. Goode and W. D. M. Hall 1976. Microfossils from a newly discovered stromatolitic iron formation in Western Australia. *Nature* **261**, 221–3.

Walter, M. R., J. H. Oehler and D. Z. Oehler 1976. Megascopic algae 1300 million years old from the Belt

Supergroup, Montana: a reinterpretation of Walcott's *Helminthoidichnites*. *J. Paleont.* **50**, 872–81.

Walton, W. R. 1955. Ecology of living benthonic Foraminifera, Todos Santos Bay, Baja California. *J. Paleont.* **29**, 952–1018.

Warrington, G. 1970. The stratigraphy and palaeontology of the 'Keuper' Series of the central Midlands of England. *Q. J. Geol Soc. Lond.* **126**, 183–223.

Werner, D. (ed.) 1977. *The biology of diatoms*. Oxford: Blackwell Scientific Publications.

West, R. G. 1968. *Pleistocene geology and biology, with especial reference to the British Isles*. London: Longmans.

West, R. G. 1971. *Studying the past by pollen analysis*. Oxford: Oxford Univ. Press.

Westphal, A. 1976. *Protozoa*. Glasgow: Blackie.

Whittaker, R. H. 1969. New concepts of kingdoms of organisms. *Science* **163**, 150–60.

Williams, D. B. and W. A. S. Sarjeant 1967. Organic-walled microfossils as depth and shoreline indicators. *Mar. Geol.* **5**, 389–412.

Wise Jr, S. W. 1973. Calcareous nannofossils from cores recovered during Leg 18, Deep Sea Drilling Project: biostratigraphy and observations of diagenesis. *IRDSDP* **18**, 569–615.

Wise Jr, S. W. and F. H. Wind 1977. Mesozoic and Cenozoic calcareous nannofossils recovered by Deep Sea Drilling Project Leg 36, drilling on the Falkland Plateau, southwest Atlantic sector of the Southern Ocean. *IRDSDP* **36**, 269–491.

Wodehouse, R. P. 1935. *Pollen grains: their structure, identification and significance in science and medicine*. New York: McGraw Hill.

Wolf, K. H. 1965. Gradational sedimentary products of calcareous algae. *Sedimentology* **5**, 1–37.

Worsley, T. R. 1973. Calcareous nannofossils: Leg 19 of Deep Sea Drilling Project. *IRDSDP* **19**, 741–50.

Wray, J. L. 1971. Algae in reefs through time. *Proc. N. Am. Paleont. Conv.* 1969, Part J, 1358–73.

Wray, J. L. 1977. *Calcareous algae*. Amsterdam: Elsevier.

Zagwijn, N. H. 1967. Ecological interpretation of a pollen diagram from Neogene Beds in the Netherlands. *RPP* **2**, 173–81.

Ziegler, W. (ed.) 1973/75/77. *Catalogue of Conodonts*. Vol. I (1973); vol. II (1975); vol. III (1977). Stuttgart, Germany: E. Schweizerbart'sche.

Systematic Index

The pages on which genera are illustrated (or on which genera and higher taxa are referred to in figure labels and captions) are given in *italic* type.

General Index

Page number in *italic* type – item appears in figure label or caption on that page.
Page number in **bold** type – item introduced and/or explained on that page.